The Tallgrass Prairie Reader

✳ ✳ ✳

A BUR OAK BOOK

Holly Carver, series editor

✳ THE ✳
TALLGRASS
PRAIRIE
READER

EDITED BY JOHN T. PRICE

University of Iowa Press, Iowa City

University of Iowa Press, Iowa City 52242

Copyright © 2014 by the University of Iowa Press

www.uiowapress.org

Printed in the United States of America

Design by April Leidig

The University of Iowa Press is a member of Green Press Initiative and is committed to preserving natural resources.

Printed on acid-free paper

ISBN: 978-1-60938-246-9 (pbk)

ISBN: 978-1-60938-310-7 (ebk)

LCCN: 2013953637

Contents

THE TWENTIETH CENTURY

✻

Acknowledgments

Like any ecosystem, the creation of a book is a collective effort, and I owe many people a debt of gratitude. First and foremost, I would like to thank Holly Carver for her vision, guidance, and patience — as well as for her courageous advocacy on behalf of the prairie and its literature. Thanks, as well, to my agent Joanne Wyckoff for her always essential advice and encouragement. At the University of Iowa Press, I wish to thank Karen Copp, Lydia Crowe, Jim McCoy, Allison Means, Amy Richard, Charlotte Wright, freelance editor Lori Vermaas, and all those at the press who have not just supported this effort, but who continue to publish books dedicated to raising awareness about tallgrass prairie and our shared natural heritage. We all owe them our thanks.

I also want to acknowledge the support of the English Department at the University of Nebraska at Omaha, and the amazing work of graduate research assistants Nathanael Tagg, Karleen Gebhardt, and Nicholas Lesiak. During the last four years, many offered their time and assistance as I sought everything from individual works of literature to author biographies to information about tallgrass ecosystems. For their essential help, I would like to thank Julene Bair, Matt Baumann, Dianne Blankenship, Thomas Dean, James Dinsmore, Pauline Drobney, Tom Montgomery Fate, Chad Graeve, Twyla Hansen, Carl Kurtz, Merloyd Lawrence, Deb Lewis, Matt Low, Susan Naramore Maher, Chris Madson, Cornelia F. Mutel, David Peterson, Glenn Pollock, Diane Quantic, Alexis Rizzuto, Barbara Robins, Linda and Robert Scarth, Steve Semken, Daryl Smith, Ned Stuckey-French, Katy Stavreva, Larry Stone, Mary Swander, Randy Williams, and the Iowa Native Plant Society.

As always, I could not have completed this project without the love, encouragement and understanding of my family, especially my wife, Stephanie, and our boys, Benjamin, Spencer, and Alden—love you, you princes of Iowa, you kings of the prairie.

Finally, I want to dedicate this book to all those who have lent their time and talents—whether as scientists, writers, artists, teachers, or volunteers—to preserving, restoring, and helping others appreciate the magnificent tallgrass prairies. You are heroes.

Introduction

It was a beautiful May morning—prairie beautiful—with lots of wind and wavy grass and cloud shadows moving across the greening hillsides. The sky seemed to be launching itself over the surrounding bluffs, soaring. It was 2009, and the occasion was the dedication of Vincent Bluff, a small prairie in western Iowa, as an official state nature preserve. A group of us had gathered at the new on-site shelter for the ceremony. This included state and local officials, scientists, area residents and volunteers, a couple of newspaper reporters, and a creative writer—me. Everyone seemed relaxed, striking up easy conversations, and pointing out various spots of interest in the surrounding prairie. It was a day to be happy.

But I was a nervous wreck.

Ever since Glenn Pollock, the preserve manager, had asked me to compose an essay for the dedication, I'd been questioning my ability to live up to the charge. By then, I'd written two books with prairie themes, had read at various environmental fundraisers, and had participated in prairie restorations elsewhere. Something just felt different about this occasion. I'm sure that was due, in large part, to the significance of the place itself. Vincent Bluff was the first urban prairie state preserve in Iowa, located on a hill in the center of Council Bluffs, where my family and I have lived for the past fifteen years (I am a lifelong resident of Iowa, and the sixth generation in my family to live in the western half of the state). Its thirty-seven acres are part of a national treasure, the Loess Hills, which rise two hundred to three hundred feet off the Missouri River floodplain, and run the entire length of Iowa's western border, spilling over into Missouri. The only other place in the world where you can find loess deposits of such dramatic height is along the Yellow River in China, near the ancient Silk Road, which delivered so much treasure to the west.

Iowa's Loess Hills are also a source of treasure. The majority of what is left of Iowa's original grasslands—a state once dominated by tallgrass prairie—can be found in these hills. As no doubt the scientists in the audience knew, pre-settlement Iowa once contained twenty-eight million acres of tallgrass, reduced now to a splotchy twenty-eight thousand (D. Smith, xviii). Of the great tallgrass prairies that once ran from central Canada to the Gulf of Mexico, and from western Indiana to eastern Nebraska, Kansas, Oklahoma, and the Dakotas, only two to three percent remain, much of it in remnants the size of Vincent Bluff or smaller. Even at Vincent Bluff, the tallgrass is found primarily on an east-facing slope, the rest being a typical (yet distinctive) Loess Hills blend of mixed- and short-grass species usually found farther west, as well as woodlands. Tallgrass prairie is, as biologist Daryl Smith has written, "the most decimated ecosystem in continental North America" (xviii). Perhaps the world.

Which is to say, the small amount of original tallgrass remaining at Vincent Bluff is a kind of miracle. It is also the result of a lot of hard work. For many years, a dedicated group of people has been laboring to restore and preserve this natural refuge, and that effort has included cooperation among state and local organizations, private donations, and thousands of hours of volunteer labor—all to save thirty-seven acres. The signs of their success could be seen in the land all around us, but there was also evidence of the ongoing challenge. Houses and yards could be seen pushing up against the border of the preserve, and if we were to walk along the top of Vincent Bluff, the full extent of its isolation would be even more obvious. In one direction we would be able to see the shopping mall and mega-grocery store and residential neighborhoods. In the other we would see Interstate 80 and the casino-lined Missouri River and the cityscape of Omaha—a metropolitan area approaching nine hundred thousand residents. Standing at the shelter, even buffeted by the wind and the chatter, I could hear the roar of semis on the highway and occasionally catch a whiff of the nearby Burger King.

I didn't know of any literature dedicated to Vincent Bluff, but all around me were competing stories about this place—its past, present, and future—as well as the story of the vast and beautiful tallgrass prairies that had once surrounded it.

Now I was being asked to contribute in my own small way to that story, but as Glenn stepped forward to introduce me at the ceremony, my strong impulse was to run for cover in the nearby invasive timber. Glenn told the audience that though he believed scientists are good at collecting and organizing data, they are not always good at making it matter to a larger public. That's what artists and writers can do, he said. They can help people care. I appreciated the sentiment, but surely, I thought, the people in attendance that morning were not in need of more reasons to care. The fact of our gathering was a direct result of their extraordinary *caring.* It was a cause I had been very late in joining. Aside from a reading I'd given for the Loess Hills Preservation Society a few years back, I had contributed very little to the effort to preserve this tiny prairie in the heart of our city.

As I walked to the front of the crowd there was friendly applause, but all I could hear was the question I was certain they were asking themselves, because I was asking it too: *Why is he here?*

LOOKING BACK, I can think of several reasons why a writer belonged at that gathering. For one, the history of the relationship between the North American grasslands and literature has always been a vexed, but significant one. Early European and Euro-American explorers, who had never seen such grasslands, had trouble describing them in their written accounts, falling back on ready-at-hand abstractions, clichés, and comparisons to other places—deserts and oceans, mainly. Or they defined them by absence, particularly the relative absence of trees as compared to eastern regions of the country. James Fenimore Cooper, who never visited the trans-Mississippi prairies, relied on the firsthand accounts of Edwin James and others to write his 1827 novel, *The Prairie.* In this adventure narrative, Cooper's forest-loving hero Natty Bumppo (or Hawkeye) declares "I often think the Lord has placed this barren belt of prairie behind the states to warn men to what their folly may yet bring the land!" (24). The immense popularity of Cooper's books encouraged Iowa citizens, during that same century, to embrace "Hawkeye" as their state nickname—forever connecting themselves with a literary character who denigrated the native prairies that, in actuality, defined their state.

Literary artists weren't alone in their treatment of the prairies and grasslands as aesthetically challenged territory. For visual artists attached to nineteenth-century expeditions, the absence of trees and mountains created serious perspective problems, as well as challenging traditional European notions of natural beauty. As art historian Joni Kinsey has pointed out, "Many artists avoided depicting pure prairies, even those committed to documenting their experiences and who in their travels encountered the scenery daily for long periods. Instead, they focused on rivers and the scenic terrain that lined them or on portraits of the region's native inhabitants" (5–6).

A causal relationship between the aesthetic treatment of prairies in popular literature and art and their ecological mistreatment is difficult to prove, but a correlative relationship seems more than possible. How far is the psychological distance between well-circulated representations of prairie as "barren" or "empty" space and public policies such as the Timber Act of 1873, which offered an additional 160 acres of land to settlers in the region if they agreed to plant 40 of those acres in trees and maintain them for 10 years? As Lisa Knopp confirms in her essay, "Far Brought," these tree-planting efforts became widespread throughout the Great Plains and Midwest, championed by powerful boosters such as Nebraska businessman and U. S. Secretary of Agriculture J. Sterling Morton. Founder of Arbor Day, Morton stated in his 1872 Fruit Address: "[Trees] make the people among whom they grow a better and more thoughtful people. If every farm in Nebraska will plant out and cultivate an orchard and a flower garden, together with a few forest trees, this will become mentally and morally the best agricultural State." Twenty years later, in 1895, the Nebraska legislature adopted "The Tree Planters State" as the official state nickname, which in 1945 they replaced with "The Cornhusker State."

Like Iowans before them, Nebraskans embraced defining metaphors that all but erased public identification with native prairies (Illinois still stands as the only "Prairie State" in the Union). Even today, just a few miles downriver from Vincent Bluff prairie, Morton's large house and orchards are carefully maintained, along with an on-site resort hotel and convention center that serves as home base for the University of Nebraska's low-residency MFA creative writing program. In the shade of those manicured and morally superior orchards, a new generation of writers is free to create art with nary a stalk of original bluestem to wield its sinful influence.

But writers, even those sympathetic to prairie landscapes, may have inadvertently promoted another kind of "emptiness." As Daryl Smith points out, the destruction of the majority of tallgrass prairies largely occurred during a seventy-year period, from 1830 to 1900. This is a geologic blink of an eye—not even a blink, a nano-blink—which corresponded with the violent disruption of tribal cultures, resulting in the loss of centuries of acquired wisdom, practical knowledge, and oral storytelling about tallgrass prairie ecosystems. These changes cleared the way for an intensive, Euro-agricultural society, but they also cleared the way for a written, more conventional form of regional literature. Travelogues and journals about the tallgrass region by European explorers and missionaries had been published as far back as the 1600s, maybe earlier. The literary cultivation of the prairies began in earnest, however, with Washington Irving's popular 1835 account of his adventures in Oklahoma in *A Tour on the Prairies*. This was, as scholar Robert Thacker acknowledges, "the first sustained literary work by an author who had traveled the prairie" (70). The creator of Rip Van Winkle and the Headless Horseman legitimized, more than any before him, the transformation of the wild prairies into a place of literary prospect. There followed a relative explosion of poetry, fiction, and nonfiction about the prairies, ranging from adventure (George Catlin and Francis Parkman Jr.) to romanticism (William Cullen Bryant and Walt Whitman) to realism (Elizabeth B. Custer and Hamlin Garland).

At the same time writers were making use of the prairies in their books, many were proclaiming its imminent extinction from the planet. The demise of the prairies—as well as the demise of the human and animal communities that had relied on them for centuries—while being acknowledged, even mourned by authors, was considered by some of them to be inevitable or already complete. "The prairies are gone," wrote former Iowan Hamlin Garland in 1893. "I held one of the ripping, snarling, breaking plows that rolled the hazel bushes and wild sunflowers under[,] . . . and so there comes into my reminiscences an unmistakable note of sadness" (*Prairie Songs*, 3). The problem with this perspective was that the prairies, although severely depleted by the end of the century, were not gone, and Garland and others may have perpetuated the idea that they were beyond help. As scholar Matthew Low observes, "The idea or image of the vanishing prairie became so commonplace so quickly that texts . . . prophesying its demise were in a certain sense self-fulfilling" (227).

Even so, by the time it was possible for an ecological literary imagination to have taken root in the region—to the depth of, say, John Muir in the Sierras (one of several environmental luminaries raised in the Midwest)—most of the original tallgrass wilderness was gone. As in more than ninety percent gone (Smith, xviii). The impact on those who witnessed the destruction of that wilderness—an almost unimaginable loss of life—must have been profound. It is often noted that during this period many of our most famous "regional" writers, such as Hamlin Garland and Willa Cather, both of whom spent their early years on the native grasslands, took off for careers in coastal cities. There were many reasons for this literary exodus, not the least of which was the fact that centers of literary culture and publishing were in the East. But I have often wondered if it also had something to do with the near disappearance of the wild landscapes of their youth. Expressions of shock concerning that dramatic environmental change can be found on the pages of these and other midwestern writers of the time. Referring to the pioneer father of Alexandra Bergson, the heroine of her novel *O Pioneers!*, Cather writes of the Nebraska prairies: "It is sixteen years since John Bergson died. . . . Could he rise from beneath it, he would not know the country under which he has been asleep. The shaggy coat of the prairie, which they lifted to make him a bed, has vanished forever" (45). Iowan and natural historian Thomas Macbride remarked in 1895: "So completely has the whole State passed beneath the plow, so quickly assumed the appearance of one vast farm, that one who thus studies the Iowa of today realizes with difficulty the strange picturesque wildness of fifty or sixty years ago" (qtd. in Mutel, 75).

Other responses were more overtly personal. William J. Haddock, an attorney and settler in the state, confessed in 1901 that the "passing of the prairie" made him "sad" and "depressed" (63, 68). As he writes in his memoir *A Reminiscence: The Prairies of Iowa and Other Notes*: "I hold that whoever has not seen the prairies has not seen the grandeur and beauty of Iowa. The contemplation of greater wealth in buildings and farm products than the pioneers saw is a poor compensation." (71). Author George Ade made a similar claim in 1922 about progress and the nearly vanished prairies and wetlands he knew as a boy in Indiana: "We have gasoline chariots now, and clothes ordered from the catalogue, but the glory of the open country has departed, save for a vivid patch here and there at some neglected corner" (48). Ohio native Sherwood Anderson, in a letter dated from the 1920s, contemplated the spiritual impact of this loss on those living in the region:

A curious notion comes often to me. Is it not likely that when the country was new and men were often alone in the fields and forests they got a sense of bigness outside themselves that has now in some way been lost? I don't mean the conventional religious thing that is still prevalent and that is nowadays being retailed to the people by the most up-to-date commercial methods, but something else. . . . Mystery whispered in the grass, played in the branches of trees overhead, was caught up and blown across the horizon line in clouds of dust at evening on the prairies.

I am old enough to remember tales that strengthen my belief in a deep semi-religious influence that was formerly at work among our people. . . . I can remember old fellows in my home town speaking feelingly of an evening spent on the big, empty plains. It has taken the shrillness out of them. They had learned the trick of quiet. It affected their whole lives. It made them significant. (qtd. in Rideout, 281)

By the turn of the century, and then well into it, the once vast prairies that were the habitation of so many young people of the pioneer generation became on their pages a realm of nostalgia, sentiment, and irretrievable youth—their own and the country's. "In fancy we wander back to the prairies and the early golden days of the state and cherish their memory," writes Haddock. "It may be called sentiment, but that is all that is left of those times now, and the recollection is dear to me" (71). Another thirty to fifty years would have to pass before an activist environmental perspective would emerge in the literature of the Midwest, in the works of such writers as Meridel Le Sueur, Aldo Leopold, John Madson, Josephine W. Johnson, and Missouri farmer and environmentalist Eugene Poirot, who wrote in 1964:

The once great prairies with their fruits and wildlife nourished our nation through its weak infancy. They nourished it again through its reckless and wasteful adolescence. The nation has now reached a maturity which should make it capable of recognizing that the prairie can no longer give that which it does not have and that as man destroys it he destroys himself. (qtd. in Samson and Knopf, 93)

Even then, consciousness of a landscape that had been almost completely destroyed—was still being destroyed—was very much in the foreground, as well as an attendant fatalism. "We drive as if going to a funeral," wrote Le Sueur as she observed the environmental disaster of the Dust Bowl in Minnesota and eastern South Dakota, "the corpse is the very earth" (266). While visiting a prairie remnant in a Wisconsin cemetery, Leopold remarked: "Few grieved when the last buffalo left Wisconsin, and few will grieve when the last Silphium [compass plant] follows him to the lush prairies of the never-never land" (54). Madson, who grew up in central Iowa, writes:

My feeling for the tallgrass prairie is like that of a modern man who has fallen in love with the face in a faded tintype. Only the frame is still real; the rest is illusion and dream. So it is with the original prairie. The beautiful face of it had faded before I was born, before I had a chance to touch and feel it, and all that I have known of the prairie is the setting and the mood—a broad sky of pure and intense light, with a sort of loftiness to the days, and the young prairie-born winds running past me from open horizons. (*Out Home*, 92)

Whether or not the extinction rhetoric of authors played a role in the destruction of tallgrass prairies, it is now mostly on the pages of their books that eyewitness accounts of that destruction—its memory, as well as the memory of the original prairie wilderness itself—can be found. Even as a literary landscape, however, the tallgrass is endangered. Since the days of Thoreau and Muir, the American environmental imagination has remained largely focused on the forests and mountains, oceans and deserts, places where some experience of wild beauty is still possible—or at least a more mainstream notion of that beauty. As Low points out, there has been an "inexcusable underexposure of the prairie in modern environmental discourse" that continues today, including within environmental literature (229). For a biome that once covered forty percent of the United States—and one that on other continents played a significant role in our evolution as a species—we're lucky if writing about grasslands of *any* kind occupies a small percentage of the pages in nature magazines and anthologies (Manning, 3). And among those limited pages, tallgrass prairie literature occupies an even smaller space. In the 2002 college edition of *The Norton Book of Nature Writing*, for example, out of the 157 pieces listed in the table of contents, ranging from 1789 to 1999, there are approximately 16 focused on grasslands, and sometimes only marginally. Of those, five authors write directly of North American tallgrass country, some focusing primarily on its wetlands and woods.

This new version of prairie "emptiness" is, I believe, adding to the challenge faced by those currently trying to restore and preserve this land, and trying to recruit others to that cause. I fear the unspoken message to young environmentalists—including writers and scholars—is that the most urgent battles for the health of our planet are to be found elsewhere, the prairies being either beyond hope or not worthy of consideration. As journalist Richard Manning puts it, "Our culture's disrespect for its grasslands has produced an environmental catastrophe. It will be the best measure of the maturing of the American environmental movement when it begins to understand and combat this destruction" (5). That effort must also take place in the hearts and minds of our citizens, the realm under the purview of literature and art. Writer Annie Dillard has argued that seeing is in part a matter of articulation, of language. The relative absence of prairie literature and writers in the American canon is in this sense a form of collective blindness. But it is even more than that. It is another kind of extinction.

Still, like the postage-stamp prairies that provide seed for restorations, the literature of the tallgrass is out there, and it matters. It certainly mattered to me and played an essential role in my own awakening to the prairies as a writer and lifelong citizen of this region. When I first sought, as a young graduate student, to understand the natural history of the tallgrass prairies during the floods in 1993—an experience I wrote about in *Not Just Any Land: A Personal and Literary Journey into the American Grasslands* and *Man Killed by Pheasant and Other Kinships*—I had very little actual tallgrass prairie to guide me. What I did have were books. Scientific identification guides, yes, and I brought them with me as I explored the grasses and flowers blooming in the ditches and cemeteries around our rural, eastern Iowa home. Equally important, however, were works of autobiographical literature that chronicled various imaginative and emotional relationships to grasslands, near and far. Although I lacked scientific knowledge about the prairie, that ignorance could to some extent be corrected. The lack of feeling for or imaginative engagement with that ecosystem was a more difficult challenge, but the impassioned voices I found in nonfiction by Paul Gruchow, Mary Swander, Linda Hasselstrom, John Madson, William Least Heat-Moon, and others helped cure that indifference.

Even then, within the grand scheme of things, my personal conversion felt insignificant, especially when set beside the larger challenge of saving the tallgrass prairies. What would it take, I wondered, for the prairie to be not only physically restored, but also restored to its rightful place at the center of our collective identity as midwesterners and Americans—as human beings?

It is a question that still needs answering.

WHICH BRINGS ME to the motivations behind assembling *The Tallgrass Prairie Reader*. When Holly Carver and I first discussed the possibility of publishing an anthology, I was pleased to have the opportunity to offer a small corrective to the literary neglect of tallgrass prairies, and to gather together, in one place, some of the most significant and memorable literary responses to those ecosystems. But then I confronted the big question all anthologists face: What to include and why?

Responding to this question began, for me, with the land itself. As many have confirmed, among grasslands ecologies, tallgrass prairies are the most damaged and in the most need of protection. Environmental historian Dan Flores states:

If the Great Plains as a whole remains pathetically underprotected ecologically, the central and southern Plains are almost entirely so. Citizens of places like Texas and Kansas are today among the most divorced of all Americans from any kind of connection with regional nature.

With midheight grassland ecology represented by existing parks on the northern Plains, however, the pressing need in the future is for large preserves in the shortgrass High Plains, and in the tallgrass prairies. (15)

As I re-read this quote, it occurred to me that something similar might be said about literary preservation. A significant anthology, *The Great Plains Reader*, edited by Diane Dufva Quantic and P. Jane Hafen, already existed, but it focused primarily on the mixed- and short-grass prairie literature of the arid plains. Another anthology— *The Great North American Prairie: A Literary Field Guide*, edited by Sara St. Antoine and intended for younger readers—included literature from the North American grasslands in general. What was needed, I believed, was a historical collection of literature from and about the tallgrass territory. There are, of course, different opinions about where the tallgrass territory begins and ends, and ecologically speaking those borders are often indefinite, porous, and shifting due to soil composition, changing moisture levels, and other environmental influences. Tallgrass communities can be found in predominantly mixed-grass country, and vice versa. And like any other label, the term "tallgrass prairie"—which has also been referred to as "true prairie"—is limited in its ability to express the complexity of that which it is supposedly describing. For the purposes of this project, however, I relied on the map researched and published by the Tallgrass Prairie Center at the University of Northern Iowa, which designates the western borders of the original tallgrass bioregion as running along the eastern edge of the Dakotas, Nebraska, and Kansas, and into Oklahoma and parts of Texas. It identifies the eastern borders primarily within Wisconsin and Illinois, with smaller communities of tallgrass located in Indiana and Ohio. Staying within these geographical boundaries meant I would not be able to include a number of grasslands authors I admire, such as Mari Sandoz, Luther Standing Bear, Linda Hasselstrom, Dan O'Brien, and Julene Bair. It also meant I wouldn't be able to include literature about some of my favorite grasslands areas, such as the magnificent Sandhills of Nebraska, a predominantly mixed-grass area that is the subject of important and eloquent nature writing by such authors as John Janovy Jr., Paul Johnsgard, Stephen R. Jones, and Lisa Dale Norton.

All too often, however, North American grasslands have been lumped together within the collective environmental imagination—*It's all just grass!* But the ecological character and history of the tallgrass does have distinct features, and (I soon discovered) a distinct kind of story to tell.

Bison, for instance. In addition to the classic bison hunting scenes found in books by widely read authors like Washington Irving and Francis Parkman Jr., I expected to find many more narrative encounters with wild herds in nineteenth-century tallgrass literature, but that wasn't the case. Accounts by explorers and missionaries in the 1600s and 1700s mention large herds, and as Lance M. Foster confirms in an essay included in this anthology, bison were an important cultural and material resource for the Iowa and other tallgrass tribes. Ecologist James J. Dinsmore, John Madson, and others have pointed out, however, that bison numbers were probably lower in the tallgrass region, in part because it was so wet and the foliage so dense. This also made it difficult for hunters on horses to chase bison at full speed across open ground—a staple of most popular literary and artistic representations of the western grasslands. Herds in the thousands could still be encountered as late as 1820 in northwestern Iowa, for example, but most of the bison had been hunted off or moved west by the time Euro-American settlement began in earnest in the mid-nineteenth century (Dinsmore 13–14). As several authors featured in this collection observed, bison bones littered portions of the tallgrass when they first ambled through them during the first half of that century. The wildly popular accounts of bison hunts were almost exclusively set in the mixed- and short-grass country, including those by Irving and Parkman. Parkman remarked that anyone crossing through the tallgrass territory should not expect to immediately encounter "the paradise of his imagination" (32–33). That would be found farther west, he promised, in the "'great American desert;' those barren wastes, the haunts of the buffalo" (33). Although bison reappear in the restoration literature of the late twentieth century—see, for instance, Louise Erdrich's and Teresa Jordan's contributions to this anthology—it is the elk, the great stag of the prairie, that is often the dominant player in nineteenth-century accounts. Indeed, in the Osage creation story that opens this collection, an elk—not a bison—brings the tallgrass prairies and savannas into existence.

Likewise, among tallgrass plant communities, big bluestem and compass plant assume starring roles, rather than the little bluestem and buffalo grass of the mixed- and short-grass regions, respectively. Also, trees. For all my earlier criticism of the American bias favoring trees over grasslands, they have always been major players in tallgrass ecologies, whether mesic woodlands and oak savannas (which grace some of the hilltops here in western Iowa), or along the ancient borders between eastern forests and prairies in Ohio, Indiana, Wisconsin, Illinois, and Missouri. In this anthology, for instance, Mark Twain and John Muir write of their boyhood experiences along the wild edges of woodland and prairie. Even the more elemental forces of wind, water, fire, and soil have distinct characteristics in tallgrass ecologies, and have shaped them in fundamental ways. George Catlin wrote about the deadly ferocity of tallgrass fires as compared to short-grass fires people could sometimes just "step over" (17). More than a century later, William Least Heat-Moon wrote about how wind, as much as anything else, created the tallgrass prairie—its plant communities, but also its stories and legends. You'll find other portraits here of rainstorms (more frequent in tallgrass country), blizzards, drought, wetlands, and "infinite" horizons, all of them major forces in tallgrass natural history and the aesthetic responses to that environment.

In selecting pieces for *The Tallgrass Prairie Reader*, another major concern was what Flores and others have observed as a disconnect between residents of this region and native ecosystems. My own experience growing up in Iowa supported this claim—I knew very few people (including me) who identified with or knew much about the tallgrass prairie environment preceding the corn and bean fields surrounding our town. That has changed, but given the extent of the destruction, the challenge remains. I wanted this anthology to reach out to those already inside a personal and environmental commitment to the tallgrass prairies, but also to those who are still trying to find the door. I wanted the selections to educate and elucidate, while also revealing as directly as possible, the impact prairie has had on the thoughts and emotions of individuals. I wanted these to be individuals who exhibited —sometimes against their will—a degree of personal connection, even intimacy with the prairie, whether positive or negative; individuals who could speak at length to the tallgrass prairie's early and ongoing power over the imagination, particularly their own; individuals who, in short, *cared* and might encourage others to do so.

Toward that end, and because of space limitations, we (the editors at the press and I) made the difficult decision to focus exclusively on auto-biographical nonfiction. This has meant leaving out important works of tallgrass literature in other genres, including the poetry of William Cullen Bryant and Carl Sandburg, both of whom wrote in earlier centuries about their encounters with large tracts of wild prairie, and also contemporary nature poets such as Twyla Hansen, whose work offers a wonderful blend of lyricism, scientific knowledge, and personal intimacy with the Nebraska grasslands. In fiction, Laura Ingalls Wilder's sympathetic writing about tallgrass prairie wildlife, including wolves, is a must-read. Also, with the exception of the Osage creation story that opens the anthology, we did not include individual examples of Native American oral stories from tall-grass country—distinct, bioregional genres that warrant more attention than they have so far received from scholars and writers. Such stories and legends are mentioned, however, in a number of the autobiographical works collected here, including by authors from tribal cultures, such as Zitkala-Ša, John Joseph Mathews, and Lance M. Foster.

Finally, although there are examples here of literary natural history, which include scientific information, this is not primarily a natural history or scientific anthology. If readers would like to begin with a narrative overview of the environmental history and ecologies of the tallgrass prairies, including the impact of Euro-American settlement, I suggest skipping ahead to John Madson's essay, "The Running Country." There are a number of other well-written, informative, and engaging books about prairie natural history and current restoration efforts. A few of my favorites are *The Emerald Horizon: The History of Nature in Iowa* by Cornelia F. Mutel, *Prairie: A Natural History* by Candace Savage, *Big Bluestem: Journey into the Tall Grass* by Annick Smith, *The Ecology and Management of Prairies in the Central United States* by Chris Helzer, *A Practical Guide to Prairie Reconstruction* (second edition) by Carl Kurtz, and *The Tallgrass Prairie Center Guide to Prairie Restoration in the Upper Midwest* by Daryl Smith, Dave Williams, Greg Houseal, and Kirk Henderson.

Within these limited literary and geographic boundaries, I've done my best to include a variety of forms, voices, and approaches—including adventure narratives, spiritual reflections, literary ethnobotany, animal portraits, "personal" natural history, childhood memoir, travel writing, humor, and reportage. Sometimes the tallgrass environment is front and center in a particular piece, sometimes the human drama seems to get most of the attention, which is yet another defining feature of the tallgrass story: the presence of people and the relatively large impact they've had, and continue to have, on native prairie habitats. In each of these works, however, the tallgrass environment—its varying degrees of physical presence or symbolic meaning, or both—influences the writer's perceptions and his or her recollections of events in significant ways. It is my hope that this variety, despite its limitations, does some honor to the ecological diversity of the tallgrass prairie itself, while providing a range of models for nature writers and students—or at the very least, something to rebel against.

One additional note on the compilation process: To keep the anthology a reasonable (and affordable) length, I have, in a number of instances, excerpted chapters and sections from larger works or made cuts within individual pieces. At times, I have also done so to reduce an overlap of information among the pieces, or to focus them more on the main theme of this anthology, the tallgrass prairie environment. In each case, I have tried to edit with extreme care, and with respect for the original artistry. Also, some of the selections already had individual titles attached to them, but other selections were (as with Twain) lifted from an unnamed chapter/section or (as with Whitman) included multiple sections, each with their own title. In these instances, I appropriated a section title or created a new one that I felt captured the theme and "spirit" of the piece. All bibliographical citations can be found at the end of each selection, with other textual and author information in the headnotes. Sometimes I've added a personal interpretation or reflection or conjecture to the headnotes, because, well, that's how every one of these pieces affected me. Personally. I hope it's the same for the reader.

In the end, this anthology offers one story, among many possible stories, of the impact tallgrass has had on the thoughts, emotions, and actions of individuals who have encountered it over the course of three centuries.

There are numerous ways to read any anthology, including jumping around to various subjects or authors of interest, but I encourage you to read this as you would a novel, starting at the beginning and progressing chronologically. Allow yourself to "experience" the tallgrass through the eyes and imaginations of these authors over time, observing and responding to the ecological and social changes as they experienced them. Begin with "A Tallgrass Commonplace," which was inspired by the popular practice of readers in earlier centuries (and more recently by William Least Heat-Moon in his book *PrairyErth*) to jot down favorite literary quotes and other scraps of knowledge in "commonplace books," often organized around particular subjects of interest. There you will find a collection of quotes from the 1600s through the 1800s by some of the last people to experience the North American prairie as a "wilderness" or "frontier"—with a few interesting predictions about the future of the region and its inhabitants. Then, beginning with Catlin, you will be immersed in more extended, nineteenth-century descriptions of personal experiences with wild prairie environments— experiences that inspired a wide range of emotions, including terror, awe, admiration, disgust, reverence, and the thrill of adventure. You will encounter the early impression of the prairie as an inexhaustible resource for food and game through hunting and cultivation, and at the same time a source of potential death from raging wildfires, blizzards, and thunderstorms. Fear and hatred of the unfamiliar—whether directed at nature or at the indigenous cultures too often referred to then as "savage"—are brutally evident. There are also those, like Eliza Woodson Farnham, who found the beauty of the prairie wilderness to be almost overwhelming, and those, like the famous Charles Dickens, who found it to be repugnant. The personal reactions to the prairie during those earlier centuries are, in short, as extreme as its weather.

Following the destruction of the majority of tallgrass prairie between 1830 and 1900, much of the writing becomes nostalgic (Muir), elegiac (Haddock), romantic (Quayle), epic and tragic (Garland), spiritual (Zitkala-Ša), and mythic (Gilmore). This reflects literary tastes at the time, but it also reflects ecological changes and the impact of those changes on those who observed and experienced them. One of the most powerful examples of this is Omaha tribal author Francis La Flesche, who chronicles his boyhood attempt to join the tribe's journey to the annual buffalo hunt out west. He is caught and, as punishment, tied to a pole in the missionary school. There a different knowledge is being imposed on him than one based on the ancient natural cycles of the grasslands—cycles that are rapidly changing, thanks to the disappearance of the free-roaming herds. Meridel Le Sueur picks up a similar theme in her portrait of the 1930s' drought, where separation from and disrespect for those cycles leads to another kind of exile from the land, forced on farmers and ranchers by one of the largest human-made natural disasters in history.

A new environmental consciousness asserts itself midcentury, as represented by writers such as John Joseph Mathews, Aldo Leopold, Josephine W. Johnson, and John Madson. During and following this period, the earlier range of responses to tallgrass—personal, spiritual, tragic, epic, etc.—is increasingly influenced by a larger ecological perspective. Restoration and preservation, informed by hard science, become major themes. The feminist perspectives in writing by Johnson and Winifred M. Van Etten in the 1960s and '70s, for instance, are intimately linked to an environmental ethic. We end in the early twenty-first century with an established awareness of tallgrass environmental history, and the need for citizens, including writers, to participate in recovery efforts. There is also a renewed awareness of the ways stories and symbols—our own and others'—can create and nurture a correspondence between the human spirit and the tallgrass prairie. A correspondence that links us to the ground beneath our feet and, as Elizabeth Dodd articulates, to the stars above our heads where, through the language of tribal cultures, the ancient prairies still call to us.

That movement from wilderness to exploration and settlement through displacement and destruction calls for healing and restoration, and what it reveals about the evolving environmental imagination is one of the central plots of this story—and the story of the tallgrass itself.

IT IS A STORY that continues to be written, on the page and in the earth.

That was evident from the crest of Vincent Bluff in May 2009, where we all walked following the dedication ceremony. Up there, we stood witness to a portion of the prairie dream restored, proof positive that our hands do not have to be, as Hamlin Garland once claimed, "all desolating" (*Boy Life*, 389). For years, volunteers had been rewriting the story of that Loess Hills prairie from the ground up, working toward an ending that will take centuries to unfold. There is so much yet to learn about prairie ecologies, and in this regard Vincent Bluff will teach us what is impossible to glean between the pages of a book. About prairies and also about ourselves. Out there, like the subterranean root systems of grasses, vital entanglements have more room to take place. Out there, knowledge can freely and irrevocably entwine with imagination, and the rituals of science can become, at once, the rituals of body and spirit. As John Morgan, the restoration scientist featured in Don Gayton's essay, "Tallgrass Dream," confirms, "There is something very subtle and very powerful that happens when you have the privilege of kneeling quietly on a piece of tallgrass prairie" (101). I couldn't agree more. When a person bends down to examine a prairie grass stem or a bizarre insect or to press their fingers into the soil, touching roots, he or she engages one of the great paradoxes of human existence: the closer we are to the earth, the more we are uplifted. That is part of the mystery of the prairie, its ongoing power to transform both "soil" and "soul"—two words separated by only one letter, the singular first-person "I" that is in all of us, seeking the happiness of which Willa Cather spoke, to "become a part of something entire[,] . . . to be dissolved into something complete and great" (*My Ántonia*, 60).

The prairie, for many of us, is that "something entire," and it is my sincere hope that this anthology will inspire you to get outside and "become a part" of the nearest stretch of prairie or grassland, however large or small. But I also hope that reading this book will leave you feeling part of the larger, continuing story of the tallgrass, no matter where you live, and that it will inspire you to add your own chapter. What form that chapter takes will be as individual as you are, and it will likely change over time. It certainly has for me. Early in my journey as a writer, I came to believe authors and other artists had a responsibility to the earth that sustained us, continues to sustain us, and that meant we had a responsibility to the prairies. That we should know them and help them, and assist others in doing the same, through our words, our art, and our actions. I still believe that, but experiences like the one I had at Vincent Bluff have taught me that fulfilling that responsibility requires more than righteous indignation, especially when it comes to the places where we live, places we often overlook as we confront global environmental challenges. What is required of us, as well, is *repentance*. Not in the sense of confessing sin and seeking forgiveness and offering reparations, though when it comes to our history on the grasslands a good dose of that is warranted. Rather, I refer to one of the ancient meanings of the word, which is "to return, as if from exile." That kind of repentance is what was required of me as a writer at Vincent Bluff, and why I think I found it so difficult. Unlike other prairies, this was right down the street from where I had lived for over a decade. This prairie was my home—a place from which, until then, I had been exiled. Now it was time to return.

That return is ongoing, always incomplete, but words had once again played an important role. I would not have been carried to Vincent Bluff that morning without them—my own words and the words of others. They carry me still, as I hope they will carry the readers of this anthology.

Closer to the prairies. Closer to home.

WORKS CITED

Ade, George. *Single Blessedness and Other Observations.* New York: Double Day, Page, 1922.

Cather, Willa. *My Ántonia.* 1918. Ontario: Broadview Literary Texts, 2003.

———. *O Pioneers!* 1913. New York: Penguin, 1989.

Catlin, George. *Letters and Notes on the Manners, Customs, and Condition of the North American Indians, Volume 2.* 1841. Minneapolis, MN: Ross and Haines, 1965.

Cooper, James Fenimore. *The Prairie.* 1827. New York: Penguin, 1964.

Dinsmore, James J. *A Country So Full of Game: The Story of Wildlife in Iowa.* Iowa City: University of Iowa Press, 1994.

Flores, Dan. "A Long Love Affair with an Uncommon Country: Environmental History and the Great Plains." *Prairie Conservation: Preserving North America's Most Endangered Ecosystem.* Ed. Fred B. Samson and Fritz L. Knopf. Washington DC: Island Press, 1996. 3–17.

Garland, Hamlin. *Boy Life on the Prairie.* New York: Harper and Brothers, 1899.

———. *Prairie Songs.* Chicago: Stone and Kimball, 1893.

Gayton, Don. *Landscapes of the Interior: Re-Explorations of Nature and the Human Spirit.* Gabriola Island, BC: New Society Publishers, 1996.

Haddock, William J. *A Reminiscence: The Prairies of Iowa and Other Notes.* Iowa City, IA: Printed for Private Circulation, 1901.

Heat-Moon, William Least. *PrairyErth: A Deep Map.* Boston: Houghton Mifflin, 1991.

Helzer, Chris. *The Ecology and Management of Prairies in the Central United States.* Iowa City: University of Iowa Press, 2010.

Kinsey, Joni. *Plain Pictures: Images of the American Prairie.* Washington, DC: Smithsonian Institution Press, 1996.

Kurtz, Carl. *A Practical Guide to Prairie Reconstruction. Second Edition.* Iowa City: University of Iowa Press, 2013.

Leopold, Aldo. *A Sand County Almanac.* 1949. New York: Ballantine Books, 1984.

LeSueur, Meridel. *North Star Country.* 1945. Lincoln: University of Nebraska Press, 1984.

Low, Matt. "To Live Only in Books: Reading and Writing Restoration." *Proceedings of the 22nd North American Prairie Conference.* Ed. by Dave Williams, Brent Butler, and Daryl Smith. Cedar Falls: Tallgrass Prairie Center, University of Northern Iowa, 2010. 226–30.

Madson, John. *Out Home.* 1979. Iowa City: University of Iowa Press, 2008.

———. *Where the Sky Began: Land of the Tallgrass Prairie.* 1982. Iowa City: University of Iowa Press, 1995.

Manning, Richard. *Grassland: The History, Biology, Politics, and Promise of the American Prairie.* New York: Viking, 1995.

Mutel, Cornelia F. *The Emerald Horizon: The History of Nature in Iowa.* Iowa City: University of Iowa Press, 2008.

Parkman, Francis, Jr. *The Oregon Trail.* 1849. New York: Oxford University Press, 1996.

Price, John T. *Man Killed by Pheasant and Other Kinships.* 2008. Iowa City: University of Iowa Press, 2012.

———. *Not Just Any Land: A Personal and Literary Journey into the American Grasslands.* Lincoln: University of Nebraska Press, 2004.

Quantic, Diane Dufva, and P. Jane Hafen, eds. *The Great Plains Reader.* Lincoln: University of Nebraska Press, 2003.

Rideout, Walter B. *Sherwood Anderson: A Writer in America, Volume 1.* Madison: University of Wisconsin Press, 2006.

Samson, Fred B., and Fritz L. Knopf, eds. *Prairie Conservation: Preserving North America's Most Endangered Ecosystem.* Washington, DC: Island Press, 1996.

Savage, Candace. *Prairie: A Natural History.* Vancouver, BC: Greystone Books, 2011.

Smith, Annick. *Big Bluestem: Journey into the Tall Grass.* Tulsa, OK: Council Oak Books, 1996.

Smith, Daryl. "Introduction: Returning Prairie to the Upper Midwest." *The Tallgrass Prairie Center Guide to Prairie Restoration in the Upper Midwest.* Ed. by Daryl

Smith, Dave Williams, Greg Houseal, and Kirk Henderson. Iowa City: University of Iowa Press, 2010. xvii–xxi.

St. Antoine, Sara, ed. *The Great North American Prairie: A Literary Field Guide.* Minneapolis, MN: Milkweed Editions, 2001.

Thacker, Robert. *The Great Prairie Fact and the Literary Imagination.* Albuquerque: University of New Mexico Press, 1989.

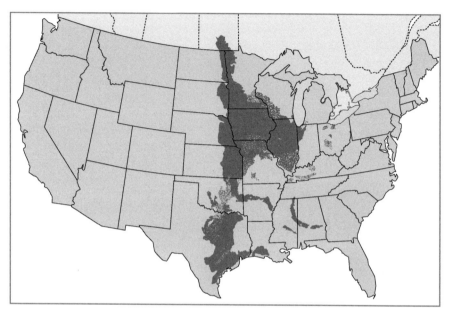

In the early nineteenth century, tallgrass prairie occupied the eastern portion of midcontinental North America between the mixed-grass prairie and the eastern deciduous forest. Today less than 2 to 3 percent of the original prairie remains. Map used courtesy of Brent Butler, Daryl Smith, and the Tallgrass Prairie Center.

This creation story was told by Black Dog (Shon-ton-ca-be), an Osage chief, to Omaha tribal member and anthropologist Francis La Flesche and ethnographer Alice C. Fletcher. It is included in their landmark study, *The Omaha Tribe* (1911), for which La Flesche and Fletcher spent years gathering firsthand information on "the customs, ceremonies, and beliefs" of the Omahas and related tribes such as the Osage following a time when, as Fletcher states in the foreword, "All that they formerly had relied on as stable had been swept away. . . . Even the wild grasses that had covered the prairies were changing" (30, 29). In this story, however, the sacred origins of the tallgrass prairies and savannas come alive again in the figure of a mighty elk that helped a people find their home.

Children of the Sun and Moon

Way beyond (an expression similar to "once upon a time") a part of the Wazha´zhe lived in the sky. They desired to know their origin, the source from which they came into existence. They went to the sun. He told them that they were his children. Then they wandered still farther and came to the moon. She told them that she gave birth to them, and that the sun was their father. She told them that they must leave their present abode and go down to the earth and dwell there. They came to the earth, but found it covered with water. They could not return to the place they had left, so they wept, but no answer came to them from anywhere. They floated about in the air, seeking in every direction for help from some god; but they found none. The animals were with them, and of all these the elk was the finest and most stately, and inspired all the creatures with confidence; so they appealed to the elk for help. He dropped into the water and began to sink. Then he called to the winds and the winds came from all quarters and blew until the waters went upward as in a mist. Before that time the winds traveled only in two directions, from north to south and then back from south to north; but when the elk called they came from the east, the north, the west, and the south, and met at a central point, and carried the water upward.

At first rocks only were exposed, and the people traveled on the rocky places that produced no plants, and there was nothing to eat. Then the waters began to go down until the soft earth was exposed. When this happened the elk in his joy rolled over and over on the soft earth, and all his loose hairs clung to the soil. The hairs grew, and from them sprang beans, corn, potatoes, and wild turnips, and then all the grasses and trees. . . .

From *The Omaha Tribe* by Alice C. Fletcher and Francis La Flesche, 1911. Reprint (Lincoln: University of Nebraska Press, 1992).

BLACK DOG

*   *   *

A Tallgrass Commonplace, 1673–1872

At first, when we were told of these treeless lands, I imagined that it was a country ravaged by fire, where the soil was so poor that it could produce nothing. But we have certainly observed the contrary; and no better soil can be found, either for corn or for vines, or for any other fruit whatever. . . . There are prairies three, six, ten and twenty leagues in length, and three in width, surrounded by forests of the same extent; beyond these, the prairies begin again, so that there is as much of one sort of land as the other. Sometimes we saw grass very short, and, at other times, five or six feet high; hemp, which grows naturally there, reaches a height of eight feet. A settler would not there spend ten years in cutting down and burning the trees; on the very day of his arrival, he could put his plow in the ground.
—Louis Jolliet, Illinois and the Mississippi River valley, 1673

The number of bison is almost beyond belief. I have seen twelve hundred killed in eight days by a single band of Savages. . . . This country is as beautiful and as fertile as Canada is lacking in these qualities.
—Nicolas de la Salle, Upper Mississippi River valley, 1680

We proceeded, continuing always to coast along the great prairies, which extend farther than the eye can reach. Trees are met with from time to time, but they are so placed that they seem to have been planted with design, in order to make avenues more pleasing to the eye than those of orchards. The base of these trees is often watered by little streamlets, at which are seen large herds of stags and hinds [female deer or elk] refreshing themselves, and peacefully feeding upon the short grass. We followed these vast plains for 20 leagues and repeated many times, "Benedicite opera Domini Domino" [Bless the Lord, ye works of the Lord].
—Fr. Claude-Jean Allouez, Illinois, 1681

3

The savage peoples who inhabit the prairies have life-long good-fortune; animals and birds are found there in great numbers, with numberless rivers abounding in fish. Those people are naturally very industrious, and devote themselves to the cultivation of the soil, which is very fertile for Indian corn. It also produces beans, squashes (both large and small) of excellent flavor, fruits, and many kinds of roots.
—Nicolas Perrot, Great Lakes region, approx. 1680–1717

We wanted then neither Powder nor Shot, and therefore we shot at random all that we met, either small birds, or Turtles, and Wood-Pigeons which were then coming from Foreign Countries in so great Numbers, that they did appear in the Air like Clouds.
—Fr. Louis Hennepin, Illinois, 1683

The golden pheasant, the quail, the partridge, the woodcock, the teeming turtle-dove, swarm in the woods and cover the open country intersected and broken by groves of full-grown forest trees which form a charming prospect which of itself might sweeten the melancholy tedium of solitude. There the hand of the pitiless mower has never shorn the juicy grass on which bisons of enormous height and size fatten. . . . Can it be thought that a land in which nature has distributed everything in so complete a manner could refuse to the hand of a careful husbandman who breaks into its fertile depths, the return which is expected of it?
—Antoine Laumet de la Mothe, Detroit region, 1701

The goodness of the country which [the Illinois and Missouri tribes] inhabit, if possible, must render life agreeable and easy to persons who, like them, are content with having the demands of nature answered, without endeavouring to increase these demands by any studied refinement in dress, equipage, or the modes of living. In short, these people, of any upon earth, seem blessed in this world: here is health and joy, peace and plenty; care and anxiety, ambition and the love of gold, and every uneasy passion seem banished from this happy region, at least to a greater degree than in almost any other part of the world.
—British Maj. Robert Rogers, 1765

A great part of the territory is miserably poor, especially that near lakes Michigan & Erie & that upon the Mississippie & the Illinois consists of extensive plains (which) have not had from appearances & will not have a single bush on them, for ages. The Districts therefore within (which) these fall will perhaps never contain a sufficient number of Inhabitants to entitle them to membership in the confederacy.
—James Monroe, in a report to Thomas Jefferson
    describing northern Illinois, 1786

Soon after we came into extensive meadows; and I was assured that those meadows continue for a hundred and fifty miles, being in winter drowned lands and marshes. By the dryness of the season they were now beautiful pastures, and here presented itself one of the most delightful prospects I have ever beheld; all the low grounds being meadow, and without wood, and all of the high grounds being covered with trees and appearing like islands; the whole scene seemed an elysium.
—Capt. Thomas Morris, Illinois, 1791

The Plains of this countrey are covered with a Leek Green Grass [big blue-stem], well calculated for the sweetest and the most norushing hay—interspersed with Cops [copses] of trees, Spreding ther lofty branchs over Pools Springs or Brooks of fine water. Groops of Shrubs covered with the most delicious froot is to be seen in every direction, and nature appears to have exerted herself to butify the Senery by the variety of flours Delicately and highly flavered raised above the Grass, which Strikes & profumes the Sensation, and amuses the mind throws it into Conjecterng the cause of So magnificent a Senerey in a Country thus Situated far removed from the Sivilised world to be enjoyed by nothing but Buffalo Elk Deer & Bear in which it abound & Savage Indians.
—William Clark, near the Missouri River, northeastern Kansas
    (now Doniphan County) July 4, 1804

The land is here the worst I had seen since I had left the banks of the Ohio; it had been gradually worse . . . and for the last two miles before we came

to Marshon's it had degenerated into natural prairies or savannas, with very little wood, and none deserving the name of timber.
—Fortescue Cuming, southwestern Ohio, 1810

On gaining a view of it, such a scene opened to us as will fall to the lot of few travelers to witness. This plain was literally covered with buffaloes as far as we could see. . . . The males were fighting in every direction, with a fury which I have never seen paralleled, each having singled out his antagonist. We judged that the number must have amounted to some thousands, and that there were many hundreds of these battles going on at the same time, some not eighty yards from us. . . . I am persuaded that our domestic bull would almost invariably be worsted in a contest with this animal, as he is inferior to him both in strength and ferocity. A shot was fired amongst them, which they did not seem to notice. Mr. Brackenridge joined me in preventing a volley being fired, as it would have been useless, and therefore wanton; for if we had killed one of these animals, I am certain the weight of his carcase in gold would not have bribed us to fetch him.
—John Bradbury, near the Missouri River,
    southeastern South Dakota, 1811

On the hills in every direction [buffalo] appeared by the thousands. Late in the evening we saw an immense herd in motion on the sides of the hills, at full speed: their appearance had something in it, which, without incurring ridicule, I might call sublime—the sounds of their footsteps, even at the distance of the two miles, resembled the rumbling of distant thunder.
—John Brackenridge, near the Missouri River,
    southeastern South Dakota, 1811

At first, we only received the impression of [the prairie's] general beauty. With longer gaze, all its distinctive features were revealed, lying in profound repose under the warm light of an afternoon's summer sun. Its indented and irregular outline of wood, its varied surface interspersed with clumps of oaks of centuries' growth, its tall grass, with seed stalks from six to ten feet high, like tall and slender reeds waving in a gentle breeze, the

whole presenting a magnificence of park-scenery, complete from the hand of Nature, and unrivalled by the same sort of scenery by European art.
—George Flower, Illinois, 1817

To the traveler, who for days traverses these prairies and barrens, their appearance is quite uninviting, and even disagreeable. He may travel from morning until night, and make good speed, but on looking around him, he fancies himself at the very spot whence he started. . . . No pleasant variety of hill and dale, no rapidly running brook delights the eye, and no sound of woodland music strikes the ear; but in their stead, a dull uniformity of prospect, "spread out immense."
—Caleb Atwater, Ohio, 1818

Some think, my Father, that you have brought all these soldiers here to take our land from us but I do not believe it. For although I am a poor simple Indian, I know that this land will not suit your farmers. If I even thought your hearts bad enough to take this land, I would not fear it, as I know there is not wood enough on it for the use of the white.
—Reported speech by Omaha Chief Big Elk to members of the
     Stephen Long expedition, eastern Nebraska, October 14, 1820

Let the reader imagine himself by the side of a rich meadow or fine grass plain several miles in diameter, decked with myriads of flowers of a most gorgeous and varied description, and he will have before his mind a pretty correct representation of one of these prairies. Nothing can surpass in richness of colour, or beauty of formation many of the flowers which are found in the most liberal profusion on these extensive and untrodden wilds.
—Rebecca Burlend, Illinois, 1830s

The Spring is any thing but what we have been taught to expect from that usually delightful season. It is a succession of rains, blows, and chills; and if the sun happen to shine, it does so gloomily, as if boding a coming storm. The whole country becomes saturated with water. . . . [But] we even contrive to pass comfortably enough the six weeks of rain, and fog, and wind

that changes the freezing winter into the warm and genial summer. We have no gradual gliding from cold to warm; it is snowy—then stormy—then balmy and delightful.
—Albert Miller Lea, on the Iowa prairies, 1836

May I not be permitted, in this place, to introduce a few reflections on the magical influence of the prairies? It is difficult to express by words the varied impressions which their spectacle produces. Their sight never wearies. . . . Doubtless there are moments when excessive heat, a want of fresh water, and other privations, remind one that life is a toil; but these drawbacks are of short duration. There is almost always a breeze over them. The security one feels in knowing that there are no concealed dangers, so vast is the extent which the eye takes in; no difficulties of road; a far spreading verdure, relieved by a profusion of variously colored flowers; the azure of the sky above, or the tempest that can be seen from its beginning to its end; the beautiful modifications of the changing clouds; the curious looming of objects between earth and sky, taxing the ingenuity every moment to rectify;—all, everything, is calculated to excite the perceptions, and keep alive the imagination. In the summer season, especially, everything upon the prairies is cheerful, graceful, and animated. . . . It is then they should be visited; and I pity the man whose soul could remain unmoved under such a scene of excitement.
—Joseph N. Nicollet, eastern South Dakota/
    western Minnesota, 1838–39

The blue heavens of Italy have tasked the inspiration of an hundred bards, and the warm brush of her own Lorraine has swept the canvas with their gorgeous transcript! But what pencil has wandered over the grander scenes of the North American prairie? What bard has struck his lyre to the wild melody of loveliness of the prairie sunset? . . . I cannot tell of the beauties of climes I have never seen; but I have gazed upon all the varied loveliness of my own fair, native land, from the rising sun to its setting, and in vain have tasked my fancy to image a fairer.
—Edmund Flagg, Illinois, 1838

The scenery of the prairie is striking and never fails to cause an exclamation of surprise. The extent of the prospect is exhilarating. The outline of the landscape is sloping and graceful. The verdure and the flowers are beautiful; the absence of shade, and consequent appearance of profusion of light, produce a gaiety which animates the beholder.
—Judge James Hall, Iowa, 1839

Eternal prairie and grass, with occasional groups of trees. [Capt. John] Frémont prefers this to every other landscape. To me it is as if someone would prefer a book with blank pages to a good story.
—Charles Preuss, 1842

We tore ourselves from the grave of our loved-one [daughter Edith], and sowing our tears along the road, went on our way toward "Big Blue." The third day we reached there, and went into a log cabin, to live with no floor, nor window, and tears were my meat and drink, day and night, until it seemed sometimes as tho' reason, could not retain her throne, unless my sorrows assuaged. No friend that seemed to understand, my sorrow. No acquaintance here, but my family—All is one vast expanse of nature, and tho' the Country is surpassingly beautiful, it is as lonely to me, as tho' I was shut up in a tomb, my heart is so sad, sad—I am glad, I'm born to die.
—Julia Louisa Lovejoy, northeast Kansas, May 6, 1854

[Father] took us all out for a ramble, out on the high prairie where we had a fine view of the country. The landscape was beautiful. The prairies look like a vast ocean stretching out until earth and sky seem to meet. We found many varieties of our cultivated flowers, growing in wild profusion. Father says we will love our new home. I already love the Nebraska prairies.
—Mollie Dorsey Sanford, near Nebraska City, May 4, 1857

The vast ocean of level prairie which lies to the west of the Red River must be seen in its extraordinary aspects, before it can be rightly valued and understood. . . . It must be seen at sunrise, when the boundless plain suddenly flashes with rose-coloured light, as the first rays of the sun sparkle

in the dew on the long rich grass, gently stirred by the unfailing morning breeze. It must be seen at noon-day . . . when each willow bush is magnified into a grove, each distant clump of aspens, not seen before, into wide forests, and the outline of wooded river banks, far beyond unassisted vision, rise to view. It must be seen at sunset, when, just as the huge ball of fire is dipping below the horizon, he throws a flood of red light, indescribably magnificent, upon the illimitable waving green, the colours blending and separating with the gentle roll of the long grass in the evening breeze, and seemingly magnified towards the horizon into the distant heaving swell of a parti-coloured sea. It must be seen, too, by moonlight, when the summits of the low green grass waves are tipped with silver, and the stars in the west disappear suddenly as they touch the earth. Finally, it must be seen at night, when the distant prairies are in a blaze, thirty, fifty, or seventy miles away; when the fire reaches clumps of aspen, and the forked tips of the flames, magnified by refraction, flash and quiver in the horizon, and the reflected lights from rolling clouds of smoke above tell of the havoc which is raging below.
—Henry Youle Hind, White Horse Plain,
    just west of Winnipeg, 1857–58

The prairie on fire all around us & no one but Elick & myself at home. It was a magnificent sight and had been, I thought, well represented in paintings that I had seen. But there was some difference to look at the real thing itself coming towards 50 tons of hay worth $20 dollars a ton on the ground or $30 at Kansas City. As we had taken the precaution to plow a few furrows away from the stacks we did not feel as uneasy as we otherwise should. But nevertheless the raging flame at every side excited us, & tonight as I am writing the horizon is lighted up at every side as if we were surrounded with furnaces and all of them were burning ore.
—George H. Hildt, Johnson County, eastern Kansas, November 1, 1857

I am sure I do not know whether we will stay here after next spring or not, —I think we will, for it seems as though Homer [Iowa] was our destiny so I try to put on a smiling face and be contented—but it is hard work some times. I do not think you need be fearful of coming in contact with "John Brownism" (by which I suppose you mean "abolitionism" and ["]opposition to southern interests") for the prairies which which—I didn't mean to write "which" twice—are our only companions are remarkably silent on the subject. [*sic*] and I will promise not to say a word on the subject, if you will only come home, and by the way [*sic*]. If you keep my mouth well stopped with novels, I believe I might consent to be come dum while you remained. Oh! ho, but the prarie chickens might croak abolitionism might they? Well, I shouldn't wonder!

—Bella Williams, age sixteen; letter to older brother
  in Alabama, January 15, 1861

The great ocean itself does not present more infinite variety than does this prairie-ocean of which we speak. In winter, a dazzling surface of the purest snow; in early summer, a vast expanse of grass and pale pink roses; in autumn too often a wild sea of raging fire. No ocean of water in the world can vie with its gorgeous sunsets; no solitude can equal the loneliness of night-shadowed prairie: one feels the stillness, and hears the silence, the wail of the prowling wolf makes the voice of solitude audible, the stars look down through infinite silence upon a silence almost as intense. This ocean has no past—time has been nought to it; and men have come and gone, leaving behind them no track, no vestige, of their presence.

—Sir William Francis Butler, near Winnipeg, 1872

THE
NINETEENTH
CENTURY

✷ ✷ ✷

Born in Vermont, Edwin James was hired in 1820, at age twenty-two, to serve as physician, botanist, geologist—and later, chief chronicler—for Major Stephen Long's western expedition. This massive operation, begun in 1819 under orders from Secretary of War John C. Calhoun, was intended to explore regions between the Mississippi and the Rocky Mountains in support of American scientific, economic, and military interests. Although suffering from scandal, sickness, supply shortages, and geographical mis-calculations, the expedition provided important and influential information about this relatively unknown territory, including the now famous desig-nation of the plains grasslands as an uninhabitable "Great American Des-ert." After retiring from government service, James settled in Iowa, where he spent the rest of his life. In this excerpt from volume 2 of his *Account of an Expedition from Pittsburgh to the Rocky Mountains* (1823), the young James describes the mirages he experienced in the tallgrass territory along the Missouri River in what is currently northwestern Missouri and into southern Nebraska and Iowa. At one point, he mistakes elk for mastodons, an extinct species Thomas Jefferson once thought might roam the wild, uncharted grasslands.

✳ ✳ ✳

Prairie Mirage

[...]

We had now ascended about eighty miles from the mouth of Grand river. The country we had passed is fertile, and presents such an intermixture of forests and grassy plains, as is extremely pleasing to the eye. Towards the north the hills became gradually more and more elevated. . . .

Leaving the immediate neighbourhood of the river, there is an ascent of several miles to the level of the great woodless plain. The bottom, and part of the sides of the vallies, are covered with trees; but in proportion to the elevation, the surface becomes more unvaried and monotonous. These vast plains, in which the eye finds no object to rest upon, are at first seen with surprise and pleasure; but their uniformity at length becomes tiresome.

For a few days the weather had been fine, with cool breezes, and broken flying clouds. The shadows of these coursing rapidly over the plain, seemed to put the whole in motion; and we appeared to ourselves as if riding on the unquiet billows of the ocean. The surface is uniformly of the description, not inaptly called *rolling*, and will certainly bear a comparison to the waves of an agitated sea. The distant shores and promontories of woodland, with here and there an insular grove of trees, rendered the illusion more complete.

The great extent of country contemplated at a single view, and the unvaried sameness of the surface, made our prospect seem tedious. We pursued our course during the greater part of the day along the same wide plain, and at evening the woody point in which we had encamped on the preceding night was yet discernible.

Nothing is more difficult than to estimate, by the eye, the distance of objects seen in these plains. A small animal, as a wolf or a turkey, sometimes appears of the magnitude of a horse, on account of an erroneous impression of distance. Three elk, which were the first we had seen, crossed our path at some distance before us. The effect of the *mirage*, together with our indefinite idea of the distance, magnified these animals to a most prodigious size. For a moment we thought we saw the mastodon of America, moving in those vast plains, which seem to have been created for his dwelling place. An animal seen for the first time, or any object with which the eye is unacquainted, usually appears much enlarged, and inaccurate ideas are formed of the magnitude and distance of all the surrounding objects; but if some well-known animal, as a deer or a wolf, comes into the field of vision so near as to be recognized, the illusions vanish, and all things return to their proper dimensions.

Soon after we had left our encampment, on one of the bright sunny mornings which occurred, when we were in the country near the sources of

EDWIN JAMES

Grand river, we discovered, as we thought, several large animals feeding in the prairie, at the distance of half a mile. These, we believed, could be no other than bisons; and after a consultation respecting the best method of surprising them, two of our party dismounted; and creeping with great care and caution, about one-fourth of a mile through the high grass, arrived near the spot, and discovered an old turkey, with her brood of half-grown young, the only animals now to be seen.

On the evening of the 20th of May, we encamped in a low, muddy bottom, overgrown with nettles and phacelias; but the only place we could find combining the three requisites—grass for our horses, and wood and water for ourselves. Here we were so tormented by the mosquitoes, harassed and goaded by the wood-ticks, that we were glad to seek relief by mounting our horses, at the earliest appearance of light on the following morning. The dew had been so heavy, that it was falling in drops from the grass and weeds where we had lain, and our blankets were dripping as if they had been exposed to a shower. We proceeded on our course about thirty miles, and encamped early in the afternoon. Having ascended Grand river nearly to the point, where we believed Field's trace must cross it, we directed our course more to the west, and had already crossed several streams running to the south, supposed to be the upper branches of the Little Platte.

The utmost uniformity prevails in the appearance of the country about the sources of the Little Platte, Nishnebottona, and other northern tributaries of the Missouri. . . .

Remains of bisons, as bones, horns, hoofs, and the like, are often seen in these plains; and in one instance, in a low swamp surrounded by forests, we discovered the recent track of a bull; but all the herds of these animals have deserted the country on this side of Council Bluff. The bones of elk and deer are very numerous, particularly about certain places, which, from the great number of tent poles, scaffolds, &c. appear to be old Indian hunting camps; and the living animals are still to be found here in plenty. As we rode along these boundless meadows, every object within several miles became visible; the smallest shrub rising a few inches above the surface of the green expanse, could be seen at a mile distant. . . .

On the evening of the 24th, we arrived on the bank of a beautiful river, at a grove of ash and cotton-wood trees. We had scarce dismounted from our horses, when a violent thunder-shower commenced; the rain fell in such torrents as to extinguish our fire, and the wind blew so violently that our blanket-tent could afford us no protection. Many large trees were blown down in the point of woods where we lay, and one fell a few yards from our camp. As the night was extremely dark, we thought the danger of moving at least equal to that of remaining where we were; and spent part of the night in the greatest anxiety, listening to the roar of the storm, and the crashing of the timber. As our horses were dispersed about the wood, we had scarce a hope they could all escape uninjured.

On the day following, after we had rode about eighteen or twenty miles, we observed the surface of the country to become suddenly hilly; and soon after were surprised by an unexpected view of the wide valley, the green meadows, and the yellow stream of the Missouri. A little after noon we encamped in a meadow on the river bottom, and by ascending one of the neighbouring bluffs; sufficiently elevated to overlook a large extent of the surrounding country, we were enabled to discover that we had arrived at the Missouri, at a point about six miles below the confluence of the great river Platte. . . .

On the 27th, we swam across Mosquito Creek; and after a ride of near thirty miles along the Missouri bottoms, encamped near the mouth of the Boyer, about six miles from the wintering place of the party. Early on the following morning we left our encampment, and were soon after cheered by the report of guns discharging at the Cantonment. The sight of the trading establishment, called Fort Lisa, gave us more pleasure than can easily be imagined, except by those who have made journies similar to ours, and have felt the deprivation of all those enjoyments which belong to the habitations of men. . . .

From chapter 5, "Journey by Land from St. Louis to Council Bluff," in volume 2 of *Account of an Expedition from Pittsburgh to the Rocky Mountains Performed in the Years 1819, 1820* by Edwin James (London: Longman, Hurst, Rees, Orme, and Brown, 1823).

✳ WASHINGTON IRVING, 1783–1859 ✳

In the autumn of 1832, internationally acclaimed author and native New Yorker Washington Irving—creator of "Rip Van Winkle" and "The Legend of Sleepy Hollow"—joined a surveying mission into what is now eastern Oklahoma, led by U. S. Commissioner on Indian Affairs Judge Henry Ellsworth and the French Creole guide and hunter Pierre Beatte. Other traveling companions included a group of military rangers under the command of Captain Jesse Bean, and their cook, Tonish. Irving had hopes of observing indigenous cultures, as well as the prairie wilderness before it disappeared. He also hoped, after having spent the last seventeen years living in Europe, to reestablish himself as an authentic American author. Although the journey itself disappointed many of Irving's expectations, the resulting book, *A Tour on the Prairies* (1835), was a commercial success and solidified Irving's American reputation. In this excerpt, set near present-day Tulsa, Irving offers readers the popular portrayal of the tallgrass prairies and woodlands as a hunter's paradise—where herds of free-roaming elk, though diminishing in number, could still be encountered—but also a place for healthy living and contemplative repose.

✳ ✳ ✳

Life on the Prairies

Having passed through the skirt of woodland bordering the river, we ascended the hills, taking a westerly course through an undulating country of "oak openings," where the eye stretched at times over wide tracts of hill and dale, diversified by forests, groves, and clumps of trees. As we were proceeding at a slow pace, those who were at the head of the line descried four deer grazing on a grassy slope, about half a mile distant. They, apparently, had not perceived our approach, and continued to graze in perfect tranquillity. A young ranger obtained permission from the Captain to go in pursuit of them, and the troop halted in lengthened line, watching him in silence. Walking his horse slowly and cautiously, he made a circuit, until a grove of wood intervened between him and the deer. Dismounting then, he left his horse among the trees, and, creeping round a knoll, was hidden from our view. We now kept our eyes intently fixed on the deer, which continued grazing, unconscious of their danger. Presently, there was the sharp report of a rifle; a fine buck made a convulsive bound and fell to the earth; his companions scampered off. Immediately our whole line of march was broken; there was a helter-skelter galloping of the youngsters of the troop, eager to get a shot at the fugitives; and one of the most conspicuous personages in the chase was our little Frenchman Tonish on his silver grey, having abandoned his pack-horses at the first sight of the deer. It was some time before our scattered forces could be recalled by the bugle, and our march resumed.

Two or three times in the course of the day we were interrupted by hurry-scurry scenes of the kind. The young men of the troop were full of excitement on entering an unexplored country abounding in game, and they were too little accustomed to discipline or restraint to be kept in order. No one, however, was more unmanageable than our Tonish. Having an intense conceit of his skill as a hunter, and an irrepressible passion for display, he was continually sallying forth, like an ill-broken hound, whenever any game was started, and had as often to be whipped back.

At length his vanity got a salutary check. A fat doe came bounding along in full sight of the whole line. Tonish dismounted, levelled his rifle, and had a fair shot. The doe kept on. He sprang upon his horse, stood up on the saddle like a posture-master, and continued gazing after the animal, as if certain to see it fall. The doe, however, kept on its way rejoicing; a laugh broke out along the line; the little Frenchman slipt quietly into his saddle, began to belabour and blaspheme the wandering pack-horses, as if they had been to blame; and for some time we were relieved from his vaunting and vapouring.

In one part of our march we came to the remains of an old Indian encampment, on the banks of a fine stream, with the moss-grown skulls of deer lying here and there about it. As we were in the Pawnee country, it was supposed, of course, to have been a camp of those formidable rovers. The Doctor, however, after considering the shape and disposition of the lodges, pronounced it the camp of some bold Delawares, who had probably made a brief and dashing excursion into these dangerous hunting-grounds.

Having proceeded some distance further, we observed a couple of figures on horseback, slowly moving parallel to us along the edge of a naked hill about two miles distant, and apparently reconnoitring us. There was a halt, and much gazing and conjecturing. Were they Indians? If Indians, were they Pawnees? There is something exciting to the imagination and stirring to the feelings, while traversing these hostile plains, in seeing a horseman prowling along the horizon. It is like descrying a sail at sea in time of war, when it may be either a privateer or a pirate. Our conjectures were soon set at rest, by reconnoitring the two horsemen through a small spy-glass, when they proved to be two of the men we had left at the camp, who had set out to rejoin us, and had wandered from the track.

Our march this day was animating and delightful. We were in a region of adventure; breaking our way through a country hitherto untrodden by white men, excepting perchance by some solitary trapper. The weather was in its perfection—temperate, genial, and enlivening; a deep blue sky, with a few light feathery clouds; an atmosphere of perfect transparency; an air pure and bland; and a glorious country spreading out far and wide in the golden sunshine of an autumnal day; but all silent, lifeless,—without a human habitation, and apparently without a human inhabitant. It was as if a ban hung over this fair but fated region. The very Indians dared not abide here, but made it a mere scene of perilous enterprise, to hunt for a few days, and then away.

After a march of about fifteen miles west, we encamped in a beautiful peninsula, made by the windings and doublings of a deep, clear, and almost motionless brook, and covered by an open grove of lofty and magnificent trees. Several hunters immediately started forth in quest of game, before the noise of the camp should frighten it from the vicinity. Our man Beatte also took his rifle, and went forth alone in a different course from the rest.

For my own part, I lay on the grass under the trees, and built castles in the clouds, and indulged in the very luxury of rural repose. Indeed, I can scarcely conceive a kind of life more calculated to put both mind and body in a healthful tone. A morning's ride of several hours, diversified by hunting incidents; an encampment in the afternoon under some noble grove on the borders of a stream; an evening banquet of venison fresh killed, roasted, or broiled on the coals; turkeys just from the thickets, and wild honey from the trees; and all relished with an appetite unknown to the gourmands of the cities. And then at night—such sweet sleeping in the open air; or waking and gazing at the moon and stars, shining between the branches of the trees!

On the present occasion, however, we had not much reason to boast of our larder. But one deer had been killed during the day, and none of that had reached our lodge. We were fain, therefore, to stay our keen appetites by some scraps of turkey, brought from the last encampment, eked out with a slice or two of salt pork. This scarcity, however, did not continue long. Before dark, a young hunter returned well laden with spoil. He had shot a deer, cut it up in an artist-like style, and, putting the meat in a kind of sack, made of the hide, had slung it across his shoulder, and trudged with it to the camp.

Not long after, Beatte made his appearance, with a fat doe across his horse. It was the first game he had brought in, and I was glad to see him with a trophy that might efface the memory of the pole-cat. He laid the carcass down by our fire, without saying a word, and then turned to unsaddle his horse; nor could any questions from us about his hunting draw from him more than laconic replies.

If Beatte, however, observed this Indian taciturnity about what he had done, Tonish made up for it by boasting of what he meant to do. Now that we were in a good hunting country, he meant to take the field; and, if you would take his word for it, our lodge would henceforth be overwhelmed with game. Luckily, his talking did not prevent his working; the doe was skilfully dissected, several fat ribs roasted before the fire, the coffee kettle replenished, and, in a little while, we were enabled to indemnify ourselves luxuriously for our late meagre repast.

The Captain did not return until late, and he returned empty-handed. He had been in pursuit of his usual game, the deer, when he came upon the tracks of a gang of about sixty elk. Having never killed an animal of the kind, and the elk being at this moment an object of ambition among all the veteran hunters of the camp, he abandoned his pursuit of the deer, and followed the newly discovered track. After some time he came in sight of the elk, and had several fair chances of a shot; but was anxious to bring down a large buck, which kept in the advance. Finding, at length, that there was danger of the whole gang escaping him, he fired at a doe. The shot took effect; but the animal had sufficient strength to keep on for a time with its companions. From the tracks of blood, he felt confident it was mortally wounded; but evening came on, he could not keep the trail, and had to give up the search until morning.

[. . .]

From chapter 14 of *A Tour on the Prairies* by Washington Irving, 1835. Reprint (Alexandria, VA: Time-Life, 1983).

✳ GEORGE CATLIN, 1796–1872 ✳

A native of Pennsylvania, George Catlin was one of the first Euro-American artists to travel extensively in the prairie regions of the upper Mississippi and Missouri Rivers during the 1830s. He sought to create a visual and written account of "vanishing" indigenous cultures, and eventually visited fifty North American tribes, producing over five hundred paintings which he displayed during tours in the United States and Europe. He also published written accounts of his travels, including his impressions of the prairies and grasslands, some of which he considered picturesque and sublime—even proposing they become the site of the first national park, almost forty years before Yellowstone was granted that distinction. In this excerpt from volume 2 of his *Letters and Notes on the Manners, Customs, and Condition of the North American Indians* (1841), set near Fort Leavenworth in northeastern Kansas, Catlin describes the beauty and terrifying power of tallgrass prairie fires.

Prairie Burning

[. . .]

T he prairies burning form some of the most beautiful scenes that are to be witnessed in this country, and also some of the most sublime. Every acre of these vast prairies (being covered for hundreds and hundreds of miles, with a crop of grass, which dies and dries in the fall) burns over during the fall or early in the spring, leaving the ground of a black and doleful colour.

There are many modes by which the fire is communicated to them, both by white men and by Indians—*par accident;* and yet many more where it is voluntarily done for the purpose of getting a fresh crop of grass, for the grazing of their horses, and also for easier travelling during the next summer, when there will be no old grass to lie upon the prairies, entangling the feet of man and horse, as they are passing over them.

Over the elevated lands and prairie bluffs, where the grass is thin and short, the fire slowly creeps with a feeble flame, which one can easily step over; where the wild animals often rest in their lairs until the flames almost burn their noses, when they will reluctantly rise, and leap over it, and trot off amongst the cinders, where the fire has past and left the ground as black as jet. These scenes at night become indescribably beautiful, when their flames are seen at many miles distance, creeping over the sides and tops of the bluffs, appearing to be sparkling and brilliant chains of liquid fire (the hills being lost to the view), hanging suspended in graceful festoons from the skies.

But there is yet another character of burning prairies, that requires another Letter, and a different pen to describe—the war, or hell of fires! where the grass is seven or eight feet high, as is often the case for many miles together, on the Missouri bottoms; and the flames are driven forward by the hurricanes, which often sweep over the vast prairies of this denuded country. There are many of these meadows on the Missouri, the Platte, and the Arkansas, of many miles in breadth, which are perfectly level, with a waving grass, so high, that we are obliged to stand erect in our stirrups, in order to look over its waving tops, as we are riding through it. The fire in these, before such a wind, travels at an immense and frightful rate, and often destroys, on their fleetest horses, parties of Indians, who are so unlucky as to be overtaken by it; not that it travels as fast as a horse at full speed, but that the high grass is filled with wild pea-vines and other impediments, which render it necessary for the rider to guide his horse in the zig-zag paths of the deers and buffaloes, retarding his progress, until he is overtaken by the dense column of smoke that is swept before the fire—alarming the horse, which stops and stands terrified and immutable, till the burning grass which is wafted in the wind, falls about him, kindling up in a moment a thousand new fires, which are instantly wrapped in the swelling flood of smoke that is moving on like a black thunder-cloud, rolling on the earth, with its lightning's glare, and its thunder rumbling as it goes.

When Ba'tiste, and Bogard, and I, and Patrick Raymond (who like Bogard had been a free trapper in the Rocky Mountains), and Pah-me-o-ne-qua (the red thunder), our guide back from a neighbouring village, were jogging along on the summit of an elevated bluff, overlooking an immense valley of high grass, through which we were about to lay our course.

"Well, then, you say you have seen the prairies on fire?" Yes. "You have seen the fire on the mountains, and beheld it feebly creeping over the grassy hills of the North, where the toad and the timid snail were pacing from its approach—all this you have seen, and who has not? But who has seen the vivid lightnings, and heard the roaring thunder of the rolling conflagration which sweeps over the *deep-clad* prairies of the West? Who has dashed, on his wild horse, through an ocean of grass, with the raging tempest at his back, rolling over the land its swelling waves of liquid fire?" What! "Aye, even so. Ask the red savage of the wilds what is awful and sublime—Ask him where the Great Spirit has mixed up all the elements of death, and if he does not blow them over the land in a storm of fire? Ask him what foe he has met, that regarded not his frightening yells, or his sinewy bow? Ask these lords of the land, who vauntingly challenge the thunder and lightning of Heaven—whether there is not one foe that travels over their land, too swift for their feet, and too mighty for their strength—at whose approach their stout hearts sicken, and their strong-armed courage withers to nothing? Ask him *again* (if he is sullen, and his eyes set in their sockets)— 'Hush!——sh!——sh!'—(he will tell you, with a soul too proud to confess —his head sunk on his breast, and his hand over his mouth)—'that's *medicine!*'"

I said to my comrades, as we were about to descend from the towering bluffs into the prairie—"We will take that buffalo trail, where the travelling herds have slashed down the high grass, and making for that blue point, rising, as you can just discern, above this ocean of grass; a good day's work will bring us over this vast meadow before sunset." We entered the trail, and slowly progressed on our way, being obliged to follow the winding paths of the buffaloes, for the grass was higher than the backs of our horses. Soon after we entered, my Indian guide dismounted slowly from his horse, and lying prostrate on the ground, with his face in the dirt, he *cried*, and was talking to the Spirits of the brave—"For," said he, "over this beautiful plain dwells the Spirit of fire! he rides in yonder cloud—his face blackens with rage at the sound of the trampling hoofs—the *fire-bow* is in his hand—he draws it across the path of the Indian, and quicker than lightning, a thousand flames rise to destroy him; such is the talk of my fathers, and the ground is whitened with their bones. It was here," said he, "that the brave son of Wah-chee-ton, and the strong-armed warriors of his band, just twelve moons since, licked the fire from the blazing wand of that great magician. Their pointed spears were drawn upon the backs of the treacherous Sioux, whose swifter-flying horses led them, in vain, to the midst of this valley of death. A circular cloud sprang up from the prairie around them! it was raised, and their doom was fixed by the Spirit of fire! It was on this vast plain of *fire-grass* that waves over our heads, that the swift foot of Mah-to-ga was laid. It is here, also, that the fleet-bounding wild horse mingles his bones with the red man; and the eagle's wing is melted as he darts over its surface. Friends! it is the season of fire; and I fear, from the smell of the wind, that the Spirit is awake!"

Pah-me-o-ne-qua said no more, but mounted his wild horse, and waving his hand, his red shoulders were seen rapidly vanishing as he glided through the thick mazes of waving grass. We were on his trail, and busily traced him until the midday-sun had brought us to the ground, with our refreshments spread before us. He partook of them not, but stood like a statue, while his black eyes, in sullen silence, swept the horizon round; and then, with a deep-drawn sigh, he gracefully sunk to the earth, and laid with his face to the ground. Our buffalo *tongues* and pemican, and marrow-fat, were spread before us; and we were in the full enjoyment of these dainties of the Western world, when, quicker than the frightened elk, our Indian friend sprang upon his feet! His eyes skimmed again slowly over the prairies' surface, and he laid himself as before on the ground.

"Red Thunder seems sullen to-day," said Bogard—"he startles at every rush of the wind, and scowls at the whole world that is about him."

"There's a rare chap for you—a fellow who would shake his fist at Heaven, when he is at home; and here, in a *grass-patch*, must make his *fire-medicine* for a *circumstance* that he could easily leave at a shake of his horse's heels."

"Not sae sure o' that, my hooney, though we'll not be making too lightly of the matter, nor either be frightened at the mon's strange octions. But, Bogard, I'll tell ye in a 'ord (and thot's enough), there's something more than odds in all this '*medicine*.' If this mon's a fool, he was born out of his own country, that's all—and if the divil iver gits him, he must take him cowld, for he is too swift and too wide-awake to be taken alive—you understond thot, I suppouse? But, to come to the plain matter—supposin that the Fire Spirit (and I go for somewhat of witchcraft), I say supposin that this *Fire Spirit* should jist impty his pipe on tother side of this prairie, and strike up a bit of a blaze in this high grass, and send it packing across in this direction, before sich a death of a wind as this is! By the *bull barley*, I'll bet you'd be after '*making medicine*,' and *taking* a bit of it, too, to get rid of the racket."

"Yes, but you see, Patrick——"

"Neever mind thot (not wishin to distarb you); and suppouse the blowin wind was coming fast ahead, jist blowin about our ears a warld of smoke and chokin us to dith, and we were dancin about a *Varginny reel* among these little paths, where the divil would we be by the time we got to that bluff, for it's now fool of a distance? Givin you time to spake, I would say a word more (askin your pardon), I know by the expression of your face, mon, you neever have seen the world on fire yet, and therefore you know nothin at all of a *hurly burly* of this kind—did ye?—did ye iver see (and I jist want to know), did ye iver see the fire in high-grass, runnin with a strong wind, about five mile and the half, and thin hear it strike into a *slash* of *dry* cane brake!! I would jist ax you that? By thuneder you niver have—for your eyes would jist stick out of your head at the thought of it! Did ye iver look way into the backside of Mr. Maelzel's Moscow, and see the flashin flames a runnin up; and then hear the poppin of the *militia fire* jist afterwards? then you have jist a touch of it! ye're jist beginnin—ye may talk about fires—but this is sich a *baste of a fire!* Ask *Jack Sanford*, he's a chop that can tall you all aboot it. Not wishin to distarb you, I would say a word more—and that is this— If I were advisin, I would say that we are gettin too far into this imbustible meadow; for the grass is dry, and the wind is too strong to make a light matter of, at this sason of the year; an now I'll jist tell ye how M'Kenzie and I were sarved in this very place about two years ago; and he's a worldly chop, and niver aslape, my word for that——hollo, what's that!"

Red Thunder was on his feet!—his long arm was stretched over the grass, and his blazing eye-balls starting from their sockets! "White man (said he), see ye that small cloud lifting itself from the prairie? he rises! the hoofs of our horses have waked him! The *Fire Spirit* is awake—this wind is from his nostrils, and his face is this way!" No more—but his swift horse darted under him, and he gracefully slid over the waving grass as it was bent by the wind. Our viands were left, and we were swift on his trail. The extraordinary leaps of his wild horse, occasionally raised his red shoulders to view, and he sank again in the waving billows of grass. The tremulous wind was hurrying by us fast, and on it was borne the agitated wing of the soaring eagle. His neck was stretched for the towering bluff, and the thrilling screams of his voice told the secret that was behind him. Our horses were swift, and we struggled hard, yet hope was feeble, for the bluff was yet *blue*, and nature nearly exhausted! The sunshine was *dying*, and a cool shadow advancing over the plain. Not daring to look back, we strained every nerve. The roar of a distant cataract seemed gradually advancing on us—the winds increased, the howling tempest was maddening behind us—and the swift-winged *beetle* and *heath hens*, instinctively drew their straight lines over our heads. The fleet-bounding antelope passed us also; and the *still swifter* long-legged hare, who leaves but a shadow as he flies! Here was no time for thought—but I recollect the heavens were overcast—the distant thunder was heard—the lightning's glare was reddening the scene—and the smell that came on the winds struck terror to my soul! The piercing yell of my savage guide at this moment came back upon the winds—his robe was seen waving in the air, and his foaming horse leaping up the towering bluff!

Our breath and our sinews, in this last struggle for life, were just enough to bring us to its summit. We had risen from a *sea of fire!* "Great God! (I exclaimed) how sublime to gaze into that valley, where the elements of nature are so strangely convulsed!" Ask not the poet or painter how it looked, for they can tell you not; but ask the *naked savage*, and watch the electric twinge of his manly nerves and muscles, as he pronounces the lengthened "hush——sh——" his hand on his mouth, and his glaring eye-balls looking you to the very soul!

I beheld beneath me an immense cloud of black smoke, which extended from one extremity of this vast plain to the other, and seemed majestically

to roll over its surface in a bed of liquid fire; and above this mighty desolation, as it rolled along, the whitened smoke, pale with terror, was streaming
and rising up in magnificent cliffs to heaven!

I stood *secure*, but tremblingly, and heard the maddening wind, which
hurled this *monster* o'er the land—I heard the roaring thunder, and saw its
thousand lightnings flash; and then I saw *behind*, the black and smoking
desolation of this *storm* of *fire!*

From volume 2 of *Letters and Notes on the Manners, Customs, and Condition of the
North American Indians* by George Catlin, 1841. Reprint (Minneapolis, MN: Ross
and Haines, 1965).

In 1842, world-famous author Charles Dickens made his first extended tour of the United States. He was twenty-nine years old, and although impressed by some of what he saw, he found much to criticize, including poor sanitary conditions in the cities, the cruel treatment of prisoners and the mentally ill, and slavery. While visiting St. Louis in April, he expressed "a great desire to see a Prairie" (175), and some locals agreed to take him to the "Looking-Glass Prairie" in St. Clair County, near the village of Lebanon on the Illinois side of the Mississippi—a river he described elsewhere as "a foul stream" of "running liquid mud" (172, 171). As this excerpt from *American Notes* (1842) attests, he was similarly unimpressed with the prairie, a scene not "to remember with much pleasure, or to covet the looking-on again, in after-life" (82–83).

✳ ✳ ✳

A Jaunt to the Looking-Glass Prairie and Back

[...]

We got over the river in due course, and mustered again before a little wooden box on wheels, hove down all aslant in a morass, with "MERCHANT TAILOR" painted in very large letters over the door. Having settled the order of proceeding, and the road to be taken, we started off once more and began to make our way through an ill-favoured Black Hollow, called, less expressively, the American Bottom.

The previous day had been—not to say hot, for the term is weak and lukewarm in its power of conveying an idea of the temperature. The town had been on fire; in a blaze. But at night it had come on to rain in torrents, and all night long it had rained without cessation. We had a pair of very strong horses, but travelled at the rate of little more than a couple of miles an hour, through one unbroken slough of black mud and water. It had no variety but in depth. Now it was only half over the wheels, now it hid the axletree, and now the coach sank down in it almost to the windows. The air resounded in all directions with the loud chirping of the frogs, who, with the pigs (a coarse, ugly breed, as unwholesome-looking as though they were the spontaneous growth of the country), had the whole scene to themselves. Here and there we passed a log hut: but the wretched cabins were wide apart and thinly scattered, for though the soil is very rich in this place, few people can exist in such a deadly atmosphere. On either side of the track, if it deserve the name, was the thick "bush;" and everywhere was stagnant, slimy, rotten, filthy water.

As it is the custom in these parts to give a horse a gallon or so of cold water whenever he is in a foam with heat, we halted for that purpose, at a log inn in the wood, far removed from any other residence. It consisted of one room, bare-roofed and bare-walled of course, with a loft above. The ministering priest was a swarthy young savage, in a shirt of cotton print like bed-furniture, and a pair of ragged trousers. There were a couple of young boys, too, nearly naked, lying idle by the well; and they, and he, and *the* traveller at the inn, turned out to look at us.

The traveller was an old man with a grey grizzly beard two inches long, a shaggy moustache of the same hue, and enormous eyebrows; which almost obscured his lazy, semi-drunken glance, as he stood regarding us with folded arms; poising himself alternately upon his toes and heels. On being addressed by one of the party, he drew nearer, and said, rubbing his chin (which scraped under his horny hand like fresh gravel beneath a nailed shoe), that he was from Delaware, and had lately bought a farm "down there," pointing into one of the marshes where the stunted trees were thickest. He was "going," he added, to St. Louis, to fetch his family, whom he had left behind; but he seemed in no great hurry to bring on these incumbrances, for when we moved away, he loitered back into the cabin, and was plainly bent on stopping there so long as his money lasted. He was a great politician of course, and explained his opinions at some length to one of our company; but I only remember that he concluded with two sentiments, one of which was, Somebody for ever; and the other, Blast everybody else! which is by no means a bad abstract of the general creed in these matters.

When the horses were swollen out to about twice their natural dimensions (there seems to be an idea here, that this kind of inflation improves their going), we went forward again, through mud and mire, and damp, and festering heat, and brake and bush, attended always by the music of the frogs and pigs, until nearly noon, when we halted at a place called Belleville.

[. . .]

From Belleville, we went on, through the same desolate kind of waste, and constantly attended, without the interval of a moment, by the same music; until, at three o'clock in the afternoon, we halted once more at a village called Lebanon to inflate the horses again, and give them some corn besides: of which they stood much in need. Pending this ceremony, I walked into the village, where I met a full-sized dwelling-house coming down-hill at a round trot, drawn by a score or more of oxen.

The public-house was so very clean and good a one, that the managers of the jaunt resolved to return to it and put up there for the night, if possible. This course decided on, and the horses being well refreshed, we again pushed forward, and came upon the Prairie at sunset.

It would be difficult to say why, or how—though it was possibly from having heard and read so much about it—but the effect on me was disappointment. Looking towards the setting sun, there lay, stretched out before my view, a vast expanse of level ground; unbroken, save by one thin line of trees, which scarcely amounted to a scratch upon the great blank; until it met the glowing sky, wherein it seemed to dip: mingling with its rich colours, and mellowing in its distant blue. There it lay, a tranquil sea or lake without water, if such a simile be admissible, with the day going down upon it: a few birds wheeling here and there: and solitude and silence reigning paramount around. But the grass was not yet high; there were bare black patches on the ground; and the few wild flowers that the eye could see, were poor and scanty. Great as the picture was, its very flatness and extent, which left nothing to the imagination, tamed it down and cramped its interest. I felt little of that sense of freedom and exhilaration which a Scottish heath inspires, or even our English downs awaken. It was lonely and wild, but oppressive in its barren monotony. I felt that in traversing the Prairies, I could never abandon myself to the scene, forgetful of all else; as I should do instinctively, were the heather underneath my feet, or an iron-bound coast beyond; but should often glance towards the distant and frequently-receding line of the horizon, and wish it gained and passed. It is not a scene to be forgotten, but it is scarcely one, I think (at all events, as I saw it), to remember with much pleasure, or to covet the looking-on again, in after-life.

We encamped near a solitary log-house, for the sake of its water, and dined upon the plain. The baskets contained roast fowls, buffalo's tongue (an exquisite dainty, by-the-way), ham, bread, cheese, and butter; biscuits, champagne, sherry; lemons and sugar for punch; and abundance of rough ice. The meal was delicious, and the entertainers were the soul of kindness and good humour. I have often recalled that cheerful party to my pleasant recollection since, and shall not easily forget, in junketings nearer home with friends of older date, my boon companions on the Prairie.

Returning to Lebanon that night, we lay at the little inn at which we had halted in the afternoon. In point of cleanliness and comfort it would have suffered by no comparison with any English alehouse, of a homely kind, in England.

[. . .]

After breakfast, we started to return by a different way from that which we had taken yesterday, and coming up at ten o'clock with an encampment of German emigrants carrying their goods in carts, who had made a rousing fire which they were just quitting, stopped there to refresh. And very pleasant the fire was; for, hot though it had been yesterday, it was quite cold to-day, and the wind blew keenly. Looming in the distance, as we rode along, was another of the ancient Indian burial-places, called The Monks' Mound; in memory of a body of fanatics of the order of La Trappe, who founded a desolate convent there, many years ago, when there were no settlers within a thousand miles, and were all swept off by the pernicious climate: in which lamentable fatality, few rational people will suppose, perhaps, that society experienced any very severe deprivation.

The track of to-day had the same features as the track of yesterday. There was the swamp, the bush, and the perpetual chorus of frogs, the rank unseemly growth, the unwholesome steaming earth. Here and there, and frequently too, we encountered a solitary broken-down waggon, full of some new settler's goods. It was a pitiful sight to see one of these vehicles deep in the mire; the axletree broken; the wheel lying idly by its side; the man gone miles away, to look for assistance; the woman seated among their wandering household goods with a baby at her breast, a picture of forlorn, dejected patience; the team of oxen crouching down mournfully in the mud, and breathing forth such clouds of vapour from their mouths and nostrils, that all the damp mist and fog around seemed to have come direct from them.

In due time we mustered once again before the merchant tailor's, and having done so, crossed over to the city in the ferry-boat: passing, on the way, a spot called Bloody Island, the duelling-ground of St. Louis, and so designated in honour of the last fatal combat fought there, which was with pistols, breast to breast. Both combatants fell dead upon the ground; and possibly some rational people may think of them, as of the gloomy madmen on the Monks' Mound, that they were no great loss to the community.

From chapter 13 of *"American Notes" and "Pictures from Italy"* by Charles Dickens, 1842 and 1846. Reprint (New York: Oxford University Press, 1987).

✳ MARGARET FULLER, 1810–1850 ✳

In May 1843, having recently stepped down as the first editor of Ralph Waldo Emerson's transcendentalist journal, *The Dial* (which published Thoreau's early writing), famed New England writer and feminist Margaret Fuller departed Buffalo on a steamer for Chicago. Although she observed that this increasingly populated area of the Midwest was no longer a pristine wilderness, an extended tour of rural De Kalb and Lee Counties in a horse-drawn wagon inspired her to write that the as yet unplowed prairies and park-like groves of timber offered settlers "the very Eden which earth might still afford"—a dramatic contrast to Dickens's negative view (122). In her account of this excursion in *Summer on the Lakes* (1844), she revels in poetic descriptions of the landscape, and shares what she has learned of local Native American history and culture, including medicinal uses of prairie plants. Although Fuller admired the self-sufficiency of farmers and their families—as this excerpt reveals—she also predicted elsewhere that "their mode of cultivation will, in the course of twenty, perhaps ten, years, obliterate the natural expression of the country" (Fuller, *Summer on the Lakes in 1843* [2007], 39).

Chicago

[. . .]

In Chicago I first saw the beautiful prairie flowers. They were in their glory the first ten days we were there—

The golden and the flame-like flowers.

The flame-like flower I was taught afterwards, by an Indian girl, to call "Wickapee;" and she told me, too, that its splendors had a useful side, for

36

it was used by the Indians as a remedy for an illness to which they were subject.

Beside these brilliant flowers, which gemmed and gilt the grass in a sunny afternoon's drive near the blue lake, between the low oakwood and the narrow beach, stimulated, whether sensuously by the optic nerve, unused to so much gold and crimson with such tender green, or symbolically through some meaning dimly seen in the flowers, I enjoyed a sort of fairyland exultation never felt before, and the first drive amid the flowers gave me anticipation of the beauty of the prairies.

At first, the prairie seemed to speak of the very desolation of dullness. After sweeping over the vast monotony of the lakes to come to this monotony of land, with all around a limitless horizon, — to walk, and walk, and run, but never climb, oh! it was too dreary for any but a Hollander to bear. How the eye greeted the approach of a sail, or the smoke of a steamboat; it seemed that any thing so animated must come from a better land, where mountains gave religion to the scene.

The only thing I liked at first to do, was to trace with slow and unexpecting step the narrow margin of the lake. Sometimes a heavy swell gave it expression; at others, only its varied coloring, which I found more admirable every day, and which gave it an air of mirage instead of the vastness of ocean. Then there was a grandeur in the feeling that I might continue that walk, if I had any seven-leagued mode of conveyance to save fatigue, for hundreds of miles without an obstacle and without a change.

But after I had rode out, and seen the flowers and seen the sun set with that calmness seen only in the prairies, and the cattle winding slowly home to their homes in the "island groves"—peacefullest of sights—I began to love because I began to know the scene, and shrank no longer from "the encircling vastness."

It is always thus with the new form of life; we must learn to look at it by its own standard. At first, no doubt my accustomed eye kept saying, if the mind did not, What! no distant mountains? what, no valleys? But after a while I would ascend the roof of the house where we lived, and pass many hours, needing no sight but the moon reigning in the heavens, or starlight falling upon the lake, till all the lights were out in the island grove of men beneath my feet, and felt nearer heaven that there was nothing but this lovely, still reception on the earth; no towering mountains, no deep tree-shadows, nothing but plain earth and water bathed in light.

Sunset, as seen from that place, presented most generally, low-lying, flaky clouds, of the softest serenity, "like," said S., "the Buddhist tracts."

One night a star shot madly from its sphere, and it had a fair chance to be seen, but that serenity could not be astonished.

Yes! it was a peculiar beauty of those sunsets and moonlights on the levels of Chicago which Chamouny or the Trosachs could not make me forget.

Notwithstanding all the attractions I thus found out by degrees on the flat shores of the lake, I was delighted when I found myself really on my way into the country for an excursion of two or three weeks. We set forth in a strong wagon, almost as large, and with the look of those used elsewhere for transporting caravans of wild beasts, loaded with every thing we might want, in case nobody would give it to us—for buying and selling were no longer to be counted on—with a pair of strong horses, able and willing to force their way through mud holes and amid stumps, and a guide, equally admirable as marshal and companion, who knew by heart the country and its history, both natural and artificial, and whose clear hunter's eye needed neither road nor goal to guide it to all the spots where beauty best loves to dwell.

Add to this the finest weather, and such country as I had never seen, even in my dreams, although these dreams had been haunted by wishes for just such an one, and you may judge whether years of dullness might not, by these bright days, be redeemed, and a sweetness be shed over all thoughts of the West.

The first day brought us through woods rich in the moccasin flower and lupine, and plains whose soft expanse was continually touched with expression by the slow moving clouds which

> Sweep over with their shadows, and beneath
> The surface rolls and fluctuates to the eye;
> Dark hollows seem to glide along and chase
> The sunny ridges,

to the banks of the Fox river, a sweet and graceful stream. We reached Geneva just in time to escape being drenched by a violent thunder shower, whose rise and disappearance threw expression into all the features of the scene.

Geneva reminds me of a New England village, as indeed there, and in the neighborhood, are many New Englanders of an excellent stamp, generous, intelligent, discreet, and seeking to win from life its true values. Such are much wanted, and seem like points of light among the swarms of settlers, whose aims are sordid, whose habits thoughtless and slovenly.

With great pleasure we heard, with his attentive and affectionate congregation, the Unitarian clergyman, Mr. Conant, and afterward visited him in his house, where almost everything bore traces of his own handy work or that of his father. He is just such a teacher as is wanted in this region, familiar enough with the habits of those he addresses to come home to their experience and their wants; earnest and enlightened enough to draw the important inferences from the life of every day.

A day or two we remained here, and passed some happy hours in the woods that fringe the stream, where the gentlemen found a rich booty of fish.

Next day, travelling along the river's banks, was an uninterrupted pleasure. We closed our drive in the afternoon at the house of an English gentleman, who has gratified, as few men do, the common wish to pass the evening of an active day amid the quiet influences of country life. He showed us a bookcase filled with books about this country; these he had collected for years, and become so familiar with the localities that, on coming here at last, he sought and found, at once, the very spot he wanted, and where he is as content as he hoped to be, thus realizing Wordsworth's description of the wise man, who "sees what he foresaw."

A wood surrounds the house, through which paths are cut in every direction. It is, for this new country, a large and handsome dwelling; but round it are its barns and farm yard, with cattle and poultry. These, however, in the framework of wood, have a very picturesque and pleasing effect. There is that mixture of culture and rudeness in the aspect of things as gives a feeling of freedom, not of confusion.

I wish it were possible to give some idea of this scene as viewed by the earliest freshness of dewy dawn. This habitation of man seemed like a nest in the grass, so thoroughly were the buildings and all the objects of human care harmonized with what was natural. The tall trees bent and whispered all around, as if to hail with sheltering love the men who had come to dwell among them.

The young ladies were musicians, and spoke French fluently, having been educated in a convent. Here in the prairie, they had learned to take care of the milk-room, and kill the rattlesnakes that assailed their poultry yard. Beneath the shade of heavy curtains you looked out from the high and large windows to see Norwegian peasants at work in their national dress. In the wood grew, not only the flowers I had before seen, and wealth of tall, wild roses, but the splendid blue spiderwort, that ornament of our gardens. Beautiful children strayed there, who were soon to leave these civilized regions for some really wild and western place, a post in the buffalo country. Their no less beautiful mother was of Welsh descent, and the eldest child bore the name of Gwynthleon. Perhaps there she will meet with some young descendants of Madoc, to be her friends; at any rate, her looks may retain that sweet, wild beauty, that is soon made to vanish from eyes which look too much on shops and streets, and the vulgarities of city "parties."

Next day we crossed the river. We ladies crossed on a little foot-bridge, from which we could look down the stream, and see the wagon pass over at the ford. A black thunder cloud was coming up. The sky and waters heavy with expectation. The motion of the wagon, with its white cover, and the laboring horses, gave just the due interest to the picture, because it seemed as if they would not have time to cross before the storm came on. However, they did get across, and we were a mile or two on our way before the violent shower obliged us to take refuge in a solitary house upon the prairie. In this country it is as pleasant to stop as to go on, to lose your way as to find it, for the variety in the population gives you a chance for fresh entertainment in every hut, and the luxuriant beauty makes every path attractive. In this house we found a family "quite above the common," but, I grieve to say, not above false pride, for the father, ashamed of being caught barefoot, told us a story of a man, one of the richest men, he said, in one of the eastern cities, who went barefoot, from choice and taste.

Near the door grew a Provence rose, then in blossom. Other families we saw had brought with them and planted the locust. It was pleasant to see their old home loves, brought into connection with their new splendors. Wherever there were traces of this tenderness of feeling, only too rare among Americans, other things bore signs also of prosperity and intelligence, as if the ordering mind of man had some idea of home beyond a mere shelter, beneath which to eat and sleep.

No heaven need wear a lovelier aspect than earth did this afternoon, after the clearing up of the shower. We traversed the blooming plain, unmarked by any road, only the friendly track of wheels which tracked, not broke the grass. Our stations were not from town to town, but from grove to grove. These groves first floated like blue islands in the distance. As we drew nearer, they seemed fair parks, and the little log houses on the edge, with their curling smokes, harmonized beautifully with them.

One of these groves, Ross's grove, we reached just at sunset. It was of the noblest trees I saw during this journey, for the trees generally were not large or lofty, but only of fair proportions. Here they were large enough to form with their clear stems pillars for grand cathedral aisles. There was space enough for crimson light to stream through upon the floor of water which the shower had left. As we slowly plashed through, I thought I was never in a better place for vespers.

[...]

From "Chicago, June 20" entry in *Summer on the Lakes in 1843* by Sarah Margaret Fuller. 1844. (Rockville, MD: Arc Manor, 2007).

✳ ELIZA WOODSON FARNHAM, 1815–1864 ✳

Born in New York State, Eliza Woodson endured a difficult childhood, which included losing her mother around age five, being separated from her siblings, and being raised by a repressive foster mother. In 1835, she reunited with her beloved sister, Mary, who had settled with her husband in rural Tazewell County, Illinois, near Peoria. There, as scholar Annette Kolodny states, the prairies offered her "a natural world . . . that seemed to compensate for the family she had lost in childhood" (Kolodny, *The Land before Her* [1984], 109). A year later Eliza married lawyer and travel writer Thomas Jefferson Farnham; the couple left the area in 1841. In New York, Eliza successfully advocated for more humane prison conditions, and later worked with the blind in Boston and farmed her late husband's ranch in Santa Cruz, California, before succumbing to tuberculosis at age forty-nine. In this excerpt from *Life in Prairie Land* (1846), she fondly recalls the borderlands between prairie and woods, where the turbulence of a spring thunderstorm—perhaps like the turbulence of her youth—gave way to "life, beauty, and joy" (70).

Spring around Prairie Lodge

The beautiful progeny of spring began now to gather around Prairie Lodge. Animate and inanimate nature teemed with the loveliest creations. The showers that had been so emphatically foretold on our arrival did not disappoint us. They fell almost daily for several weeks, and were generally accompanied by lightning and thunder, such as the dwellers in the east have no conception of. Nothing of the kind can be more magnificent, unless it be the marshalling of the same storms on the vast plains farther west, where they are said to be even more terrific. They come more generally toward evening, and not unfrequently continue till near morning. Nothing can exceed the rapidity with which they gather after the first signal is given. A little cloud not larger than a man's hand rises on the horizon, and in fifteen minutes the earth is deluged, and the pealing heavens seem on fire. There are few showers here unaccompanied by the most striking electric phenomena: sometimes the whole arch is lighted by a continuous flickering glare, rent occasionally by a more intense vein. The thunder roll is ceaseless, with such lightning! The deep peals that accompany the brighter flashes only strike with a more appalling tone. At other times the whole vault is filled with a darkness that seems ponderable, till a mighty flash rends the pall and searches the very soul. It is gone, and the solid earth trembles under the mighty concussion. Again darkness, as if eternal night had come, wraps the scene till the flame leaps forth with a more blinding glare than before, and a crash follows that seems to shatter the foundation of the world. The third or fourth signal is followed by the storm, which breaks through the sable rack as if half the ocean had been lifted from its bed and were wandering in the upper air. In an inconceivably short space of time the plains around you are deluged, so that every succeeding flash is reflected from innumerable little pools as if you were in the midst of a shallow lake broken by islands of sedge and grass. I never appreciated the sublime power of the elements till I witnessed these storms. They are one of the most glorious features of the country.

Their effect was heightened too by contrast with the scenes which followed them. The vast expanse of country over which they ranged was in a few hours after as quiet and smiling as if the upper elements had dispensed only peace and sunshine from the first hour of creation. And beauty born of these awful warrings stole over every rolling height and into every green glade in our landscape. The swelling bud, the unfolding leaf and flower followed in the path of their majestic progress, making rich and beautiful what had before been desolate and wintry. The spirit that had all the night perhaps raved with such fearful and angry power, seemed, when the bright and peaceful morning came, to have borne a magician's wand after his wrath, and kindled life, beauty, and joy on the plains it had threatened to devastate.

[...]

Spring around Prairie Lodge

On a spring morning before sunrise, if you are in the vicinity where grove and prairie meet, the air resounds with a peculiar noise, between the whistle of the quail and the hoarse blowing of the night-hawk, but louder than either. You inquire what it is, and are told it is the prairie cocks greeting the opening day.

Spring morning on the prairies! I wish I could find language that would convey to the mind of the reader an adequate idea of the deep joy which the soul drinks in from every feature of this wonderful scene! If he could stand where I have often stood, when the rosy clouds were piled against the eastern sky, and the soft tremulous light was streaming aslant the dewy grass, while not a sound of life broke on the ear, save the wild note just mentioned, so much in harmony with the whole of visible nature, he would feel one of the charms which bind the hearts of the sons and daughters of this land.

We are within the borders of a little grove. Before us stretches a prairie; boundless on the south and east, and fringed on the north by a line of forest, the green top of which is just visible in a dark waving line between the tender hue of the growing grass and the golden sky. South and east as far as the eye can stretch, the plain is unbroken save by one "lone tree," which, from time immemorial, has been the compass of the red man and his white brother. The light creeps slowly up the sky; for twilight is long on these savannahs. The heavy dews which the cool night has deposited glisten on the leaves and spikes of grass, and the particles, occasionally mingling, are borne by their own weight to the earth. The slight blade on which they hung recovers then its erect position, or falls into its natural curve, with a quick but gentle motion, that imparts an appearance of life to that nearest you, even before the wind has laid his hand on the pulseless sea beyond. A vast ocean, teeming with life; redolent of sweet odors! It yields no sound save the one which first arrested our attention, and this is uttered without ceasing. It is not the prolonged note of one, but the steady succession of innumerable voices. It comes up near you and travels on, ringing more and more faintly on the ear, till it is returned by another line of respondents, and comes swelling in full chorus, stronger and nearer, till the last seems to be uttered directly at your feet.

But the light is gaining upon the grey dawn. Birds awaken in the wood behind us, and salute each other from the swinging branches. Insects begin their busy hum. And now, the sun has just crowded his rim above a bank of gorgeous clouds, and pours a flood of dazzling light across the grassy main. Each blade becomes a chain of gems, and, as the light increases, and the breath of morning shakes them, they bend, and flash, and change their hues, till the whole space seems sprinkled with diamonds, rubies, emeralds, amethysts, and all precious stones. Nothing can be conceived more beautiful or joyous than such a scene at this hour. The contiguous wood conveys an idea of home, such as you have borne from the forest-clad states of the east. It is a refuge from the vastness which oppresses the mind, because it can never wholly compass it. You rejoice, you exult in the friendly presence of the trees; not because they afford you a grateful retreat from the ardent sun; not because they adorn your rude dwelling; not because they promote the growth of fruit and flowers; not even because they congregate the dear little birds about your home; but because they afford the natural and familiar alternative to which the mind recurs when it is weary of the majesty which lies beyond them. You have sat under them in childhood; you have swept the fragments from the little spaces among their roots and carpeted them with moss, and festooned them with the wild flowers which nodded near. You have peopled these magic palaces with fairies, and felt a joy which words can never tell, in dreaming how happy the little beings might be where nothing is visible to their tiny eyes but exquisite beauty, and no sound falls on their small ears but the melodies of growing life. You have listened to the winds, sighing plaintively through the boughs, and felt your soul grow fit for companionship with all things whatsoever that are beautiful and lovely. And now your heart turns fondly to these tall tenants of the plain as to elder brothers, and for a moment you look coldly on the naked expanse beyond. But stop! the sun is fairly up. The flashing gems have faded from the grass tops; the grouse has ceased his matin song; the birds have hailed the opening day, and are gaily launching from the trees: the curtain which has hung against the eastern sky is swept away, and the broad light pours in resistless. The wind comes coursing gently up from the far distance, bending the young herbage, and bearing to your senses sweet sounds and odors, nursed on the unsullied breast of Nature.

The tenants of the farm-yard are now astir; the cows are milked, and all the animals whose services the farmer does not call to aid his labors, are dismissed to ramble in the boundless pasture. The generous oxen are summoned to the yoke, and the labor of the day commences. If I have lingered long over this revel of nature, a spring morning on the prairies, with the grouse be all the blame!

[...]

From chapter 9 of *Life in Prairie Land* by Eliza Woodson Farnham, 1846. Reprint (New York: Arno Press, 1972).

Born into a wealthy Boston family, Francis Parkman Jr. resisted pressure to become a lawyer, and instead pursued his interests in historical research and writing, for which he became well known. In March 1846, at the age of twenty-three, he departed on a western journey that took him to Westport (Kansas City), then along the Oregon Trail through Nebraska, into Wyoming, and back through Colorado and Kansas. The resulting book, *The Oregon Trail* (1849), remains one of the most revealing accounts of the American frontier during a time of major western migration. Parkman's narrative includes graphic descriptions of bison hunts on the Great Plains, detailed (though often disdainful and disrespectful) portraits of tribal people and cultures, and accounts of his own physical ailments during the journey, from which he never fully recovered. In this excerpt, he describes his party's experiences traveling through the tallgrass prairies near Fort Leavenworth, Kansas—a region which, he warns readers, is not the "paradise" (31) they may have imagined.

＊ ＊ ＊

Jumping Off

[. . .]

By sunrise on the twenty-third of May we had breakfasted; the tents were levelled, the animals saddled and harnessed, and all was prepared. '*Avance donc!* get up!' cried Delorier from his seat in front of the cart. Wright, our friends' muleteer, after some swearing and lashing, got his insubordinate train in motion, and then the whole party filed from the ground. Thus we bade a long adieu to bed and board, and the principles of Blackstone's Commentaries. The day was a most auspicious one; and yet

Shaw and I felt certain misgivings, which in the sequel proved but too well founded. We had just learned that though R—— had taken it upon him to adopt this course without consulting us, not a single man in the party was acquainted with it; and the absurdity of our friend's high-handed measure very soon became manifest. His plan was to strike the trail of several companies of dragoons, who last summer had made an expedition under Colonel Kearney to Fort Laramie, and by this means to reach the grand trail of the Oregon emigrants up the Platte.

We rode for an hour or two, when a familiar cluster of buildings appeared on a little hill. 'Hallo!' shouted the Kickapoo trader from over his fence, 'where are you going?' A few rather emphatic exclamations might have been heard among us, when we found that we had gone miles out of our way, and were not advanced an inch toward the Rocky Mountains. So we turned in the direction the trader indicated; and with the sun for a guide, began to trace a 'bee-line' across the prairies. We struggled through copses and lines of wood; we waded brooks and pools of water; we traversed prairies as green as an emerald, expanding before us for mile after mile; wider and more wild than the wastes Mazeppa rode over:

> 'Man nor brute,
> Nor dint of hoof, nor print of foot,
> Lay in the wild luxuriant soil;
> No sign of travel; none of toil;
> The very air was mute.'

Riding in advance, as we passed over one of these great plains, we looked back and saw the line of scattered horsemen stretching for a mile or more; and far in the rear, against the horizon, the white wagons creeping slowly along. 'Here we are at last!' shouted the Captain. And in truth we had struck upon the traces of a large body of horse. We turned joyfully and followed this new course, with tempers somewhat improved; and toward sunset encamped on a high swell of the prairie, at the foot of which a lazy stream soaked along through clumps of rank grass. It was getting dark. We turned the horses loose to feed. 'Drive down the tent-pickets hard,' said Henry Chatillon, 'it is going to blow.' We did so, and secured the tent as well as we could; for the sky had changed totally, and a fresh damp smell in the wind warned us that a stormy night was likely to succeed the hot clear day. The prairie also wore a new aspect, and its vast swells had grown black and sombre under the shadow of the clouds. The thunder soon began to growl at a distance. Picketing and hobbling the horses among the rich grass at the foot of the slope, where we encamped, we gained a shelter just as the rain began to fall; and sat at the opening of the tent, watching the proceedings of the Captain. In defiance of the rain, he was stalking among the horses, wrapped in an old Scotch plaid. An extreme solicitude tormented him, lest some of his favorites should escape, or some accident should befall them; and he cast an anxious eye toward three wolves who were sneaking along over the dreary surface of the plain, as if he dreaded some hostile demonstration on their part.

On the next morning we had gone but a mile or two, when we came to an extensive belt of woods, through the midst of which ran a stream, wide, deep, and of an appearance particularly muddy and treacherous. Delorier was in advance with his cart; he jerked his pipe from his mouth, lashed his mules, and poured forth a volley of Canadian ejaculations. In plunged the cart, but midway it stuck fast. Delorier leaped out knee-deep in water, and by dint of *sacres* and a vigorous application of the whip, he urged the mules out of the slough. Then approached the long team and heavy wagon of our friends; but it paused on the brink.

'Now my advice is —,' began the Captain, who had been anxiously contemplating the muddy gulf.

'Drive on!' cried R——.

But Wright, the muleteer, apparently had not as yet decided the point in his own mind; and he sat still in his seat on one of the shaft-mules, whistling in a low contemplative strain to himself.

'My advice is,' resumed the Captain, 'that we unload; for I'll bet any man five pounds that if we try to go through, we shall stick fast.'

'By the powers, we shall stick fast!' echoed Jack, the Captain's brother, shaking his large head with an air of firm conviction.

'Drive on! drive on!' cried R—— petulantly.

'Well,' observed the Captain, turning to us as we sat looking on, much edified by this by-play among our confederates, 'I can only give my advice, and if people won't be reasonable, why they won't, that's all!'

Meanwhile, Wright had apparently made up his mind; for he suddenly began to shout forth a volley of oaths and curses, that, compared with the French imprecations of Delorier, sounded like the roaring of heavy cannon after the popping and sputtering of a bunch of Chinese crackers. At the same time, he discharged a shower of blows upon his mules, who hastily dived into the mud, and drew the wagon lumbering after them. For a moment the issue was dubious. Wright writhed about in his saddle, and swore and lashed like a madman; but who can count on a team of half-broken mules? At the most critical point, when all should have been harmony and combined effort, the perverse brutes fell into lamentable disorder, and huddled together in confusion on the farther bank. There was the wagon up to the hub in mud, and visibly settling every instant. There was nothing for it but to unload; then to dig away the mud from before the wheels with a spade, and lay a causeway of bushes and branches. This agreeable labor accomplished, the wagon at length emerged; but if I mention that some interruption of this sort occurred at least four or five times a day for a fortnight, the reader will understand that our progress towards the Platte was not without its obstacles.

We travelled six or seven miles farther, and 'nooned' near a brook. On the point of resuming our journey, when the horses were all driven down to water, my homesick charger Pontiac made a sudden leap across, and set off at a round trot for the settlements. I mounted my remaining horse, and started in pursuit. Making a circuit, I headed the runaway, hoping to drive him back to camp; but he instantly broke into a gallop, made a wide tour on the prairie, and got past me again. I tried this plan repeatedly, with the same result: Pontiac was evidently disgusted with the prairie; so I abandoned it, and tried another, trotting along gently behind him, in hopes that I might quietly get near enough to seize the trail-rope which was fastened to his neck, and dragged about a dozen feet behind him. The chase grew interesting. For mile after mile I followed the rascal, with the utmost care not to alarm him, and gradually got nearer, until at length old Hendrick's nose was fairly brushed by the whisking tail of the unsuspecting Pontiac. Without drawing rein, I slid softly to the ground; but my long heavy rifle encumbered me, and the low sound it made in striking the horn of the saddle startled him; he pricked up his ears, and sprang off at a run. 'My friend,' thought I, remounting, 'do that again, and I will shoot you!'

Fort Leavenworth was about forty miles distant, and thither I determined to follow him. I made up my mind to spend a solitary and supperless night, and then set out again in the morning. One hope, however, remained. The creek where the wagon had stuck was just before us; Pontiac might be thirsty with his run, and stop there to drink. I kept as near to him as possible, taking every precaution not to alarm him again; and the result proved as I had hoped; for he walked deliberately among the trees, and stooped down to the water. I alighted, dragged old Hendrick through the mud, and with a feeling of infinite satisfaction picked up the slimy trail-rope, and twisted it three times round my hand. 'Now let me see you get away again!' I thought, as I remounted. But Pontiac was exceedingly reluctant to turn back; Hendrick too, who had evidently flattered himself with vain hopes, showed the utmost repugnance, and grumbled in a manner peculiar to himself at being compelled to face about. A smart cut of the whip restored his cheerfulness; and dragging the recovered truant behind, I set out in search of the camp. An hour or two elapsed, when, near sunset, I saw the tents, standing on a rich swell of the prairie, beyond a line of woods, while the bands of horses were feeding in a low meadow close at hand. There sat Jack C—, cross-legged, in the sun, splicing a trail-rope, and the rest were lying on the grass, smoking and telling stories. That night we enjoyed a serenade from the wolves, more lively than any with which they had yet favored us; and in the morning one of the musicians appeared, not many rods from the tents, quietly seated among the horses, looking at us with a pair of large gray eyes; but perceiving a rifle levelled at him, he leaped up and made off in hot haste.

I pass by the following day or two of our journey, for nothing occurred worthy of record. Should any one of my readers ever be impelled to visit the prairies, and should he choose the route of the Platte, (the best, perhaps, that can be adopted,) I can assure him that he need not think to enter at once upon the paradise of his imagination. A dreary preliminary, protracted crossing of the threshold, awaits him before he finds himself fairly upon the verge of the 'great American desert;' those barren wastes, the haunts of the buffalo and the Indian, where the very shadow of civilization lies a hundred leagues behind him. The intervening country, the wide and fertile belt that extends for several hundred miles beyond the extreme frontier, will probably answer tolerably well to his preconceived ideas of the prairie; for this it is from which picturesque tourists, painters, poets and novelists, who have seldom penetrated farther, have derived their conceptions of the whole region. If he has a painter's eye, he may find his period of probation not wholly void of interest. The scenery, though tame, is graceful and pleasing. Here are level plains, too wide for the eye to measure; green undulations, like motionless swells of the ocean; abundance of streams, followed through all their windings by lines of woods and scattered groves. But let him be as enthusiastic as he may, he will find enough to damp his ardor. His wagons will stick in the mud; his horses will break loose; harness will give way, and axle-trees prove unsound. His bed will be a soft one, consisting often of black mud, of the richest consistency. As for food, he must content himself with biscuit and salt provisions; for strange as it may seem, this tract of country produces very little game. As he advances, indeed, he will see, mouldering in the grass by his path, the vast antlers of the elk, and farther on, the whitened skulls of the buffalo, once swarming over this now deserted region. Perhaps, like us, he may journey for a fortnight, and see not so much as the hoof-print of a deer; in the spring, not even a prairie-hen is to be had.

Yet, to compensate him for this unlooked-for deficiency of game he will find himself beset with 'varmints' innumerable. The wolves will entertain him with a concerto at night, and skulk around him by day, just beyond rifle-shot; his horse will step into badger-holes; from every marsh and mud-puddle will arise the bellowing, croaking and trilling of legions of frogs, infinitely various in color, shape and dimensions. A profusion of snakes will glide away from under his horse's feet, or quietly visit him in his tent at night; while the pertinacious humming of unnumbered mosquitoes will banish sleep from his eyelids. When thirsty with a long ride in the scorching sun over some boundless reach of prairie, he comes at length to a pool of water, and alights to drink, he discovers a troop of young tadpoles sporting in the bottom of his cup. Add to this, that all the morning the sun beats upon him with a sultry, penetrating heat, and that, with provoking regularity, at about four o'clock in the afternoon, a thunder-storm rises and drenches him to the skin. Such being the charms of this favored region, the reader will easily conceive the extent of our gratification at learning that for a week we had been journeying on the wrong track! . . .

From chapter 4 of *The Oregon Trail* by Francis Parkman Jr., 1849. Reprint (New York: Oxford University Press, 1999).

Outside of what he says in his own writing, little can be found about the biography of this often quoted author, or whether in fact he ever existed—Van Tramp may be a pseudonym for one or several people. Perhaps a reader of this anthology will help shed light on the mystery. What is known, however, is that in 1858, "John C. Van Tramp" compiled a massive collection of history, lore, and firsthand accounts of the western regions of the country, *Prairie and Rocky Mountain Adventures*, and given the number of editions through 1870, it must have been popular. In this excerpt from the 1870 edition, he/they share a story told by a hunter in Illinois about his dog, Phil, and a child who got lost in the tallgrass—a common fear expressed by many living near the wild prairies, including Laura Ingalls Wilder.

✳ ✳ ✳

Phil and the Lost Boy

[. . .]

A western hunter gives the following graphic description of his favorite dog, and the adventures of a day:

Phil was a *setter* dog, whose peculiar gift or talent is to trace out birds and stand and mark them, until the hunter comes up to shoot them as they rise, and then to retrieve them for him. But Phil was no common one, I assure you. Of course he could scent a bird at any reasonable distance, and follow its track through the tall prairie grass with unerring certainty; could distinguish at once the track of a prairie chicken or a plover from a hawk or a bittern, and was never known to follow or set the latter, or retrieve them when shot, unless bidden. He was a handsome dog, too, with fine hair, white and brown in spots; with long fringes upon his legs and tail; a hazel eye, long face, and a head that would do credit to a canine statesman or philosopher. His soft silky ears, hanging smoothly down, giving full prominence to the bump (much prized by hunters, though unknown to Combe) of prairie-chickenetiveness. Many a time have I taken down my fowling-piece, slung on my game-bag, while he was upsetting stool, chair and stand in the exuberance of his joy at seeing the well-known preparations for a hunt. Then he would start and run and bark at anything, or nothing, roll over in the grass and then spring to his feet again, to entice me to the prairie or the field. But all these rude demonstrations of joy were stilled in a moment when we had reached the ground where the game might be expected. He then commenced his serious business. No voice is now heard from him, he takes no notice of me except to mark the direction which I take, but with a steady run he courses zig-zag across the field, his tail in continual motion with a rolling swing. Now he stops suddenly, pauses a moment as if to assure himself that he is not mistaken, and then goes on less rapidly. He has scented game; he no longer swings his tail, no longer pursues a devious course, but with a steady, quiet motion, step by step he follows up the scent cautiously, slower and slower; and now he stops. Look at him! It were worth a painter's while to picture him, though few could do him justice. He stands mute and motionless as a statue, his right leg raised and folded at the knee, his tail rigid and straight as an iron bar, his body drawn forward; no motion—you scarce perceive that he breathes. But it is clear that it is not the posture of repose. His earnest look, his keen eye gazing intensely forward at the spot where the bird has cowered, and every muscle held firmly to its trust. He no longer looks for his master or heeds his presence, or even hears his voice; every thought, every faculty, every nerve feels but one impulse, and obeys one power. Phil has made a "point."

But it was not for his skill in hunting that I most valued him. He had that talent in common with his race; but he had others not often found in a setter. They all know how to track and set birds—it is part of their natures—but they rarely know aught more. They can hunt birds, but that is the extent of their capacity. One is often surprised, astonished even, at their sagacity in this matter, while they betray such extreme dulness in every other. They are dogs of one idea; every other faculty seems to be dwarfed to make a prodigy of this. Their whole power, their whole intelligence, seems concentrated in this one point, and no wonder that it is brilliant; but they have no general knowledge, or even the mind to acquire it. But this was not the case with Phil. He was behind none of them in this particular branch, while he was before them in every other. He had general intelligence. He was not a professor merely, Phil was a philosopher. He had ideas not pertaining to his own department of bird hunting.

You could tell him of other things, and he knew when he understood you; and he would let you know it, not only by doing what you wished, but by his looks, eyes, everything. But I will tell you one of his doings, and you can judge if he does not deserve my praise. But I find I must do this, if ever, at another time; for it will lead me so far into the prairie, where so many things must be explained to enable you to understand me, that this already long letter would be extended beyond all reasonable limits.

But now, before you can fully understand the story of Phil, you must have some good idea of a prairie. But how to give you this, I know not. There is no describing them. They are like the *ocean*, in more than one particular; but in none more than in this: the utter impossibility of producing any just impression of them by description. They inspire feelings so unique, so distinct from anything else, so powerful, yet vague and indefinite, as to defy description, while they invite the attempt. Nothing but the ocean compares with the prairie, in its impression on the mind; and like the ocean, it is impossible to tell in what its distinctive character consists; unless it be their vastness, the want of anything on which the eye can rest, and say that there the prairie or the ocean ends. I think it must be this; for every other feature about them I have seen change, and leave them the prairie still. I have seen them, in the mid-winter, covered with snow; a white waste, cold and bleak, *so* white that the sky looked strangely blue, almost black, above them, shutting down on them far, far inside their viewless limits. Then, again, I have seen them covered with green verdure, blooming rich with flowers (not in stinted patches like those sweet spots we know in childhood, where some opening in the forest shade lets the warm sunlight in), but by acres, some in curved belts, circling the round knolls; others stretching for miles along the devious wanderings of some water-course; here, with red flaunting flowers crowning the hill-top; there, a few yards of blue-bells marking some latent spring; and here, a small still lake covered with the white lotus floating on its water so close as to leave scant room for the Ibis, with plumage white as their flowerets, to stand among them.

And again I have seen the prairies, when the first winter's frost fell upon them, their green verdure changed to a light yellow, almost white; the tall dry grass lying flat and motionless, waiting the careless hand of some hunter, or the lightning's flash, to give them to the flames. The wild deer, no longer sheltered by the grass, standing out boldly on the hill-top, their light forms of beauty backed by the blue sky, watching, for hours, the verdureless prairie, waiting until the evening's shade invites them far away to the burr-oaks to feed upon the acorns. The wolf cowering beside the small mound, raised by the gopher for a home, or by the surveyor's landmark; or, conscious of discovery, skulking away to some reedy marsh, gazing back at times with a sneaking look of mingled cowardice and cruelty. The cranes stalking on the prairie, or, in wide circles, cleaving the still air, higher and higher, until their large forms seem dwindled to a speck scarce larger than the golden plover that hurries by so near on its swift wing.

And then again I have seen them on fire, when the bright sunlight dimmed the flames, while their smoke rolled up and on over hill and hollow till the whole sky was darkened. And then I have watched until night came on and the whole scene was changed. The pillar of cloud had become the pillar of fire. There was fire in every form, from the small torch-light made by the tuft of slough-grass, to acres flaming from the long blue-joint on the river bottom. Flames everywhere, now moving slowly on where the sweet grass had enticed the wild flock of deer to crop the herbage close, while the soft night wind just gave it life enough to lick up one by one the few scant leaves still left—now stopped by an old Indian trail, until some loose leaf or bending stem of grass led it across the track to pursue its slow and silent course, now rushing before the wild west wind with a speed that outstrips the wolf and almost overtakes the deer; with loud-hummed roar climbing the hill-side and down the valley unchecked by the dividing stream, and passing all barriers in its fiery course. Here and there staying its speed among the short silk grass that belts some large cane-marsh, while on each side, like the wings of an army marching with quick step while the centre halts, it rushes on, surrounding the whole flat, and then crossing the narrow barrier, on every side seizing the dry reeds and cane, and gathering strength as they draw closer the red circle of their forces; going up at last in one triumphant flash of flames, dying themselves on the last conquered spot, and leaving the eye free amid the sudden gloom, to gaze once more on the far distant fires, miles away, skirting the farthest verge of the horizon like day's first burst of light.

And then, once more, have I seen them after the fire had swept then leaving them verdureless and black—so black as to weary and pain the eye almost as did their white dress in winter.

But all these changes, and more which I have seen, are but so many different phases of the same scene, no one of them, or all of them describe it; it would be the prairie without them. Their vastness, their solitude, the soberness which they inspire—and in this again they resemble the ocean, for who ever saw one new to the scene laugh on the sea-shore? A thousand minor features make up the picture which would tire in description, and yet without them all description fails to be correct. I will name but two of them, the surface of the ground and its covering. The first is best described by the term rolling hillocks or ridges, varying from two to ten yards in hight, irregular, with round basins or long troughs between them, presenting a sky line closely resembling the ocean when a strong wind has suddenly changed its course, breaking the continuity of the swells. And the surface so described is covered over, everywhere without a spot of naked earth, with grass, and much of it of great growth; grass covering acre after acre, mile after mile, with one unvaried interminable green. This grass is from two feet to two yards in hight, varying with the soil and species. This refers to the wild prairie away from the cultivated farms. You will perceive at once the difficulty of keeping a straight course across the prairies. I have been "lost" more times in one year on the prairies, than in twenty spent in the woods of Western New York when in their wildest state. Not two weeks since I spent an hour on the prairies within half a mile of home waiting for the stars to come out to guide me. Some time ago a German woman came to my place long before the sun was up, asking help, or rather the help of Phil. Her little boy had wandered and was lost. You know what being lost *in the woods* means, but for a child, that is nothing, is safety itself, when compared with being lost on the prairies. Two within my own knowledge, within as many years, have wandered; one fell a prey to the wolves, and one was never heard of more. You will not wonder at this when you reflect on the description I have given and shall give you. A child of five years old can see over the grass only occasionally, and then with no extensive view. There are no trees to guide, no fences to restrain their steps, but foot-paths enough to mislead them, trails made by Indian or buffalo, loading from one distant ford or woodland to another. And then the sparse settlement makes every course but the right one fatal. These present so many dangers as to render the night and wolf superfluous perils. I strove in vain to explain to the woman that my dog was not a blood-hound but a bird-dog, that he would follow no human footsteps but my own, that I feared he could not be made to follow her boy's track. But she could not or would not believe but that Phil would follow and do anything I told him, and I almost repented having said anything to check for a moment the illusion of hope in the wretched mother's breast. You know that it was not said to save myself the trouble of going with her; I should of course have gone with her at any rate. But she had heard a great deal of my dog, and had seen him track, she told me, the little snipe and plover, whose whole foot was not so large as one of Hanka's toes; and with true womanly tact she reminded me how months before she had gone to show me where a wild turkey had crossed the prairie, and how she had seen Phil take up the sporr and follow it, recounting with earnest interest all the difficulties he had overcome; how the bird flew over the narrow brook, leaving him no track to follow, how he ran up and down the stream to search for it, and then swam over and scoured the prairie on the other side until he found the track once more. I listened with an aching heart, for I knew the difficulties far better than she could or would. I was soon ready to follow her, and on the way she told me that her little boy had been playing before the door while she went to carry their dinner to the men folks on the prairie. That when she came back he was gone; that she ran over the prairie to seek him, and called him until the men heard her and came to her help; that before nightfall their few nighbors, men and women, joined them in the search; how the dark night came but no child; how she and her husband had wandered through its gloom, calling the boy, and making noises to scare the wild beasts from the place, and how she had left before the first light of morning to come for me. She told me all this while hurrying along at a speed which tested even a hunter's stride, fresh as I was from the night's rest. We reached her house as the first light of the morning began to spread over the premises. It was a small board building of such size as the boards' length would make, on the very out edge of the cultivated country. The sides of the house were banked up, except the doorway, with coarse prairie turf a foot in thickness to the bottom of the small window, on the south a narrow footpath led from the door down a sloping bank to a shallow well, dug near the slough at the bottom. A wagon, plow, and a few more farming tools lay scattered round, and in the house a scanty supply of household goods. At the door lay a small pair of wooden shoes which Hanka had thrown off while at play. A small but unfenced spot was cultivated near the house, while north and east might be seen other cottages like it, scattered here and there at intervals, and on the south and west the limitless prairie, without a tree or shrub, far as the eye could see. But why draw a picture that will not distinguish this cottage or spot from a hundred others on the broad prairie? And now began my almost hopeless task of teaching a setter in one lesson the trade of a blood-hound.

My plans were soon laid. I threw aside my hunting coat, set up my gun, and taking some of the boy's clothing, tried to make Phil understand what I wished him to do. He would smell of them because I told him, but without interest or intelligence, and would then turn and look at the gun as if expecting me to take it up again. I left it, however, and called him out of the house. I was glad to see him smell of the small wooden shoes lying by the door, though this he did of course.

The boy had now been gone some eighteen hours, and no scent of his footsteps could be hoped for near the house, even if Phil could be made to know that I wanted him to follow them. They had searched the day before the grounds around the house, and the foot-paths leading to the neighbors. I determined, therefore, at once to strike off into the prairie. Phil followed me, looking wistfully back at times, at the house where I had left my gun. We had left the house a mile or more, when calling Phil, I tried once more to make him understand my object. He would smell of the little sock which I had brought with me, look wistfully in my face, as if to search out my meaning. He would then start off in one direction, looking back to see if I approved of that. I would call him back and make him again smell the child's sock, but it seemed useless; he would be off again another way, looking back to see if *that* was right, and being called back again, looked perplexed and discouraged, and walked slowly by my side. The neighbors meanwhile scattered far and near in the almost hopeless search—hopeless, for the boy might have wandered many miles, and we knew that we might pass within a dozen yards of him, in the tall prairie grass, without knowing he was there. But the poor mother clung to me and Phil, with a sinking heart, however, for she could not but observe that he was not searching for her lost treasure. And thus we wandered on hour after weary hour. Time after time I endeavored to make Phil understand me, but in vain. Once he ran to me, looking bright and glad, and when I showed him the boy's stocking he eagerly took it in his mouth and walked proudly, with head erect, as if to say, "Now I understand you want me to carry it." In spite of self-control, my face must have betrayed my disappointment, for he dropped his head and tail, and slowly brought me back the sock, which I took, but at the same time caressed him and walked slowly on. At length he stops again, snuffs the ground, looks pleased, hurries this way and that to catch a warmer scent, looks up with bright eyes at me, then runs slowly, as nosing the ground. We follow him, and on my part for the first time with hope, it *might* be he had at last caught my meaning. But then again he might be following the track of game, and this was the most natural supposition. But no, he is scenting up a tall weed, too high for a bird to touch; it cannot be deer, for their sharp hoofs would have left a print on the sod which would not escape my eye; nor wolf, for Phil has not the angry look, the glaring eye, and lips drawn up to show his white tusks ready for his foe, features which the wolf's scent always gives him. But on he goes, scenting every tuft of grass, or now unheeded prairie flower, pausing at some, and snuffing a long slow breath, with eyes half-closed lest light should interfere with the one sense on which he relies. The mother is close by me, asking every moment "Is he tracking Hanka? will he find Hanka?" I dare not say yes, for I am not certain, but I have never seen him move so after any kind of game, and I know his varied movements when pursuing each. But the track is not warm, whatever made it, for he stops, now turns round and stops again, then takes a wider circle and comes round to the same point again; "he is at fault." He makes another effort on a wider circle still, and is yet at fault. He now gives one sharp cry of angry vexation, and then turns suddenly and retraces his own footsteps, following at a fast run his back track, several hundred yards. Stops, scents the ground, catches the trail and follows over the track once more, cautious and slowly, to within a few rods of first fault—and then turns off with cheerful steps. He has recovered the trail and runs briskly on, but soon checks himself and turns half round, as if on second thought he would examine a weed he had just passed. I examined it too, and there, on the dry rough stem of the resin-weed, hung a few shreds of blue cotton. The mother saw me looking at them, and then ran forward and seized the precious relic, "It was Hanka's, I knew it was Hanka's!" I thought so too, for the color is such as no Yankee has yet imitated with success. But Phil has breathed on it, and she has handled it, and I cannot judge how long it has hung there. But she is calling her friends to come in. In the meanwhile Phil has got the start of us and we hurry on to overtake him, but cautiously avoid the track he follows, lest he might be at fault again and have to retrace his steps.

How intently the mother watched Phil's movements, but happily without the fear which troubled me, who could understand his difficulties far better than she could. But he is going steadily on now, not fast, and I have much trouble to keep the impatient mother from outstripping him, and soiling the trail. The crowd gather, one by one, after us from the prairie. Keeping them at a distance as well as might be, we follow close by Phil, watching his every movement. He's working gloriously, but on a faint trail. He understands the matter now, and has all our excitement. With his mouth open, lest the too strong draught of air through his nose should blunt the delicacy of its nerves, he tracks for hours the wanderings of that child. And now the last doubt as to the character of the track is removed, for just before us, in an old Buffalo trail, is a child's track. I hastily put my foot over it to hide it from the mother's sight, for fear her eagerness might interface with Phil, our only hope and guide. But the effort was vain, for she noticed the movement, and, darting forward, saw another track. I stopped her before she could reach it, and while she is crying, almost screaming, "'Tis Hanka's sporr, 'tis Hanka's sporr; mein kint, mein kint!" I examined with a hunter's eye and care the track. It is a child's foot-print, beautifully moulded in the soft dust of the Buffalo trail. It was made long after the sun was up, and the dew gone, as the dust was dry when the foot pressed it, for, although smooth, it has not the coherence of dust, pressed and dried afterwards. The slightest breath disturbs it, and the slow-worm, which has made the only trail across it, has scarcely crawled ten yards beyond along the Buffalo path, which it is painfully pursuing with dull, tortuous movement. Yet it was clear that for some hours the sun had shone upon that foot-mark, and it might be miles must be passed before we could overtake the foot that made it, unless stayed by sleep or exhaustion. The task was not easy, for the boy had taken the Buffalo path. I cautioned the crowd to keep back at least a stone's throw, and hurried on to overtake Phil, busy in that most difficult and delicate operation, following a track over dry dust. But he was working well. Cheerful and confident, swinging his fringed tail around with its widest sweep, dodging his head from side to side of the narrow path to catch the scent left on the green herbage at its edge, where the boy's clothes or hands had chanced to touch. Phil and his master were both excited, and the scene was enough to excite any one. There, on the wide prairie, in the bright sunshine, the deep blue sky above and the green earth beneath, bending alike to meet at the horizon. The ancient path we were treading, made long years ago by the large buffalo and the pursuing Indian, both banished now to the Far West, withering before the pale face of their common enemy. The trail now leading over the low hill-tops; now down their gentle slopes to the low grounds, skirting the marsh, then rising up again. And then the game we were pursuing—not to kill but to save—richer than the finest fur or proudest antler that dwell on the green deserts; for it was the dearest treasure of two human hearts, the richest gem of a prairie home. But Phil has stopped by a large gopher-mound, near the hilltop, where the grass is shortest, the mother and myself beside him. The boy had been on the hillock, doubtless, to look out for home. Vain hope! No sign of human habitation or human handiwork can be seen from here. He had turned round and round upon it, but could catch no sight of any particular object. Campbell's last man was scarcely more hopelessly alone. He had sat down to rest him, perhaps to weep; for I could see the print of his heels half way down the small earthen hillock. But he had left; and Phil, having snuffed for many rods along the trail, in vain, now came running back, and taking a narrow circle round the hillock and recovering the track, starts off in a new direction. Fortunately, now the track leads through the green grass, and Phil follows swiftly, so quickly as to render needless my caution to the crowd; for we have left them far behind, and none but the mother and myself keep up with Phil. He leads us down the hill to a small brook, where the boy had gone to drink. We could see where his small feet had struggled in the marsh, and where he had knelt down, both hands were printed in the soft soil. From here, the trail turned back again towards the high ground and the distanced crowd. But now Phil stops a moment, and his whole manner changes. He no more noses the ground, following the various windings of the track; but, with head erect and neck stretched out, marches straight forward, with steady gait and gaze. He no longer heeds the track, for he can scent the boy where he lies hid. I noted the change at once, and knew its meaning, but dared not tell the mother. She observed it soon, and cried out that Phil had left off hunting; but in an instant, recollecting to have seen him retrieve, cried out "He has found him!" "he has found mein kint!" and, rushing past us, in an instant more, I heard the boy's scream of fright, and her wild cry of joy.

We were soon with her, and Phil seemed almost disposed to dispute her right to the child, but joined most heartily in her exultation, leaping upon me, running to the boy, as he lay in his mother's arms, rubbing his nose on his face and hands, then racing away again to greet, with boisterous mirth, each new-comer to the group.

We were now on our way home, laughing and shouting, a joyous troop. I led the way; Phil followed me close, except at times, when he went back to look after the boy, carried in the strong men's arms, by turns, with his mother watching beside him.

I left them at the end of three miles, and struck across the prairie for my home, some five miles distant, and reached it at nightfall, tired with my day's adventures.

From *Prairie and Rocky Mountain Adventures; or, Life in the West* by John C. Van Tramp (Columbus, OH: Segner and Condit, 1870).

Born on Long Island, New York, Walt Whitman worked as a printer, teacher, newspaper editor, Union army nurse, government clerk, freelance journalist, and poet. The publication of *Leaves of Grass* in 1855 (and its subsequent editions) eventually established him as one of America's preeminent authors. In *Specimen Days* (1882), a collection of autobiographical pieces, he recounts his journey west in 1879–80, which took him through Ohio, Illinois, Missouri, Kansas, and all the way to Denver—regions that had become considerably more accessible to tourists due to the advance of the railroad. In this excerpt from the 1887 edition he reflects on the Great Plains and prairies as "America's Characteristic Landscape" (232) and as a distinct, ongoing source of that nation's ideals and creative expression.

✳ ✳ ✳

America's Characteristic Landscape

At a large popular meeting at Topeka—the Kansas State Silver Wedding, fifteen or twenty thousand people—I had been erroneously bill'd to deliver a poem. As I seem'd to be made much of, and wanted to be good-natured, I hastily pencill'd out the following little speech. Unfortunately, (or fortunately,) I had such a good time and rest, and talk and dinner, with the U. boys, that I let the hours slip away and didn't drive over to the meeting and speak my piece. But here it is just the same:

> "My friends, your bills announce me as giving a poem; but I have no poem—have composed none for this occasion. And I can honestly say I am now glad of it. Under these skies resplendent in September beauty—amid the peculiar landscape you are used to, but which is

new to me—these interminable and stately prairies—in the freedom and vigor and sane enthusiasm of this perfect western air and autumn sunshine—it seems to me a poem would be almost an impertinence. But if you care to have a word from me, I should speak it about these very prairies; they impress me most, of all the objective shows I see or have seen on this, my first real visit to the West. As I have roll'd rapidly hither for more than a thousand miles, through fair Ohio, through bread-raising Indiana and Illinois—through ample Missouri, that contains and raises everything; as I have partially explor'd your charming city during the last two days, and, standing on Oread hill, by the university, have launch'd my view across broad expanses of living green, in every direction—I have again been most impress'd, I say, and shall remain for the rest of my life most impress'd, with that feature of the topography of your western central world—that vast Something, stretching out on its own unbounded scale, unconfined, which there is in these prairies, combining the real and ideal, and beautiful as dreams.

"I wonder indeed if the people of this continental inland West know how much of first-class *art* they have in these prairies—how original and all your own—how much of the influences of a character for your future humanity, broad, patriotic, heroic and new? how entirely they tally on land the grandeur and superb monotony of the skies of heaven, and the ocean with its waters? how freeing, soothing, nourishing they are to the soul?

"Then is it not subtly they who have given us our leading modern Americans, Lincoln and Grant?—vast-spread, average men—their foregrounds of character altogether practical and real, yet (to those who have eyes to see) with finest backgrounds of the ideal, towering high as any. And do we not see, in them, foreshadowings of the future races that shall fill these prairies?

"Not but what the Yankee and Atlantic States, and every other part—Texas, and the States flanking the south-east and the Gulf of Mexico—the Pacific shore empire—the Territories and Lakes, and the Canada line (the day is not yet, but it will come, including Canada entire)—are equally and integrally and indissolubly this Nation, the *sine qua non* of the human, political and commercial New World. But this favor'd central area of (in round numbers) two thousand miles square seems fated to be the home both of what I would call America's distinctive ideas and distinctive realities."

[. . .]

America's Characteristic Landscape

Speaking generally as to the capacity and sure future destiny of that plain and prairie area (larger than any European kingdom) it is the inexhaustible land of wheat, maize, wool, flax, coal, iron, beef and pork, butter and cheese, apples and grapes—land of ten million virgin farms—to the eye at present wild and unproductive—yet experts say that upon it when irrigated may easily be grown enough wheat to feed the world. Then as to scenery (giving my own thought and feeling,) while I know the standard claim is that Yosemite, Niagara falls, the upper Yellowstone and the like, afford the greatest natural shows, I am not so sure but the Prairies and Plains, while less stunning at first sight, last longer, fill the esthetic sense fuller, precede all the rest, and make North America's characteristic landscape.

Indeed through the whole of this journey, with all its shows and varieties, what most impress'd me, and will longest remain with me, are these same prairies. Day after day, and night after night, to my eyes, to all my senses —the esthetic one most of all—they silently and broadly unfolded. Even their simplest statistics are sublime.

Earth's Most Important Stream

The valley of the Mississippi river and its tributaries, (this stream and its adjuncts involve a big part of the question,) comprehends more than twelve hundred thousand square miles, the greater part prairies. It is by far the most important stream on the globe, and would seem to have been marked out by design, slow-flowing from north to south, through a dozen climates, all fitted for man's healthy occupancy, its outlet unfrozen all the year, and its line forming a safe, cheap continental avenue for commerce and passage from the north temperate to the torrid zone. Not even the mighty Amazon (though larger in volume) on its line of east and west—not the Nile in Africa, nor the Danube in Europe, nor the three great rivers of China, compare with it. Only the Mediterranean sea has play'd some such part in history, and all through the past, as the Mississippi is destined to play in the future. By its demesnes, water'd and welded by its branches, the Missouri, the Ohio, the Arkansas, the Red, the Yazoo, the St. Francis and others, it already compacts twenty-five millions of people, not merely the most peaceful and money-making, but the most restless and warlike on earth. Its valley, or reach, is rapidly concentrating the political power of the American Union. One almost thinks it *is* the Union—or soon will be. Take it out, with its radiations, and what would be left? From the car windows through Indiana, Illinois, Missouri, or stopping some days along the Topeka and Santa Fe road, in southern Kansas, and indeed wherever I went, hundreds and thousands of miles through this region, my eyes feasted on primitive and rich meadows, some of them partially inhabited, but far, immensely far more untouch'd, unbroken—and much of it more lovely and fertile in its unplough'd innocence than the fair and valuable fields of New York's, Pennsylvania's, Maryland's or Virginia's richest farms.

Prairie Analogies—The Tree Question

The word Prairie is French, and means literally meadow. The cosmical analogies of our North American plains are the Steppes of Asia, the Pampas and Llanos of South America, and perhaps the Saharas of Africa. Some think the plains have been originally lake-beds; others attribute the absence of forests to the fires that almost annually sweep over them—(the cause, in vulgar estimation, of Indian summer.) The tree question will soon become a grave one. Although the Atlantic slope, the Rocky mountain region, and the southern portion of the Mississippi valley, are well wooded, there are here stretches of hundreds and thousands of miles where either not a tree grows, or often useless destruction has prevail'd; and the matter of the cultivation and spread of forests may well be press'd upon thinkers who look to the coming generations of the prairie States.

Mississippi Valley Literature

Lying by one rainy day in Missouri to rest after quite a long exploration —first trying a big volume I found there of "Milton, Young, Gray, Beattie and Collins," but giving it up for a bad job—enjoying however for awhile, as often before, the reading of Walter Scott's poems, "Lay of the Last Minstrel," "Marmion," and so on—I stopp'd and laid down the book, and ponder'd the thought of a poetry that should in due time express and supply the teeming region I was in the midst of, and have briefly touch'd upon. One's mind needs but a moment's deliberation anywhere in the United States to see clearly enough that all the prevalent book and library poets, either as imported from Great Britain, or follow'd and *doppel-gang'd* here, are foreign to our States, copiously as they are read by us all. But to fully understand not only how absolutely in opposition to our times and lands, and how little and cramp'd, and what anachronisms and absurdities many of their pages are, for American purposes, one must dwell or travel awhile in Missouri, Kansas and Colorado, and get rapport with their people and country.

Will the day ever come—no matter how long deferr'd—when those models and lay-figures from the British islands—and even the precious traditions of the classics—will be reminiscences, studies only? The pure breath, primitiveness, boundless prodigality and amplitude, strange mixture of delicacy and power, of continence, of real and ideal, and of all original and first-class elements, of these prairies, the Rocky mountains, and of the Mississippi and Missouri rivers—will they ever appear in, and in some sort form a standard for our poetry and art? (I sometimes think that even the ambition of my friend Joaquin Miller to put them in, and illustrate them, places him ahead of the whole crowd.)

Not long ago I was down New York bay, on a steamer, watching the sunset over the dark green heights of Navesink, and viewing all that inimitable spread of shore, shipping and sea, around Sandy hook. But an intervening week or two, and my eyes catch the shadowy outlines of the Spanish peaks. In the more than two thousand miles between, though of infinite and paradoxical variety, a curious and absolute fusion is doubtless steadily annealing, compacting, identifying all. But subtler and wider and more solid, (to produce such compaction,) than the laws of the States, or the common ground of Congress or the Supreme Court, or the grim welding of our national wars, or the steel ties of railroads, or all the kneading and fusing processes of our material and business history, past or present, would in my opinion be a great throbbing, vital, imaginative work, or series of works, or literature, in constructing which the Plains, the Prairies, and the Mississippi river, with the demesnes of its varied and ample valley, should be the concrete background, and America's humanity, passions, struggles, hopes, there and now — an *eclaircissement* as it is and is to be, on the stage of the New World, of all Time's hitherto drama of war, romance and evolution — should furnish the lambent fire, the ideal.

[...]

Upon Our Own Land

"Always, after supper, take a walk half a mile long," says an old proverb, dryly adding, "and if convenient let it be upon your own land." I wonder does any other nation but ours afford opportunity for such a jaunt as this? Indeed has any previous period afforded it? No one, I discover, begins to know the real geographic, democratic, indissoluble American Union in the present, or suspect it in the future, until he explores these Central States, and dwells awhile observantly on their prairies, or amid their busy towns, and the mighty father of waters. A ride of two or three thousand miles, "on one's own land," with hardly a disconnection, could certainly be had in no other place than the United States, and at no period before this.

From *Specimen Days in America* by Walt Whitman (London: Walter Scott, 1887).

Born into a wealthy Michigan family, Elizabeth Bacon met and eventually married Union General George Armstrong Custer in 1864. During his subsequent posts in the Great Plains, she insisted on joining him, often "tenting" for extended periods of time. In 1876, he and hundreds of his soldiers were killed after attacking a large group of Lakota, Northern Cheyenne, and Arapaho warriors near the Little Bighorn River in eastern Montana—a devastating military defeat for which he was criticized by President Ulysses S. Grant and many others. Elizabeth took it upon herself to redeem her husband's public reputation by writing three popular memoirs about their life and adventures together. Despite the obvious (and very debatable) idolization of the general in these books, they offer realistic portrayals of the Great Plains and prairie environments. In this excerpt from *Boots and Saddles* (1885), she describes a deadly blizzard near Yankton, South Dakota, where in April 1873 the Seventh Calvary was posted. Elizabeth and her African American friend and cook, Mary, heroically scrambled to save the lives of those caught in the storm, while George labored to recover from an illness."

✳ ✳ ✳

A Blizzard

After so many days in the car, we were glad to stop on an open plain about a mile from the town of Yankton, where the road ended.

The three chief considerations for a camp are wood, water, and good ground. The latter we had, but we were at some distance from the water, and neither trees nor brushwood were in sight.

The long trains were unloaded of their freight, and the plains about us seemed to swarm with men and horses. I was helped down from the Pullman car, where inlaid woods, mirrors, and plush surrounded us, to the ground, perfectly bare of every earthly comfort. The other ladies of the regiment went on to the hotel in the town. The general suggested that I should go with them, but I had been in camp so many summers it was not a formidable matter for me to remain, and fortunately for what followed I did so. The household belongings were gathered together. A family of little new puppies, some half-grown dogs, the cages of mocking-birds and canaries, were all corralled safely in a little stockade made of chests and trunks, and we set ourselves about making a temporary home. The general and a number of soldiers, composing the headquarters detail, were obliged to go at once to lay out the main camp and assign the companies to their places. Later on, when the most important work was done, our tents were to be pitched. While I sat on a chest waiting, the air grew suddenly chilly, the bright sun of the morning disappeared, and the rain began to fall. Had we been accustomed to the climate we would have known that these changes were the precursors of a snow-storm.

When we left Memphis, not a fortnight before, we wore muslin gowns and were then uncomfortably warm; it seemed impossible that even so far north there could be a returned winter in the middle of April. We were yet to realize what had been told us of the climate—that there were "eight months of winter and four of very late in the fall." On the bluffs beyond us was a signal-station, but they sent us no warning. Many years of campaigning in the Indian Territory, Kansas, Colorado, and Nebraska, give one an idea of what the weather can do; but each new country has its peculiarities, and it seemed we had reached one where all the others were outdone. As the afternoon of that first day advanced the wind blew colder, and I found myself eying with envy a little half-finished cabin without an enclosure, standing by itself. Years of encountering the winds of Kansas, when our tents were torn and blown down so often, had taught me to appreciate any kind of a house, even though it were built upon the sand as this one was. A dug-out, which the tornado swept over, but could not harm, was even more of a treasure. The change of climate from the extreme south to the far north had made a number of the men ill, and even the superb health of the general had suffered. He continued to superintend the camp, however, though I begged him from time to time as I saw him to give up. I felt sure he needed a shelter and some comfort at once, so I took courage to plan for myself. Before this I had always waited, as the general preferred to prepare everything for me. After he had consented that we should try for the little house, some of the kind-hearted soldiers found the owner in a distant cabin, and he rented it to us for a few days. The place was equal to a palace to me. There was no plastering, and the house seemed hardly weatherproof. It had a floor, however, and an upper story divided off by beams; over these Mary and I stretched blankets and shawls and so made two rooms. It did not take long to settle our few things, and when wood and water were brought from a distance we were quite ready for house-keeping, except that we lacked a stove and some supplies. Mary walked into the town to hire or buy a small cooking-stove, but she could not induce the merchant to bring it out that night. She was thoughtful enough to take along a basket and brought with her a little marketing. Before she had come within sight of our cabin on her return, the snow was falling so fast it was with difficulty that she found her way.

Meanwhile the general had returned completely exhausted and very ill. Without his knowledge I sent for the surgeon, who, like all of his profession in the army, came promptly. He gave me some powerful medicine to administer every hour, and forbade the general to leave his bed. It was growing dark, and we were in the midst of a Dakota blizzard. The snow was so fine that it penetrated the smallest cracks, and soon we found white lines appearing all around us, where the roof joined the walls, on the windows and under the doors. Outside the air was so thick with the whirling, tiny particles that it was almost impossible to see one's hand held out before one. The snow was fluffy and thick, like wool, and fell so rapidly, and seemingly from all directions, that it gave me a feeling of suffocation as I stood outside. Mary was not easily discouraged, and piling a few light fagots outside the door, she tried to light a fire. The wind and the muffling snow put out every little blaze that started, however, and so, giving it up, she went into the house and found the luncheon-basket we had brought from the car, in which remained some sandwiches, and these composed our supper.

The night had almost settled down upon us when the adjutant came for orders. Knowing the scarcity of fuel and the danger to the horses from exposure to the rigor of such weather after their removal from a warm climate, the general ordered the breaking of camp. All the soldiers were directed to take their horses and go into Yankton, and ask the citizens to give them shelter in their homes, cow-sheds, and stables. In a short time the camp was nearly deserted, only the laundresses, two or three officers, and a few dismounted soldiers remaining. The towns-people, true to the unvarying western hospitality, gave everything they could to the use of the regiment; the officers found places in the hotels. The sounds of the hoofs of the hurrying horses flying by our cabin on their way to the town had hardly died out before the black night closed in and left us alone on that wide, deserted plain. The servants, Mary and Ham, did what they could to make the room below-stairs comfortable by stopping the cracks and barricading the frail door. The thirty-six hours of our imprisonment there seems now a frightful nightmare. The wind grew higher and higher, and shrieked about the little house dismally. It was built without a foundation, and was so rickety it seemed as it rocked in a great gust of wind that it surely would be unroofed or overturned. The general was too ill for me to venture to find my usual comfort from his re-assuring voice. I dressed in my heaviest gown and jacket, and remained under the blankets as much as I could to keep warm. Occasionally I crept out to shake off the snow from the counterpane, for it sifted in between the roof and clapboards very rapidly. I hardly dared take the little phial in my benumbed fingers to drop the precious medicine for fear it would fall. I realized, as the night advanced, that we were as isolated from the town, and even the camp, not a mile distant, as if we had been on an island in the river. The doctor had intended to return to us, but his serious face and impressive injunctions made me certain that he considered the life of the general dependent on the medicine being regularly given.

During the night I was startled by hearing a dull sound, as of something falling heavily. Flying down the stairs I found the servants prying open the frozen and snow-packed door, to admit a half dozen soldiers who, becoming bewildered by the snow, had been saved by the faint light we had placed in the window. After that several came, and two were badly frozen. We were in despair of finding any way of warming them, as there was no bedding, and, of course, no fire, until I remembered the carpets which were sewed up in bundles and heaped in one corner, where the boxes were, and which we were not to use until the garrison was reached. Spreading them out, we had enough to roll up each wanderer as he came. The frozen men were in so exhausted a condition that they required immediate attention. Their sufferings were intense, and I could not forgive myself for not having something with which to revive them. The general never tasted liquor, and we were both so well always we did not even keep it for use in case of sickness.

I saw symptoms of that deadly stupor which is the sure precursor of freezing, when I fortunately remembered a bottle of alcohol which had been brought for the spirit-lamps. Mary objected to using the only means by which we could make coffee for ourselves, but the groans and exhausted and haggard faces of the men won her over, and we saw them revive under the influence of the fiery liquid. Poor fellows! They afterwards lost their feet, and some of their fingers had also to be amputated. The first soldier who had reached us unharmed, except from exhaustion, explained that they had all attempted to find their way to town, and the storm had completely overcome them. Fortunately one had clung to a bag of hard-tack, which was all they had had to eat. At last the day came, but so darkened by the snow it seemed rather a twilight. The drifts were on three sides of us like a wall. The long hours dragged themselves away, leaving the general too weak to rise, and in great need of hot, nourishing food. I grew more and more terrified at our utterly desolate condition and his continued illness, though fortunately he did not suffer. He was too ill, and I too anxious, to eat the fragments that remained in the luncheon-basket. The snow continued to come down in great swirling sheets, while the wind shook the loose window-casings and sometimes broke in the door. When night came again and the cold increased, I believed that our hours were numbered. I missed the voice of the courageous Mary, for she had sunk down in a corner exhausted for want of sleep, while Ham had been completely demoralized from the first. Occasionally I melted a little place on the frozen window-pane, and saw that the drifts were almost level with the upper windows on either side, but that the wind had swept a clear space before the door. During the night the sound of the tramping of many feet rose above the roar of the storm. A great drove of mules rushed up to the sheltered side of the house. Their brays had a sound of terror as they pushed, kicked, and crowded themselves against our little cabin. For a time they huddled together, hoping for warmth, and then despairing, they made a mad rush away, and were soon lost in the white wall of snow beyond. All night long the neigh of a distressed horse, almost human in its appeal, came to us at intervals. The door was pried open once, thinking it might be some suffering fellow-creature in distress. The strange, wild eyes of the horse peering in for help, haunted me long afterwards. Occasionally a lost dog lifted up a howl of distress under our window, but before the door could be opened to admit him he had disappeared in the darkness. When the night was nearly spent I sprang again to the window with a new horror, for no one, until he hears it for himself, can realize what varied sounds animals make in the excitement of peril. A drove of hogs, squealing and grunting, were pushing against the house, and the door which had withstood so much had to be held to keep it from being broken in.

It was almost unbearable to hear the groans of the soldiers over their swollen and painful feet, and know that we could do nothing to ease them. To be in the midst of such suffering, and yet have no way of ameliorating it; to have shelter, and yet to be surrounded by dumb beasts appealing to us for help, was simply terrible. Every minute seemed a day; every hour a year. When daylight came I dropped into an exhausted slumber, and was awakened by Mary standing over our bed with a tray of hot breakfast. I asked if help had come, and finding it had not, of course, I could not understand the smoking food. She told me that feeling the necessity of the general's eating, it had come to her in the night-watches that she would cut up the large candles she had pilfered from the cars, and try if she could cook over the many short pieces placed close together, so as to make a large flame. The result was hot coffee and some bits of the steak she had brought from town, fried with a few slices of potatoes. She could not resist telling me how much better she could have done had I not given away the alcohol to the frozen men!

The breakfast revived the general so much that he began to make light of danger in order to quiet me. The snow had ceased to fall, but for all that it still seemed that we were castaways and forgotten, hidden under the drifts that nearly surrounded us. Help was really near at hand, however, at even this darkest hour. A knock at the door, and the cheery voices of men came up to our ears. Some citizens of Yankton had at last found their way to our relief, and the officers, who neither knew the way nor how to travel over such a country, had gladly followed. They told us that they had made several attempts to get out to us, but the snow was so soft and light that they could make no headway. They floundered and sank down almost out of sight, even in the streets of the town. Of course no horse could travel, but they told me of their intense anxiety, and said that fearing I might be in need of immediate help they had dragged a cutter over the drifts, which now had a crust of ice formed from the sleet and the moisture of the damp night air. Of course I declined to go without the general, but I was more touched than I could express by their thought of me. I made some excuse to go up-stairs, where, with my head buried in the shawl partition, I tried to smother the sobs that had been suppressed during the terrors of our desolation. Here the general found me, and though comforting me by tender words, he still reminded me that he would not like any one to know that I had lost my pluck when all the danger I had passed through was really ended.

[. . .]

From *Boots and Saddles; or, Life in Dakota with General Custer* by Elizabeth B. Custer (New York: Harper and Brothers, 1885).

✳ MARK TWAIN, 1835-1910 ✳

Samuel Clemens, popularly known as Mark Twain, grew up in the Mississippi River town of Hannibal, Missouri, and as a young man worked as a riverboat pilot—an experience he wrote about in *Life on the Mississippi* (1883). Although he lived most of his life away from his home state, and traveled across the globe as an international celebrity, his childhood experiences of the river and its natural surroundings were at the heart of a number of his most popular works, including *The Adventures of Tom Sawyer* (1876) and *Adventures of Huckleberry Finn* (1885). In this excerpt from his posthumously published *Autobiography*, an expanded version of which was reissued in 2010, Clemens recollects in rich detail his boyhood experiences along the borders of prairie and woodland while visiting his uncle's farm.

My Uncle's Farm

Beyond the road where the snakes sunned themselves was a dense young thicket, and through it a dim-lighted path led a quarter of a mile; then out of the dimness one emerged abruptly upon a level great prairie which was covered with wild strawberry plants, vividly starred with prairie pinks, and walled in on all sides by forests. The strawberries were fragrant and fine, and in the season we were generally there in the crisp freshness of the early morning, while the dew-beads still sparkled upon the grass and the woods were ringing with the first songs of the birds.

[. . .]

The country schoolhouse was three miles from my uncle's farm. It stood in a clearing in the woods, and would hold about twenty-five boys and girls.

We attended the school with more or less regularity once or twice a week, in summer, walking to it in the cool of the morning by the forest paths, and back in the gloaming at the end of the day. All the pupils brought their dinners in baskets—corn dodger, buttermilk and other good things—and sat in the shade of the trees at noon and ate them. It is the part of my education which I look back upon with the most satisfaction. My first visit to the school was when I was seven. A strapping girl of fifteen, in the customary sunbonnet and calico dress, asked me if I "used tobacco"—meaning did I chew it. I said, no. It roused her scorn. She reported me to all the crowd, and said—

"Here is a boy seven years old who can't chaw tobacco."

By the looks and comments which this produced, I realized that I was a degraded object; I was cruelly ashamed of myself. I determined to reform. But I only made myself sick; I was not able to learn to chew tobacco. I learned to smoke fairly well, but that did not conciliate anybody, and I remained a poor thing, and characterless. I longed to be respected, but I never was able to rise. Children have but little charity for each other's defects.

As I have said, I spent some part of every year at the farm until I was twelve or thirteen years old. The life which I led there with my cousins was full of charm, and so is the memory of it yet. I can call back the solemn twilight and mystery of the deep woods, the earthy smells, the faint odors of the wild flowers, the sheen of rain-washed foliage, the rattling clatter of drops when the wind shook the trees, the far-off hammering of wood-peckers and the muffled drumming of wood-pheasants in the remotenesses of the forest, the snap-shot glimpses of disturbed wild creatures skurrying through the grass,—I can call it all back and make it as real as it ever was, and as blessed. I can call back the prairie, and its loneliness and peace, and a vast hawk hanging motionless in the sky, with his wings spread wide and the blue of the vault showing through the fringe of their end-feathers. I can see the woods in their autumn dress, the oaks purple, the hickories washed with gold, the maples and the sumachs luminous with crimson fires, and I can hear the rustle made by the fallen leaves as we plowed through them. I can see the blue clusters of wild grapes hanging amongst the foliage of the saplings, and I remember the taste of them and the smell. I know how the wild blackberries looked, and how they tasted; and the same with the pawpaws, the hazelnuts and the persimmons; and I can feel the thumping rain, upon my head, of hickory nuts and walnuts when we were out in the frosty dawns to scramble for them with the pigs, and the gusts of wind loosed them and sent them down. I know the stain of blackberries, and how pretty it is; and I know the stain of walnut hulls, and how little it minds soap and water; also what grudged experience it had of either of them. I know the taste of maple sap, and when to gather it, and how to arrange the troughs and the delivery-tubes, and how to boil down the juice, and how to hook the sugar after it is made; also how much better hooked sugar tastes than any that is honestly come by, let bigots say what they will. I know how a prize watermelon looks when it is sunning its fat rotundity among pumpkin vines and "simblins;" I know how to tell when it is ripe without "plugging" it; I know how inviting it looks when it is cooling itself in a tub of water under the bed, waiting; I know how it looks when it lies on the table in the sheltered great floor-space between house and kitchen, and the children gathered for the sacrifice and their mouths watering; I know the crackling sound it makes when the carving knife enters its end, and I can see the split fly along in front of the blade as the knife cleaves its way to the other end; I can see its halves fall apart and display the rich red meat and the black seeds, and the heart standing up, a luxury fit for the elect; I know how a boy looks, behind a yard-long slice of that melon, and I know how he feels; for I have been there. I know the taste of the watermelon which has been honestly come by, and I know the taste of the watermelon which has been acquired by art. Both taste good, but the experienced know which tastes best. I know the look of green apples and peaches and pears on the trees, and I know how entertaining they are when they are inside of a person. I know how ripe ones look when they are piled in pyramids under the trees, and how pretty they are and how vivid their colors. I know how a frozen apple looks, in a barrel down cellar in the winter time, and how hard it is to bite, and how the frost makes the teeth ache, and yet how good it is, notwithstanding. I know the disposition of elderly people to select the specked apples for the children, and I once knew ways to beat the game. I know the look of an apple that is roasting and sizzling on a hearth on a winter's evening, and I know the comfort that comes of eating it hot, along with some sugar and a drench of cream. I know the delicate art and mystery of so cracking hickory nuts and walnuts on a flatiron with a hammer that the kernels will be delivered whole, and I know how the nuts, taken in conjunction with winter apples, cider and doughnuts, make old people's old tales and old jokes sound fresh and crisp and enchanting, and juggle an evening away before you know what went with the time. I know the look of Uncle Dan'l's kitchen as it was on privileged nights when I was a child, and I can see the white and black children grouped on the hearth, with the firelight playing on their faces and the shadows flickering upon the walls, clear back toward the cavernous gloom of the rear, and I can hear Uncle Dan'l telling the immortal tales which Uncle Remus Harris was to gather into his book and charm the world with, by and by; and I can feel again the creepy joy which quivered through me when the time for the ghost story of the "Golden Arm" was reached—and the sense of regret, too, which came over me, for it was always the last story of the evening, and there was nothing between it and the unwelcome bed.

I can remember the bare wooden stairway in my uncle's house, and the turn to the left above the landing, and the rafters and the slanting roof over my bed, and the squares of moonlight on the floor, and the white cold world of snow outside, seen through the curtainless window. I can remember the howling of the wind and the quaking of the house on stormy nights, and how snug and cosy one felt, under the blankets, listening; and how the powdery snow used to sift in, around the sashes, and lie in little ridges on the floor, and make the place look chilly in the morning, and curb the wild desire to get up—in case there was any. I can remember how very dark that room was, in the dark of the moon, and how packed it was with ghostly stillness when one woke up by accident away in the night, and forgotten sins came flocking out of the secret chambers of the memory and wanted a hearing; and how ill chosen the time seemed for this kind of business; and how dismal was the hoo-hooing of the owl and the wailing of the wolf, sent mourning by on the night wind.

I remember the raging of the rain on that roof, summer nights, and how pleasant it was to lie and listen to it, and enjoy the white splendor of the lightning and the majestic booming and crashing of the thunder. It was a very satisfactory room; and there was a lightning rod which was reachable from the window, an adorable and skittish thing to climb up and down, summer nights, when there were duties on hand of a sort to make privacy desirable.

I remember the 'coon and 'possum-hunts, nights, with the negroes, and the long marches through the black gloom of the woods, and the excitement which fired everybody when the distant bay of an experienced dog announced that the game was treed; then the wild scramblings and stumblings through briars and bushes and over roots to get to the spot; then the lighting of a fire and the felling of the tree, the joyful frenzy of the dogs and the negroes, and the weird picture it all made in the red glare—I remember it all well, and the delight that every one got out of it, except the 'coon.

I remember the pigeon seasons, when the birds would come in millions, and cover the trees, and by their weight break down the branches. They were clubbed to death with sticks; guns were not necessary, and were not used. I remember the squirrel-hunts, and prairie-chicken hunts, and wild turkey hunts, and all that; and how we turned out, mornings, while it was still dark, to go on these expeditions, and how chilly and dismal it was, and how often I regretted that I was well enough to go. A toot on a tin horn brought twice as many dogs as were needed, and in their happiness they raced and scampered about, and knocked small people down, and made no end of unnecessary noise. At the word, they vanished away toward the woods, and we drifted silently after them in the melancholy gloom. But presently the gray dawn stole over the world, the birds piped up, then the sun rose and poured light and comfort all around, everything was fresh and dewy and fragrant, and life was a boon again. After three hours of tramping we arrived back wholesomely tired, overladen with game, very hungry, and just in time for breakfast.

From volume 1 of *Autobiography of Mark Twain* by Mark Twain, edited by Harriet Elinor Smith (Berkeley: University of California Press, 2010).

THE
TWENTIETH
CENTURY

✳ ✳ ✳

The son of Omaha Chief Iron Eye, La Flesche was born and raised on the tribal reservation in eastern Nebraska. Educated in a Presbyterian mission school overlooking the Missouri River, he went on to earn a law degree and, as an anthropologist working with the Smithsonian's Bureau of American Ethnology, studied and recorded the traditional customs of several tribes, including the Omahas and Osages. The Omahas had long thrived in the tallgrass prairies and woodlands along the Missouri River, but as with many indigenous tribes, they came under extreme pressure to assimilate, a painful process La Flesche exposed through his writing. In this moving chapter from his memoir, *The Middle Five* (1900), La Flesche describes his unsuccessful attempt as a boy to join fellow tribal members as they journey west for the annual bison hunt—a prairie ritual soon to disappear, along with the wild herds.

✷ ✷ ✷

A Runaway

Vacation had come, and the Indians were about to start on their annual summer buffalo hunt. Some of the scholars were to accompany their parents, and others, after a brief home visit, were to return to the school and continue their studies while the tribe was away.

In the three villages there was great hurry and bustle in every family. Pack saddles were brought out of the caches where they had lain through the winter. The task of mending them fell to the older people of the household, while the younger folk busied themselves in retrimming their more ornate trappings. Goods not necessary for the journey were stored away, and the dwellings were made ready for the long absence.

At last there remained but one day before the time set for the departure of the tribe. In the afternoon I bade my parents goodby, and reluctantly returned to the school. Quite a number of the boys and girls had already come back, among them Lester and Warren. Brush had not left the school, so on my arrival I received from the three boys the usual greetings we accorded each other when one returned after an absence. We four paced the long front porch, arm in arm, for a while, and then went and sat down in the shade of a tree.

"Where is Edwin?" asked Brush; "isn't he coming back?"

"No," I replied; "his mother wanted him to; but his father didn't want to leave him behind, so he's going on the hunt."

"He'll have lots of fun," said Warren; "I wish I could go!"

The next morning, immediately after breakfast, Brush borrowed the superintendent's spy-glass, and we went to a high point whence we could watch the movements of the people in the village nearest the school. We took turns in looking through the glass. Already the head of the great caravan had gone behind the first hill, but my family had not yet started. We looked toward Edwin's house, and saw that the people were just moving. It was a wonderful sight to us, the long procession on the winding trail, like a great serpent of varied and brilliant colors. At last I saw my father mount a horse and move forward, the rest of the family followed him, and I watched them until they finally disappeared beyond the green hills. It was nearly noon when the end of the line went out of sight.

While the movements were going on in the village, we could hear the neighing of horses, the barking of dogs, and the hum of voices, but now there was a stillness in the deserted village which brought upon us a sense of loneliness that was hard to overcome. We slowly returned to the Mission and ate our noonday meal without speaking. There seemed to be a general depression among the remaining pupils at the school. A silence pervaded all the surroundings which made each boy wish to retire from the other and to be alone.

At breakfast, the next morning, there was the same sense of stillness; even the superintendent and the teachers at their table seemed to be homesick, and they passed the dishes to each other in silence. The reading of the

Scriptures and the prayer of the superintendent was in a tone that added to the gloominess which had taken possession of our simple little souls.

As we were slowly marching out of the dining-room, when the worship was over, the superintendent stopped Brush and said to him:

"I want you to go after the mail this morning; go on horse-back so as to get back soon. I have some work for you to do this afternoon. Take Dolly, and use the large saddle; the other one needs mending."

"Let's go down to the spring," said Lester to Warren and me.

So while Brush went to the barn to saddle up, we three went to the spring and sat under an elm that stood near by.

"Say, boys, I'm going to the hunt!" said Lester, startling us with the sudden announcement; "I heard that two families down at the Wood-eaters' village can't get away for two days yet, and I'm going down there so I can go with them. The Omahas always wait on the Wa-tae (Elkhorn River), for those that are last."

"If you're going, I'm going too," spoke up Warren; "I don't want to stay here."

"If you two go, I'm going!" I exclaimed.

"All right, let's all go then," said Lester, rising. "We must hurry up; some one might see us!"

We followed a narrow path that led through a ravine just beyond the spring. We were in the greatest excitement; every little sound aroused within us the fear of detection, and we frequently sought for a hiding place, while we carefully avoided all well-beaten paths. Silently we plodded our way through the bushes until we came to a hill where there were no trees, then we ran as fast as our legs could carry us for another wooded place.

We stopped a moment when passing to take a look at the village. Silence prevailed. Not a living thing was astir. Three whirlwinds chased each other along the winding paths between the houses, making funnel-shaped dust clouds as they sped on.

"The ghosts have entered the village," said Lester, in our own language, and in a melancholy tone; "they always do that as soon as the living leave their houses!"

Entering the ravine for which we were making, we continued our journey. The nettle weeds caused us much suffering, for we were barefooted, and wore short trousers. We came to an opening; before us lay the road to the Agency; we looked cautiously around, then started to cross it to go into another ravine that headed toward the big village, when the snorting of a horse was heard with startling distinctness.

"Quick! quick! get down!" exclaimed Lester in a loud whisper, as he dropped into the gully of the old abandoned wagon-road.

Warren and I followed hastily, pulling the tall grass over us. We heard the footsteps of the horse come nearer and nearer to our hiding place. It stopped and reached its head down, and began to nibble the grass under which I lay concealed. I looked up through a slight opening, and, behold! there on the horse sat Brush with one leg thrown over the pommel of the saddle, busily reading a book. I could see the boy's eyes and his lips moving as he read, and at times it seemed as though his eyes were looking right into mine. I was in great suspense while the horse stood there, but at length Brush picked up the reins and urged Dolly on. As soon as he disappeared at the bend of the road, we rose and darted across and ran down to the ravine.

We entered the big village of sod houses through which we had to pass. Here, too, we felt the sense of desolation that pervaded even the hills around. Somewhere from the midst of these peculiar dwellings came the doleful howl of a stray dog, the only sound that broke the stillness of the place. What sensations my companions experienced upon hearing the melancholy wail of that deserted beast I do not know; but, like the rapid advance of a fire over the prairie, a thrill that made the very roots of my hair creep vibrated through my body. Involuntarily we paused to listen; the long-drawn moan came to a close, and the ghostly echoes carried on the sound as though to mock the lost creature.

"Let's run!" exclaimed Lester, in a frightened tone; "let's get away from here!"

And so we sped on until, all out of breath, we were far beyond the limits of the village.

[. . .]

We trudged along to Me-chah´-pe's house. The family had gathered about an outside fire, and were eating their supper in the dusk. Upon our coming into the light of the fire we were recognized; the mother and grown daughter greeted us with exclamations of surprise and sympathy; while the father and the two sons glanced at each other with expressions of amusement. A place was assigned us in the circle, and soon we were busily engaged with the simple fare placed before us by the good and hospitable mother.

"Why do parents when they go away leave their children at the House of Teaching, I wonder?" commented the woman, as she apportioned the food for us. "Some people show no signs of affection for their sons and daughters until they sicken and die; then they tear their hair and rend the air with their loud wails. It is well enough while the parents are at home that they should place the young ones in the care of the White-chests; but, when going on a long journey like this, they should take the children with them."

By the side of every Indian house stands a raised platform made of poles, elevated upon posts, some seven or eight feet high, planted firmly in the ground. This platform is used for drying corn and squash, at the time of harvesting; but, through the summer when the people are at home, the young men and boys take possession of it, for sleeping in the open air. As weariness began to be felt, one by one, the family arose, and, without formality, each sought his place of rest. We school-boys and the sons of Me-chah´-pe repaired to the platform, climbing the "stairs" made of a single log, with notches cut in it for steps.

This was the first night I had ever spent out of doors. The novel experience, and the excitements of the day, filled my mind with strange speculations, and I lay awake long after my companions had gone to sleep. Now and then, I heard the chatter of birds and the whirring of their wings, as they flew by far above me, and I wondered if they could see in the darkness. The roar of the river filled the still air, and the crash of a tree uprooted by the current sent its echoes far and wide; then the sounds about me grew to faint murmurings, until I was conscious of them no more.

When I awoke, the dawn was coming, and the stars were beginning to turn pale. There was a gentle stir in the tent near by; a tall man came out, and his shadowy form passed from view into the slowly rising mist. A woman moved noiselessly to the fire-place, and, bending over, began to gather the embers together, blowing them to life with her breath. The gray streak along the horizon slowly turned to a rosy hue; here and there the birds began drowsily to peep and twitter, then, when the sun shot its rays through the heavens, a thousand voices burst into rapturous song.

My companions awoke, and one by one we climbed down the rude ladder to the ground.

When we gathered for breakfast, the mother, as she helped the food, asked, "Where is Na-zhé-de-ah?" (Lester.)

Warren and I looked at each other; neither of us could explain his absence.

"Call him," said the good woman, addressing her son; "we must hurry, the sun is up!"

No response came to the young man's call. It was evident that Lester had slipped away before any one was awake.

Breakfast over, Me-chah´-pe and his sons saddled and packed the horses, while the wife and daughter gathered the various utensils. Warren and I tried to make ourselves useful by holding up the packs with our shoulders, as they were being placed on the horses.

Me-chah´-pe looked at Warren, then at me, as he shouldered his rifle, and said, "I am sorry that I have not enough horses for all of us to ride. You see those I have are heavily burdened; so we will have to do as our fathers did, take one step forward, then another, and keep stepping forward until we get to the place where we are going. Are we ready? Here we go!"

And we did go,—horses, dogs, and all. Soon we were joined by the man of the lonely sod house and his family, and together we made quite a cavalcade as we went up hill and down hill, and up hill and down hill again. By and by, we reached a long ridge, called by the Indians "the tortuous ridge," which zigzagged in a westerly direction, and along it lay the hunting trail.

The sun grew hot; Warren and I were drenched with perspiration as we plodded on. Every now and then Me-chah´-pe gave us an encouraging word, when we showed signs of lagging. We were determined to keep on, for were we not going to a buffalo hunt! The heat increased. The dogs did not now chase each other and run after birds as when we started out, but let loose their tongues and panted, keeping close to the shadows of the horses. On we all trudged, while the one baby slept on its mother's back, its little head rocking from side to side with the motion of her steps.

As we reached an elevated point on the ridge, Me-chah´-pe shaded his face with his hand and scanned the horizon. Far ahead of us his experienced eye caught sight of an object, like a mere speck. He pointed it out to us, saying, "There's somebody coming."

Warren and I looked at each other in alarm, and then kept our eyes on the speck, which grew larger and larger as the distance between it and us lessened.

"The horse looks like one of your father's," said Me-chah´-pe to me. "I think it is some one looking for you!"

My heart sank when I recognized the horse as father's, and the rider as my uncle, and, for the first time in my life, I was not glad to meet him.

Warren and I were captured, and there was no escape. We tried to be brave when Me-chah´-pe shook hands with us, as his party moved westward; but we were far from happy when, ignominiously mounted on father's horse, one behind the other, we followed my uncle, who walked so rapidly that the animal had to trot now and then to keep up. The road over which we had so laboriously travelled on our outward way was soon retraced, and the sun still high when my uncle, who had wandered all night in search of us, turned us over to Gray-beard.

It was thought best to punish us; so Warren was taken to the top of the house and locked up in the attic, where he was to reflect upon the wrong he had committed in running away. But I am quite sure he thought more about the devil and the ghosts in that horrid place than of anything else.

As for me, I was marched to the dinning-room, placed with my back to one of the posts, and my arms brought around it and tied; then I was left alone in this uncomfortable position,—to repent.

The afternoon was close and hot; the windows and doors were open, but the place was very quiet. Now and then I heard the cry of a bird, or the laughter of the happy wren. The time seemed very long as I stood there, with my arms thrown back around the post and my hands tied so that I could not defend myself against the flies that attacked my bare feet. A rooster came to the back door and entered the dining-room. He shied on discovering me; but, as I did not move, he began picking in the cracks of the floor. He spied my toe, looked at it curiously, turning his head from side to side, then stretched his neck and gave it a dab. I was in no mood to be amused by his actions, so I sent him flopping and squawking under the table. Recovering from his surprise, he ran around, sprang on the table, then on the sill of the open window, tossed up his head, flapped his wings, gave a lusty crow, and hopped out.

Immediately I saw eight little fingers hook themselves on the outer edge of the window-sill, and a head with black hair held back by a rubber comb rise higher and higher until two bright eyes gazed right into mine. The head disappeared, and shortly after a little figure cautiously approached the door, looked all around, and then came up to me. It was Rosalie. Her bright smiling face threw a sunbeam into my gloomy little heart. Without saying a word she wiped the perspiration from my face with the corner of her apron; then she went away softly in the direction of the kitchen. Soon she returned with a tin cup having in it bits of ice. She took a lump and put it in my mouth, then stood looking in my face. After a while, she said, "I like you, don't I?"

"'M h'm!" I assented with my mouth closed, nodding my head.

"When we get big, we're going to be married, ain't we?"

"'M h'm!" again I answered.

"We won't send our children to this horrid old place, will we?"

"'M 'm! 'M 'm!" I replied with emphasis, shaking my head and stamping the floor.

The little sweet-heart, seeing that the flies troubled my ankles, went out and came back with a linden branch and brushed away the pests. I slid to the floor and sat down with my legs stretched out. Rosalie dropped down too, and sat whisking away the flies.

Gradually things took on queer shapes, and the sounds seemed to come from afar; there was a moment of confusion and then,—I found myself on a wide prairie. Heavy clouds were swiftly approaching; the thunder rolled long and loud, and the lightning darted hither and thither. Off in the distance I saw a forest. I pushed toward it with all my strength so as to take shelter before the storm should come upon me; but as I labored on there crept over me a consciousness of a weight upon my back which, hitherto, I had not noticed. It retarded my progress, and from time to time I was obliged to stop and give a little spring to shift the burden higher up. A cry of terror came from the thing I was carrying; then I knew it was little Rosalie. I tried to speak words of encouragement to her, but my strength was fast failing. Great drops of rain fell, and the wind drove the dust into my face, blinding me. I tottered on with my load, but the timber was still far away. A vivid flash, a deafening crash, and I fell to the ground with a cry. I tried to rise, but my legs and arms were as though dead.

With a start I opened my eyes. The room was darkened; there was a great commotion; all through the house, windows were being rapidly closed and the doors swung to with a bang. A terrific storm had arisen, and the building was in danger of destruction. Rosalie lay asleep with her head resting on my knees.

From *The Middle Five: Indian Schoolboys of the Omaha Tribe* by Francis La Fleshe, 1900. Reprint (Lincoln: University of Nebraska Press, 1978).

Born near Belfast, Ireland, William J. Haddock immigrated with his parents to Philadelphia in 1849, before moving to eastern Iowa in 1856. After briefly working on farms and at a sawmill, he studied and practiced law, taught school, served as the superintendent of schools for Johnson County, and was eventually appointed circuit court judge—all before the age of forty-one. His wife, Emily, was equally accomplished, being the first woman to receive a bachelor of law degree from the University of Iowa. As a young man in the 1850s, living northeast of Iowa City and in Cedar Rapids, Haddock had a chance to observe large expanses of unplowed tallgrass prairie, which awed him with its size, beauty, and natural diversity. Fifty years later, as he reveals in his 1901 memoir, those prairies had been almost completely destroyed because of agricultural and urban development. In this excerpt, Haddock expresses the shock he experienced in the face of those changes, while contemplating how the loss of the prairies will impact his home state and its citizens.

The Passing of the Prairies

[. . .]

Time has worked great changes on the prairies since I first saw them. They were gradually diminished by the encroachments of the settlers. The residents on the barrens would secure a prairie tract as near to their dwellings as practicable and cultivate it as a feeder to their home farms. This was well enough for a time. But when the pine lumber supply became more abundant through improved shipping facilities, bolder settlers went out on the prairies and made homes there. The prairie homes increased in number and multiplied in all parts of the state. The open prairies diminished in this way at a marvelous rate, till finally, in the thickly settled parts of the state the prairies, as such, entirely disappeared.

"Prairie," today, to the average citizen, is only a name. To him it means a fair farm and comfortable buildings and shady groves and orchards on the land that once was the prairie. He has no sentiment about the change and cannot understand it. He never saw the prairie, and hence never saw Iowa during its Golden Days.

The old distinguishing features of the prairie days are passing or are gone. Indian summer, beautiful and balmy in the past, is no longer seen in Iowa. There may be some compensation for the loss of other features of our former times, but there is none in this. Indian summer was a blessing, with nothing now to take its place. We are to that extent poor indeed.

The prairie fires are gone, but that is more reasonable. Iowa has now no surplus grass to destroy. The farmers have no hay to burn. They cannot afford to waste the crop for a fall and spring conflagration. The great flocks and herds of Iowa consume the whole grass and hay crop of the state. Fat beeves are sold and shipped in train loads, fed largely on the product which was heretofore wasted annually by prairie fires.

The prairie chicken one would like to see still in the land, but that cannot be. Its day has passed. Its doom was not hastened by the hunters. Its fate was sealed by what we term the march of civilization. It was a timid bird and raised its brood in the security of the prairie. It was pitiful to see it clinging to its old home in spite of the plow and the reaper. It is not a migratory bird that could take to new fields on the destruction of its old haunts. It did what it could to exist under the new dispensation, but the struggle was too great for the handsome bird and it has well nigh faded from Iowa forever. We have no compensatory feature to record for its loss.

Of late years the English sparrow gained a foothold in the state. There is no compensation for anything in that. It is an outcropping of dudish idiocy, in its first introduction, in the aping of everything English, "you know." This bird follows city developments and is a songless epicure that revels in fresh horse manure. It is a loathsome object compared with the beauty of the wild prairie hen.

The passing of the prairie made me sad. Every drive or trip I made told of the change. I thought of the first prairie I had ever seen, when I rode (in Jeanie Deans' words) "sae willyard a powny" over it. That was long ago and I must expect great changes. I still wondered what that prairie was like now and if it had been effected by the spirit of the times. I concluded to go and see it. I half hoped to see the wide expanse of green flowering prairie again, and see the plover, with his long legs, walking around as lord of the whole domain. I took a buggy and a driver, this time, and went out on a sunny morning, well on in June. I rather looked to bring home some sweet-williams, prairie-pinks, and wild roses for my trouble. But they were gone. We miss the wealth of flowers yielded by both prairies and barrens of "lang syne" in the springtime of the year. We can now scarcely find by hunting far and wide samples of the old time plants. They are nearly of the past. Civilization has been too much for them. They left before the march of our modern progress as if they had been sentient as well as things of beauty. We have in their stead the bluegrass and white clover of the east. These were carried here by the "movers" who traveled with the "Prairie Schooner" as it was called. The name was applied to a certain class of white, canvas-covered wagons with feed trough swung at the end. They carried their own hay, which often contained blue-grass and white clover and in feeding the seed was scattered around their camping places. A start of this kind was enough for a neighborhood. When I first saw Iowa I saw no blue-grass nor white clover. They eventually became thick enough and many years afterwards I sent lots of this seed, raised on my own land, to farmers on western prairies to sow on well pastured tracts. These pasturage plants are some compensation for the loss of the bright spontaneous flowers of long ago.

I went through Scott township on my way out and reached the south road leading to the scene of "lang syne." As I began to catch a view of the location, I thought there must be some mistake. I could not exactly locate the site I had formerly ridden over. Even if the prairie was pretty well settled, I assumed that there would be open tracts still with some of the former glory about it. I was satisfied that the settlers there would have small frame houses and shiplap barns, and a place for hogs under the corn cribs. I expected to find a few small logs or tree tops in the front yard near the door with an ax sticking in a chunk of wood there, just as it was left after cutting the morning stove wood.

When I got well in sight of that prairie I was on an elevation and the scene before me was astonishing. It was picturesque and pleasing. It was covered as far as the eye could reach with groves of timber. The trees were large and of numerous varieties. Large, comfortable barns and dwelling houses nestled among these groves. There has been some enchantment at work. I would drive over the scene and investigate. I drove well out on what had been the prairie. Good roads and bridges instead of mere trails; no sloughs to wade or creeks to jump there. I drove to the east and back along the north toward the west and the same appearance of comfort and wealth was everywhere. All was changed from the wild, lonely grandeur of 1856 to a great cultivated garden spot with rich, comfortable homes on it. There were fruit trees and evergreens in plenty and grape-vines and small fruit in abundance, too.

On a sunny slope facing the south I found a settler with 240 acres of land. He had grand groves, sheltering him with a wealth of pines and many other varieties of evergreens, for both protection and ornament. His lawns, gardens and yard must have contained at least five acres. The "wind breaks," as he called those fine groves, contained twice as much. I found this old settler and made friends with him and was invited in and shown around. His girls gave us a lunch and fine rich strawberries, of which his gardens contained bushels. We had hot coffee prepared on a gas stove. I was told that it was gasoline vapor gas. These dainty little maidens said it was "very nice" and that the plant furnished gas for the whole house as well as for the cooking stove. We went out to see his cornfields in which his boys were cultivating the corn. These young men were riding as comfortably as in a buggy. Each had two fat horses and a "riding plow." I asked the young fellows if they went to see the girls in that rig, as it looked like that kind of thing. They laughed and said that it would be of no use to go in that way; that the girls had to be cultivated very differently. They had a top buggy apiece, if that helped it at all. We came back by the house and found one of the girls at the piano working off "When Johnny comes marching Home again" in fine style. The other one was working a sewing machine. Their mother was not at home. She was at a P. P. Club, about a mile north. This meant Prairie Pink Club and it was devoted to the "Philosophy of Sociology." I did not know what that meant, the P. P.'s, I think did not either. But what of that?

It gave great opportunities for gossip and for working up essays, from the encyclopaedias, and reviewing the social economy of society from the days of the Mound-builders down to the advent of the latest Paris bonnet. What more could any mortal want? The girls had not reached that point yet. They were devoted to a very different philosophy.

This was prairie farming, was it? Cooking on gas stoves; working corn on buggy plows; living on ripe strawberries and the fat of the land. This kind of life was a picnic. This was not the kind of thing that I was used to in my agricultural days. The prairies were gone, but let them go if this is what takes their place. Still I miss and regret the departure of their old time splendor.

I wondered what my favorite prairie at Cedar Rapids was like after all these years. There must be trees there now of great size, and the settlers enjoying their shade, in splendor and comfort. I determined to go and see it and to go at once. So we drove back through these luxurious prairie homes to Oasis. I there took the cars for Cedar Rapids. This far surpassed footing it as of yore. But we must put on style nowadays to keep pace with the stylish agriculturist of the prairies.

Late that evening we reached Cedar Rapids, crossing the river lower down than I expected, and a little south of the town. I took a street car for a hotel. That ride in an electric car struck me as a great contrast with what I had found when there first. Then I was independent and walked to a log tavern where I spent a night. Now I take a comfortable car, name a splendid hotel to the conductor and I am let off at its very door. A waiter comes out, takes my grip and shows me in. Everything is in readiness and I am waited on right royally, just as if they had been expecting me, as Samantha Allen said.

I was out early next morning and tried to find my log hotel, but it was gone, as well as every structure that I recollected. I found instead splendid blocks of fine buildings and miles of well paved streets. There seemed to be no end of business places. I took an electric car and rode out in the suburbs on the bluffs to the north of town and found fine palatial residences everywhere. Cedar Rapids is an amazing place now. I noticed that three great through rail roads had depots right in the town. Some of these roads have their main offices and shops there, too.

I wandered up the railroad and finally reached the river and was surprised at the amount of water I saw there. There was the Cedar River dam which supplied power for manufacturing purposes. I saw that activity and energy had taken the place of the quiet repose of long ago. I walked slowly back to my hotel and concluded to visit my beautiful prairie with the western sun shining on the long grass and see it at its best. I told the landlord to order a buggy for me. It was soon announced as a "turnout." I turned out too, instructing the driver to cross the river and drive over the prairie westward. When I got on the bridge and looked for the prairie I was stunned. I saw before me one grand panorama of city buildings covering that entire area that I had traversed with so much pleasure when it was an open prairie. I drove on the streets all through that territory, but did not find a vestige of former days. The houses were in solid blocks and detached buildings, for any and all purposes incident to city life. This was settling up a prairie with a vengeance! I went back depressed with the "turnout." But what did it matter? We must look for the prairies to pass and be dedicated to new uses and purposes and the occupants be the happier and the better for that. I returned to Iowa City, to the law office and to work again in good heart.

Time rolled on, as time will, until it has rolled into several years since my last search for the bright, flower covered prairies. The day of such enchanting scenes is past. They and their former glory are nothing now but a recollection. Still my great interest in them has not waned even in these changed conditions. They are teeming garden spots now and I like to hear from the settlers on them.

The summer of 1901 was dry and there were loud complaints of the crop failure in consequence. I wrote to a friend living on the bottom land prairie in Woodbury County, on one of the branches or tributaries of the Little Sioux River, and asked for a report of his season's crop. After a good deal of delay he answered. I quote from his letter. He says, "I am so stiff and tired after the day's work that I can hardly sleep at night. That is why I did not write sooner. The trouble is I could not get 'help' enough to do the work. I will take things more comfortably now that the corn is all cribbed." "Our crop was good for this season. We have about 4000 bushels of this year's corn in crib, and about 2000 bushels of last year's corn in crib and it is now selling in our nearest shipping point at 57 cents per bushel. I have about 700 bushels of wheat in bin worth now 62 cents per bushel. I have 40 head of cattle and 60 head of hogs and 9 head of horses, and I do not owe any man one dollar. We are all enjoying good health. So you see we have no kick coming." That is the kind of a letter to give us a fair notion of the prairies in the fall of the past dry summer. This man's surplus stock and crop, just now, on the 10th day of December, 1901, when he wrote, was worth at least $5000, without touching the proper outfit to run the farm. Wherever we find the prairies they are always ahead. We get "no kick" there. There is little use in fretting at the passing of these superb lands under the plow, if that is the way it works. If any of the settlers in the timber or on the barrens in the "lang syne" could have figured out that much capital ahead he would have sold out, gone to town and taken to banking, issued wild-cat money galore and when the game was up gone back to the farm again.

Iowa as a state was always great. It is greater now, since its prairies are broken up and settled, than ever before. The comparative importance of a state is usually estimated by its resources and population and the wealth and the intelligence and comfort of its people. Gauged in this way, Iowa is pre-eminently the chief of states. Its population is nearly two and one-half millions today which makes it a power in the union. It was not over one-fifth of what it now is when I first saw it. We now have over 8500 miles of railroads in the state. When I came, there were 55 miles only and part of that sunk in the mud out of sight. The fight in political lines just now is how to make the most out of these roads by taxation. There was no trouble of that sort when I first saw Iowa. The anxiety then was to get what road they had to the surface again.

The settlers who selected this state in which to make their homes were great as a people and carried that element into every branch and function of the state. The course of such a people in such a country will ever be onward and will know no check for ages yet to come.

Although Iowa is a beautiful state today it was more beautiful in the early days. The green rolling prairies were its glory. The wild blossoming bushes on their borders made them enchanting in the gay springtime of the year. All this had to give place to the wants and necessities of the settlers as the state was developed. In fancy we wander back to the prairies and the early golden days of the state and cherish their memory. It may be called sentiment, but that is all that is left of those times now, and the recollection is dear to me.

I hold that whoever has not seen the prairies has not seen the grandeur and beauty of Iowa. The contemplation of greater wealth in buildings and farm products than the pioneers saw is a poor compensation for missing the early charms of the country.

We rejoice that no distinction can be made as to the patriotism or loyalty of the great family of Iowa citizens, and that all can join in heartfelt admiration of the state and hurrah for Iowa.

From *A Reminiscence: The Prairies of Iowa and Other Notes* by William J. Haddock (Iowa City, IA: Printed for Private Circulation, 1901).

Born in Parkville, Missouri, William A. Quayle attended Baker University in Kansas where he was later appointed professor of Greek language and then university president. An ordained Methodist minister, he left his academic position to serve churches in Kansas City, Indianapolis, and Chicago, and was elected a bishop of the church in 1908. A lifelong bibliophile, he amassed a large collection of rare books, while also authoring numerous books on literature, social theory, religion, nature, and other themes. In this complex, wide-ranging selection from *The Prairie and the Sea* (1905), Quayle begins with reflections on a remnant he owned in eastern Kansas—"kept to prairie for prairie's sake" (51)—and on the animals and plants that act as an expression of the personality, spirit, and even emotions of the prairie. He ends with a story of family loss that reveals, perhaps, his own deeply personal reasons for considering himself a "son of the prairies" (55).

The Prairie

[. . .]

You must not be in the prairie; but the prairie must be in you. That alone will do as qualification for biographer of the prairies. As Tennyson first had Ulysses and his sea, drunk like quaffing wine, and then began his trumpeting until you saw gray Ulysses and his mariners and saw the dim lights on receding rocks and heard the deep moan round with many voices and felt the mystery of this man and sea, so he who tells the prairie mystery must wear the prairie in his heart. . . .

I own a square mile of prairie. Or possibly this may be stronger language than the facts justify. I leave the reader to judge. I have given my autograph for a square mile of prairie. This method of purchase appeals to my imagination. I dwell fondly upon it. The customary way of buying ground is to pay cash for it. This seems to me crude and plebeian. Anybody can buy for cash. There is nothing creditable to character in that class of transaction. Anybody with a dollar in hand can buy a dollar's worth of commodities. But when for a solemn mile of prairie, a four-square block of God's out-of-doors, with the height of the sky above it and the depth of the world beneath it, and the radiancy of dawns and sunsets shed over it, and the dim dawn of dusks enfolding it like a blessed compassion,—a mile east, a mile west, a mile south, a mile north,—and all the time to be tramping on your own grass and breathing air brewed on your own ground and lifting head into your own sky and gazing at your own firmament, bless me this *is* plutocracy! And then to take one's own hand in congratulation, remembering that all this is held in fee simple without the cost of a postage stamp, simply by the execution of an autograph,—why, this method of purchase is as unique as the prairie itself.

[. . .]

But to stand about the center of this section of prairie and to look and breathe! I think that if I did this often I would sprout wings. I know I could crow; and I would not put it past me to cackle. But it is exhilarant to own your own prairie grass and prairie air; and to tramp on one and in the other—this is kingly. On glorious nights of gloomy dusk—without a silvery moon, but vigiled by the stars—to watch the Pleiades blink, and to feel the wind stream from far and hidden spaces past me as it hails from beyond the confines of the world, blown over infinite spaces of nameless seas and from the mountain land of those dim stars and to feel the wondering eyes of those stars, homes of these winds, searching my face on this tranced night,—this is delight keen as wine, rapturous like love. O, this square mile of prairie is an intoxicant to the soul!

The prairie-hawk is the brown prairie arrow. The Indian arrows are all broken or lost or archaic, mere curling of blue smoke to tell that once a camp-fire burnt here. And the one arrow left to cleave swift way over the prairie and through the sky is the prairie-hawk. And his is the speed vagrant, but terrific, which should embody the spirit of the prairie when set to aggressive winged motion. His flight has the notionateness of prairie winds, and the sudden detour as of a change of mind, a leap straight on, and then a notionate, abrupt change in direction, as if he had just bought wings and were out trying what sort of wings they were. I have seen him lurch along a bleak winter's russet landscape as if he were joking with it, in patches of torn flight, utterly erratic and utterly engaging. The freedom of the fields, the prairie-fields, is on him. His flight is, in ordinary, low. He loves his prairies and would keep close against their breast. They mother him.

[. . .]

The jack-rabbit—him we will not forget seeing the prairie can not forget him nor he the prairie. Thither he hastes; this he loves; and to see him lying flat in the brown grasses, long ears forgetful of alert erectness, falling along his shoulders as if the wind had blown them so, and to see him standing alert, listening ("aures erecti" as says our friend Vergil), ready for a leap; then to see him give those wild prairie bounds, as if spurred forward, not by fear, but by delight of the long, brown-like sea roads, ready for fleet running; and his racing is as if the tumble-weed made bounds, lurching to the jest of fleet prairie winds. His going is spasmodic like the blowing of prairie winds. He could not wear a pedometer. He is prairie-begotten and is as lithe as a lynx, and as eager as a gust of March wind blowing Spring back to the world. He is the voiceless swiftness of the prairie.

The prairie-wolf, his name inclines me to him. He, too, is a lover of the prairie. Wolf he is, sullen and whelpish. His swinging gallop, with head thrown back waggishly over his shoulder, is free as the blowing of winter winds. His lair is the prairie-paved sky. He is not moral. He cares for no works on ethics. He looks out for No. 1, in which lucrative employment both on and off the prairie, many are engaged. . . . Many is the night when I have lain awake listening to the eery barking of the prairie-wolf. At the first it is weird. I may have only been dreaming; but that cry appeals to me as the expression of the weirdness of the prairies, their strange unknowableness. This wolf bark is like the laughter of a child maniac, repetitional, meaningless, remorseless, a laughter without joy in or behind it. The cry is a wandering voice of the prairie levels; disappearing and reappearing among the billows of a rolling prairie, but is mirthless, insistent, uncanny. Through the still hours of nights, quiet as the quiet stars, I have heard this invasion of unbridled wild voices mixing unmusical screeches in sullen, joyless chorus;

they were the prairie's stretches breaking into a vagabond song, a meaningless bacchanalian revel. . . . The wolf is careless of any man; and his lope, than which nothing could be less routine or more care-free, less stilted, less an acquisition, or more an extemporaneous procedure, — is the heedlessness of the prairies, the needlessness of wings, the playing with the ground as if it were a jest, with waggish head thrown over the shoulder as to insult your laggard speed. The wolf-leap is the prairie in cruel motion, not creeping like feline hypocrisy, but the vagabond swing of a wild, elastic delight in the unfenced wonder of the prairie. The wolf is a prairie child.

And the prairie-chicken is child of the sullen winter grasses — dappled brown like a winter prairie field, so that when this wild thing lies close along the grass, an expert eye might forgive itself for not beholding it, until the wild thing leaps from its neighborly wild grasses and whirs away, brown-blown rags against a gray sky, and is as if the brown prairie had found wings. I have watched this singularity of flight; have seen the companies crowded in great multitudes or only a few survivors of what had once been great flocks; and their movement is like the free moods of far prairie winds' lurches of flight across the sky in a moment. A blur of wings — a brown battle rush and wildness that knows not man nor his peace measures — and they are gone. . . . When winter clouds lower and the brown prairies swirl to the anger of the sullen prairie winds, to see and hear the flutter and wings of a flock of prairie-chickens is to have brown prairies slip the leash of the earth and take to the sky.

And the buffalo, — who shall write his epitaph or rehearse his story? He is beast of the prairie. To see that burly figure, that huge bronzed head protruding from prairie or picture or cast in bronze, is to make a certificate of America. The buffalo belongs here. He is the might of the prairie; and like the prairie he is vanishing. I like not to think of the pathos of his evanishment, as I like not to think of the sadness of disappearing grasses dying out for lack of room, and from invasion of the grasses of civilization; but to read in Irving or Inman or Greeley or Richardson, or any other, of those vast herds of trampling bronzed beasts, with heads bent low, their prodding horns stooped for battle, their massed legions, their prodigious onset, their trampling as if the thunder had dropped from the sky and had begun to trample on the solid world, the trembling ground, fairly oscillant to the hammering hoofs,—these are the native rough-riders of the prairie: these are the children of angered prairies, fleet of foot, furious of onward going, ruthless as death, grim as fate, a hot breath as of the spirit of the wild prairie. The prairie a-foot and angered and battle mooded—this is the buffalo. For the buffalo was a sullen, laughterless, frigid occupant of the horizon, as stationed there by the prairie to keep intruders out. . . . I know not anything more expressive of America. And he is a prairie figure. The woods he courted not; but the long, green-gray, brown, or green grass-grown, wind-swept and wind-billowed prairie-stretches set against the turquoise hedges of remote skies, these he loved,—there he flourished and rejoiced. The buffalo is the monstrous prairie become perambulant, the interior of a continent heaving into prodigious and portentous battle charge.

[. . .]

And the meadow-lark! I have a quarrel with whoever named him. He is ill-named; and that is unfair dealing with man or bird. This jaunty bird is not meadow-lark; he is prairie-lark. A meadow is civilized; and the lark is not civilized. Meadows with tame grass grow too eagerly, and mature too precociously, for this bird of the dappled yellow breast to nest his young and get them ready for their lifelong voyage across the fields. Blue grass, clover, alfalfa, timothy are all unfitted for this prairie-nester. But prairie grasses, to which this sweet bird is native, grow slowly enough to allow all gentle larks to be graduated from their nests or ever a sickle clatters across the hayfield. The prairie-lark! Now that is delicious. Now I know why his garments are mottled like the hawk's and the prairie-chicken's; now I know why his breast is yellow. 'T is watching for the dawn so long as to have caught and held the earliest yellow rays of Summer on his breast when morning spilled its light through twenty million dewdrops on his heart. Now I know why his voice is haunting, and why it haunts me so. It is the laughter of the prairie. It is the prairie gladness lifted to a song.

[. . .]

All the prairie has are grass and flowers. Trees belong not to the prairies. Is that strange? How strange? Shrubs belong not to the prairies, save here and there a clump of sumacs cluster like sheep about a gentle shepherd. How often, when a child, a child of the prairie, have I gone out barefoot and alone to go wild-strawberrying upon the prairies! Nobody told me I was going to poetry or picking poetry. Maybe nobody knew. I did not then, though I do now. But out on the lifting prairie hills, free, free, free, with the prairie wind blowing full in my face and pushing me rudely as to get rid of my intrusion, though I heeded not, thinking it was jesting, as indeed it was; but the prairie wind piped with its prairie orchestra and the sky gave the prairies room; and the clouds fled from the winds affrighted, and the straying breeze came in sudden gusts with lulls of quiet; and the grass answered to each touching of the wind more mobile than a sea-wave, and would at a gust bend low as to let the wind, so evidently in a hurry, get past. And the curlew's call; and the prairie-lark—that is his name, I will call him that—lit, like a winged song, in the green pools of grass and gave his lyric to the wind as a love-token; and I, the little lad with bare feet and freckled hands and freckled face and holding tight to the tin berry-pail and with wild sense of wideness and of freedom—albeit as a child all undefined; for children define nothing. They are too busy living. Definitions come when we have gotten into the back bays of life. I called to the winds and tried to answer the prairie-lark, and could not, seeing he has monopoly of his own singing, and will not lend his lute even to a freckled boy out wild-strawberrying,—and called aloud, not knowing why I called, and danced with the wild bee and waded knee-deep (a small boy's knee-deep) in prairie grasses, and felt the quiver of the green grass-blades about my ankles and naked knees, and ran or puttered, as my mood turned, toward a group of sumacs, if I could see one, across the quivering plain, and was rarely disappointed in finding wild strawberries in their shadow. And the sumac thickets on the prairies are plowed under; and the strawberry beds are lost; and the prairie is a field of corn; but those sumacs cast shadows in my heart; and the wild strawberries crimson my fingers and my lips; and the tang of those winds and voices when I was a boy is on me; and the lark flirts from the bunch of grasses where his little folks are growing wings, and tosses his song out like a rollicking voice of the prairie winds,—and where am I? Yes, I know now,—I am strawberrying out in the prairies with bare legs and freckled face, and girt round by miles of unfenced prairie and thousands of miles of prairie winds and an ecstasy of wideness.

And to look at a sky-full of prairie, and never a tree or bush on all of it, unless a palmy clump of sumacs camped there! How shall we explain this? We can not. Some things are to be named as a child is. Nobody knows why prairies are treeless and shrubless. That is one of the prairie mysteries. When planted here, trees and shrubs grow and prosper. There are no hostilities of soil or breath; but they did not grow there of themselves. That is all. Grass owned the prairie. How often I have let my eyes drift like a cloud across long miles of prairie, and not a sight of any shrub from sky to sky! The wild grass had its way. A clump of wild weeds, clouds of wild flowers, but not a bush in sight; for so is the prairie sacred to the grass. And with wide stretching prairies the eye grows a glutton. It knows not the meaning of enough. I have, many's the day, swept swiftly across prairies fifty miles an hour, watching every inch of the journey, loving every inch of the journey, drinking it in as the thirsty earth does dew but never having sufficiency, not to say satiety. I become exhilarant. I become bacchanal. The wideness, the delight, the freedom fairly inebriate the spirit. The air is fresh and keen: the wind scurries like Indian riders: the grasses lean down to kiss the earth, who is their mother, and lift again to catch the wind's caress, and answer to it in fitful allegiancy; then race madly like a thing gone mad; then in a moment, without visible provocation, quiet, till the calm is like the calm of the high heavens between dim stars. Who will articulate, for those who know it not or feel it not, this drench of delight, this rapture of living?

[. . .]

The prairie is the sea of the land. The ancestors of this hand that writes were sea-born and seafaring folks for nameless generations; and this son of theirs never saw the sea until he was man full grown. He heard that somber minstrelsy only in his dreams by dark and day, but was reared mid-prairie and has found that, as a fact in his spiritual biography, the prairie took with him the place of the ancestral sea. Prairie and sea plant no other hedgerows than the sky. Both billow out into the universe. This is a great, strange presence,—this intimation of the infinite, this feeling that your journey leads you into space, that if your feet would walk to the horizon verge they would thereafter journey out into the sky. My opinion holds that this feeling is more visible in prairies than seas, the reason being, probably, that sea and sky are both amethyst, and sea melts into sky. They seem not two but one. All appears terraqua; but with the prairie, its chrysophrase is strikingly, beautifully, and I will add gloriously, contrasted with the sky amethyst, so that here are two lands apparent on which the path is set. The prairie path leads to the sky path; the paths are one: the continents are two; and you make your journey from the prairies to the sky. You tramp across the chrysophrase into the amethyst. O journey from the near into the far, O journey from the swaying green into the becalmed, illimitable blue, O journey from the earth with prairie-wind a-blowing in the face into the far, dim spaces level as the seashore, where, by imperceptible tilt of landscape, I am led at last up to the purple hilltops of eternity, where blow unhindered and forever the winds of God,—O journey, journey!

The spacious prairie is helper to a spacious life. Mountains shut us in; prairies let us out. Mountains are barrier builders; prairies are barrier destroyers. Prairies make level roadways for the soul to walk, and invite outward, outward to the sky, which invitation is passionate and eloquent beyond describing. Prairies lead into the sky! Had you learned that, my heart? They aid to grow a roomy life. Big thoughts are nurtured here, with little friction. The world does not seem great, but is great. Goals seem suggestions and not destinations. Room, room! On the prairies you may stand tiptoe and your uplifted finger-tips have no fear that they will touch the sky, and you may have and feel no lack of room. "No hindrance" would appear a legitimate motto for the stately prairies; and the motto is sublime.

Who writes about the grass with all the poetry of it? No one has expressed that poetry. Ruskin has approximated being its voice more adequately than anybody else; but I think the prairies will die without grass finding a voice. Its democracy may be against it. John J. Ingalls celebrated blue grass; but blue grass is civilized. Prairie grass is barbarian. One has been taken to school: the other knows not the meaning of a schoolhouse door. The one is conventional: the other is free as birds. The one belongs to dooryards and pastures: the other to the spaces where winds are grown and storms begin. One is reproducible: the other, uprooted, dies. This is the pathos of the prairie, that, once turned over by the plow, prairie grasses die. Some other vegetation grows on the sod where prairie was. Grass does not. Prairie grass plowed up is eliminated. It grows but once. I have a plot of prairie kept to prairie for prairie's sake; and no one can dig a plowshare into that sod. Sacred it shall remain to prairie grass. If you uproot a pine you can plant another, or cut down an oak you can plant an acorn, and so of fruit or flower. Violets have seeds. Prairie grasses have only roots, so that, when once the prairie grasses die, no cunning art known to the husbandman can ever coax their radiant greens into life again. Was I not right? Is not this the pathos of the prairies? They die out like the buffalo. Where civilization digs deep with its spade, forests readily rise to reassert lost ascendency. . . . But prairies have no art of resurrection. They, like the brokenhearted, have no to-morrow. And if prairie grasses cling with a tenacious persistence along the field edges, lay this disposition not to obstinacy but to love of life. 'T is the wild grasses' last chance. Sunflowers can seed themselves across a landscape, but prairie grasses can only creep inches where roots in the dark soil reach out. A voice calling for help, "Let not the prairie grasses die. Keep a plot for memory, for nature." Green, vivid prairie for remembrance! Remembrance of the morning of the world! And to me alfalfa, millet, timothy, white clover, red clover, I love them, and faithfully, but all appeal in lesser passion than the prairie grass. They belong to to-day: this, to yesterday. They are the here: these, the there. They can not live without civilization: these can not well live with it. Day was when the land was theirs alone. Besides, prairie grass is so beautiful. Blue grass (and I do not offend it) is demure; but prairie grass is vivid, as if God had just dyed it. Essential surprise is on its face, the wide wonder of a face just waking in the dewy morn. Prairie grass never seems to know anybody. It forgets faces, or, what I suppose is more accurate, does not recognize them. The prairies belong to the sky, and do not, in their nomad vocabulary, know the meaning of a face. They creep where their king is; they journey toward the sky.

I can no more get enough of a wide prairie than I can of a sunrise. I can sit for months of days watching the level stretches and never feel a sense of enoughness, not to say satiety. I know no limit. Owning the landscape, that is the prairie. March, march, march, what is on the march? Why, silly friend, the grass is on the march.

[. . .]

When a little babe and held against a mother's gentle breast, this writer voyaged across the prairies from Fort Leavenworth to the Rockies. And what a voyage that must have been! Though he, a babe, had no knowledge of the strange witchery of his journey, to sail across a prairie-level sea, with a mother's arms for a hammock, and a mother's bosom for a pillow, and a mother's singing for a sea lullaby,—that was a voyage to make all sea-goings uninteresting as a twice-told tale. But he, poor little lad, knew naught of this divine poetry, but across the long green levels steadily sailed,—tossed on the stranded billows, this prairie schooner sailed. Instead of tritons to blow the sail, there were oxen to pull it. But this was a brave voyage. The crew were my father and my mother; and I, the little tyke, was sole passenger. And afterward, when I was but a little lad, only old enough to remember a very, very little, I sailed back across these identical prairies; only now, I sailed motherless; and the crew was my father, I still being sole passenger. No arms nor breast nor lily wonder of a mother's face, nor star-shining of a mother's eyes, nor the whole world of tenderness of a mother's hugging arms, but the bearded sea-going man, crew of the prairie schooner, and the little passenger. . . . And the voyaging—nothing besides the crew and passenger—and sailing on steadily with the merest dalliance of motion; and camping at night and noon and we cooked our supper and breakfast with a fire of buffalo chips; and the lad would fall asleep watching his father attending the beasts, and would find himself sleeping in the night under the schooner cover, and would look out and find the stars shining far but brilliant in the prairie sky, and would waken his father to ask if all were well, and if God were there, and where was mother, mother?—and being told by the father, whose voice, as he now recalls, choked at the saying, that God was here, and mother was out past and beyond the stars; and she was with God there, as God was with us here; but she was watching her little lad;—then the lonely little lad would sob a little, and fall asleep on his father's heart! He recollects! And now the father and mother are both with God there; and their lad is with God here. But the voyaging across the prairie in the white prairie schooner was the invasion of his soul by the prairie, and has staid with him through years, and will stay with him all the years that are to come. He is, by all pre-emption, a son of the prairies.

The Prairie

Irving made mention of the loneliness of the prairies. He is quite right. They are lonely, with a loneliness for which tears are no alleviation. Whittier, with an inaccuracy born of ignorance, talks of the prairie moaning like a broken heart. This is far away from truth; sea-waves moan: prairies never moan. By no license of interpretation can you so construe them. Prairie grasses swish. In that sound heard at night is something weird as a dead man seen alone by moonlight. I have lain all night long, many's the night, listening to the weird voices of the grass,—I know nothing comparable with it. No fields of oats, wheat, or corn can be mentioned with it. Corn rustles; grass swishes; winter grass heard by dark, when the winds brew tempests, will put your spirit in frame for believing "The Rime of the Ancient Mariner" and the poems of Edgar Allan Poe. Loneliness, wide as a sky, will grip your spirit. A moan will lurch from your lips. You will feel the tragedy of space. The being shut out in a universe alone, forsaken, solitary, in the unpathed spaces where no stars light their lamps nor any angels ever happen by nor any guidance is afforded and where a compass would be a terrestrial bauble because its needle would find no pole star,—such loneliness will drench you, as I have been drenched with the falling waves of ragged seas. The swish of the grass, the long reaching of the darkness, spaces laying hold on you like an iron hand, spaces speaking to you in a husky whisper, fearful as battle, frightful as death! The lonely prairies, with the rush of the tireless wind and the swish of the tawny grasses and the last touch of loneliness, the "kye-yi" of the prairie-wolf tossing his wild cry for the winds to carry where they will,—loneliness, thy other name, thy one true synonym, is prairie.

Or to watch prairies by moonlight. Not to have poetry of light poured, as among the hills, into a bowl, but on the wide prairies to distill as an atmosphere from marge to marge. Not so many nights ago I chanced to stand upon a prairie when the moon was full and very silvery. I was as one drenched in a silver haze, a halo such as angels wear about their brows when they are visitants to man. A-near the prairies were rapturous in the light which concealed while it revealed: further the silver shone above the nodding grasses and, far out on the last marge of prairie and of sky, the moonlight enveloped the landscape like a recollection of life. Nothing to hinder the moonlight. No shadow cast. Only silvery light sown to wide spaces where night-winds wakened the nodding grasses with petulant hands, petulant yet caressing. No shadow, only light; and the calling of the wind in whispers to the nodding grasses, "Wake, O wake! We are come: we, moonlight and the summer wind: wake!" and the prairies lift up their lips for the kisses of the moonlight and the wind, and then fall back into a happy sleep, shone through with happy dreams. And moonlight and the prairie winds fall a-wooing each other till the dawn.

From *The Prairie and the Sea* by William A. Quayle (Cincinnati, OH: Jennings and Graham, 1905).

✳ JOHN MUIR, 1838–1914 ✳

Founder of the Sierra Club, and considered one of the luminaries of American environmentalism, John Muir spent much of his youth in southern Wisconsin, where his family emigrated from Scotland in 1849. After studying botany and geology at the University of Wisconsin, Muir eventually followed his passion for exploration and the study of nature to the Sierra Nevada Mountains of California, where his writing and activism was instrumental in establishing Yosemite National Park. Muir's belief in the ecological and spiritual integrity of nature, for its own sake, increasingly drew him into conflict with those who saw nature as existing primarily for human use. Although he found the writing process arduous, Muir published numerous essays, articles, editorials, and books over the course of his lifetime, including *The Story of My Boyhood and Youth* (1913), in which he recalls exploring the Wisconsin mix of oak savannas, wetlands, and wet-mesic prairies surrounding his family's Fountain Lake farm, now a National Historic Landmark. In this excerpt, Muir gives a sense of why, despite severe treatment by his father, he would later refer to his time there as "the happiest days and scrap portions of my life" (Gifford, *John Muir: His Life and Letters* [1996], 113).

A New World

[. . .]

To this charming hut, in the sunny woods, overlooking a flowery glacier meadow and a lake rimmed with white water-lilies, we were hauled by an ox-team across trackless carex swamps and low rolling hills sparsely dotted with round-headed oaks. Just as we arrived at the shanty, before we had time to look at it or the scenery about it, David and I jumped down in a hurry off the load of household goods, for we had discovered a blue jay's nest, and in a minute or so we were up the tree beside it, feasting our eyes on the beautiful green eggs and beautiful birds, — our first memorable discovery. The handsome birds had not seen Scotch boys before and made a desperate screaming as if we were robbers like themselves, though we left the eggs untouched, feeling that we were already beginning to get rich, and wondering how many more nests we should find in the grand sunny woods. Then we ran along the brow of the hill that the shanty stood on, and down to the meadow, searching the trees and grass tufts and bushes, and soon discovered a bluebird's and a woodpecker's nest, and began an acquaintance with the frogs and snakes and turtles in the creeks and springs.

This sudden plash into pure wildness — baptism in Nature's warm heart — how utterly happy it made us! Nature streaming into us, wooingly teaching her wonderful glowing lessons, so unlike the dismal grammar ashes and cinders so long thrashed into us. Here without knowing it we still were at school; every wild lesson a love lesson, not whipped but charmed into us. Oh, that glorious Wisconsin wilderness! Everything new and pure in the very prime of the spring when Nature's pulses were beating highest and mysteriously keeping time with our own! Young hearts, young leaves, flowers, animals, the winds and the streams and the sparkling lake, all wildly, gladly rejoicing together!

Next morning, when we climbed to the precious jay nest to take another admiring look at the eggs, we found it empty. Not a shell-fragment was left, and we wondered how in the world the birds were able to carry off their thin-shelled eggs either in their bills or in their feet without breaking them, and how they could be kept warm while a new nest was being built. Well, I am still asking these questions. When I was on the Harriman Expedition I asked Robert Ridgway, the eminent ornithologist, how these sudden flittings were accomplished, and he frankly confessed that he did n't [*sic*] know, but guessed that jays and many other birds carried their eggs in their mouths; and when I objected that a jay's mouth seemed too small to hold its eggs, he replied that birds' mouths were larger than the narrowness of their bills indicated. Then I asked him what he thought they did with the eggs while a new nest was being prepared. He didn't know; neither do

I to this day. A specimen of the many puzzling problems presented to the naturalist.

We soon found many more nests belonging to birds that were not half so suspicious. The handsome and notorious blue jay plunders the nests of other birds and of course he could not trust us. Almost all the others—brown thrushes, bluebirds, song sparrows, kingbirds, hen-hawks, nighthawks, whip-poor-wills, woodpeckers, etc.—simply tried to avoid being seen, to draw or drive us away, or paid no attention to us.

We used to wonder how the woodpeckers could bore holes so perfectly round, true mathematical circles. We ourselves could not have done it even with gouges and chisels. We loved to watch them feeding their young, and wondered how they could glean food enough for so many clamorous, hungry, unsatisfiable babies, and how they managed to give each one its share; for after the young grew strong, one would get his head out of the door-hole and try to hold possession of it to meet the food-laden parents. How hard they worked to support their families, especially the red-headed and speckledy woodpeckers and flickers; digging, hammering on scaly bark and decaying trunks and branches from dawn to dark, coming and going at intervals of a few minutes all the live-long day!

We discovered a hen-hawk's nest on the top of a tall oak thirty or forty rods from the shanty and approached it cautiously. One of the pair always kept watch, soaring in wide circles high above the tree, and when we attempted to climb it, the big dangerous-looking bird came swooping down at us and drove us away.

We greatly admired the plucky kingbird. In Scotland our great ambition was to be good fighters, and we admired this quality in the handsome little chattering flycatcher that whips all the other birds. He was particularly angry when plundering jays and hawks came near his home, and took pains to thrash them not only away from the nest-tree but out of the neighborhood. The nest was usually built on a bur oak near a meadow where insects were abundant, and where no undesirable visitor could approach without being discovered. When a hen-hawk hove in sight, the male immediately set off after him, and it was ridiculous to see that great, strong bird hurrying away as fast as his clumsy wings would carry him, as soon as he saw the little, waspish kingbird coming. But the kingbird easily overtook him, flew just a few feet above him, and with a lot of chattering, scolding notes kept diving and striking him on the back of the head until tired; then he alighted to rest on the hawk's broad shoulders, still scolding and chattering as he rode along, like an angry boy pouring out vials of wrath. Then, up and at him again with his sharp bill; and after he had thus driven and ridden his big enemy a mile or so from the nest, he went home to his mate, chuckling and bragging as if trying to tell her what a wonderful fellow he was.

This first spring, while some of the birds were still building their nests and very few young ones had yet tried to fly, father hired a Yankee to assist in clearing eight or ten acres of the best ground for a field. We found new wonders every day and often had to call on this Yankee to solve puzzling questions. We asked him one day if there was any bird in America that the kingbird could n't [sic] whip. What about the sandhill crane? Could he whip that long-legged, long-billed fellow?

"A crane never goes near kingbirds' nests or notices so small a bird," he said, "and therefore there could be no fighting between them." So we hastily concluded that our hero could whip every bird in the country except perhaps the sandhill crane.

We never tired listening to the wonderful whip-poor-will. One came every night about dusk and sat on a log about twenty or thirty feet from our cabin door and began shouting "Whip poor Will! Whip poor Will!" with loud emphatic earnestness. "What's that? What's that?" we cried when this startling visitor first announced himself. "What do you call it?"

"Why, it's telling you its name," said the Yankee. "Don't you hear it and what he wants you to do? He says his name is 'Poor Will' and he wants you to whip him, and you may if you are able to catch him." Poor Will seemed the most wonderful of all the strange creatures we had seen. What a wild, strong, bold voice he had, unlike any other we had ever heard on sea or land!

A near relative, the bull-bat, or nighthawk, seemed hardly less wonderful. Towards evening scattered flocks kept the sky lively as they circled around on their long wings a hundred feet or more above the ground, hunting moths and beetles, interrupting their rather slow but strong, regular wing-beats at short intervals with quick quivering strokes while uttering keen, squeaky cries something like *pfee, pfee*, and every now and then diving nearly to the ground with a loud ripping, bellowing sound, like bull-roaring, suggesting its name; then turning and gliding swiftly up again. These fine wild gray birds, about the size of a pigeon, lay their two eggs on bare ground without anything like a nest or even a concealing bush or grass-tuft. Nevertheless they are not easily seen, for they are colored like the ground. While sitting on their eggs, they depend so much upon not being noticed that if you are walking rapidly ahead they allow you to step within an inch or two of them without flinching. But if they see by your looks that you have discovered them, they leave their eggs or young, and, like a good many other birds, pretend that they are sorely wounded, fluttering and rolling over on the ground and gasping as if dying, to draw you away. When pursued we were surprised to find that just when we were on the point of overtaking them they were always able to flutter a few yards farther, until they had led us about a quarter of a mile from the nest; then, suddenly getting well, they quietly flew home by a roundabout way to their precious babies or eggs, o'er a' the ills of life victorious, bad boys among the worst. The Yankee took particular pleasure in encouraging us to pursue them.

Everything about us was so novel and wonderful that we could hardly believe our senses except when hungry or while father was thrashing us. When we first saw Fountain Lake Meadow, on a sultry evening, sprinkled with millions of lightning-bugs throbbing with light, the effect was so strange and beautiful that it seemed far too marvelous to be real. Looking from our shanty on the hill, I thought that the whole wonderful fairy show must be in my eyes; for only in fighting, when my eyes were struck, had I ever seen anything in the least like it. But when I asked my brother if he saw anything strange in the meadow he said, "Yes, it's all covered with shaky fire-sparks." Then I guessed that it might be something outside of us, and applied to our all-knowing Yankee to explain it. "Oh, it's nothing but lightnin'-bugs," he said, and kindly led us down the hill to the edge of the fiery meadow, caught a few of the wonderful bugs, dropped them into a cup, and carried them to the shanty, where we watched them throbbing and flashing out their mysterious light at regular intervals, as if each little passionate glow were caused by the beating of a heart. Once I saw a splendid display of glow-worm light in the foothills of the Himalayas, north of Calcutta, but glorious as it appeared in pure starry radiance, it was far less impressive than the extravagant abounding, quivering, dancing fire on our Wisconsin meadow.

Partridge drumming was another great marvel. When I first heard the low, soft, solemn sound I thought it must be made by some strange disturbance in my head or stomach, but as all seemed serene within, I asked David whether he heard anything queer. "Yes," he said, "I hear something saying *boomp*, *boomp*, *boomp*, and I'm wondering at it." Then I was half satisfied that the source of the mysterious sound must be in something outside of us, coming perhaps from the ground or from some ghost or bogie or woodland fairy. Only after long watching and listening did we at last discover it in the wings of the plump brown bird.

The love-song of the common jack snipe seemed not a whit less mysterious than partridge drumming. It was usually heard on cloudy evenings, a strange, unearthly, winnowing, spiritlike sound, yet easily heard at a distance of a third of a mile. Our sharp eyes soon detected the bird while making it, as it circled high in the air over the meadow with wonderfully strong and rapid wing-beats, suddenly descending and rising, again and again, in deep, wide loops; the tones being very low and smooth at the beginning of the descent, rapidly increasing to a curious little whirling storm-roar at the bottom, and gradually fading lower and lower until the top was reached. It was long, however, before we identified this mysterious wing-singer as the little brown jack snipe that we knew so well and had so often watched as he silently probed the mud around the edge of our meadow stream and springholes, and made short zigzag flights over the grass uttering only little short, crisp quacks and chucks.

The love-songs of the frogs seemed hardly less wonderful than those of the birds, their musical notes varying from the sweet, tranquil, soothing peeping and purring of the hylas to the awfully deep low-bass blunt bellowing of the bullfrogs. Some of the smaller species have wonderfully clear, sharp voices and told us their good Bible names in musical tones about as plainly as the whip-poor-will. *Isaac, Isaac*; *Yacob, Yacob*; *Israel, Israel*; shouted in sharp, ringing, far-reaching tones, as if they had all been to school and severely drilled in elocution. In the still, warm evenings, big bunchy bullfrogs bellowed, *Drunk! Drunk! Drunk! Jug o' rum! Jug o' rum!* and early in the spring, countless thousands of the commonest species, up to the throat in cold water, sang in concert, making a mass of music, such as it was, loud enough to be heard at a distance of more than half a mile.

Far, far apart from this loud marsh music is that of the many species of hyla [tree frog], a sort of soothing immortal melody filling the air like light.

We reveled in the glory of the sky scenery as well as that of the woods and meadows and rushy, lily-bordered lakes. The great thunderstorms in particular interested us, so unlike any seen in Scotland, exciting awful, wondering admiration. Gazing awe-stricken, we watched the upbuilding of the sublime cloud-mountains,—glowing, sun-beaten pearl and alabaster cumuli, glorious in beauty and majesty and looking so firm and lasting that birds, we thought, might build their nests amid their downy bosses; the black-browed storm-clouds marching in awful grandeur across the landscape, trailing broad gray sheets of hail and rain like vast cataracts, and ever and anon flashing down vivid zigzag lightning followed by terrible crashing thunder. We saw several trees shattered, and one of them, a punky old oak, was set on fire, while we wondered why all the trees and everybody and everything did not share the same fate, for oftentimes the whole sky blazed. After sultry storm days, many of the nights were darkened by smooth black apparently structureless cloud-mantles which at short intervals were illumined with startling suddenness to a fiery glow by quick, quivering lightning-flashes, revealing the landscape in almost noonday brightness, to be instantly quenched in solid blackness.

But those first days and weeks of unmixed enjoyment and freedom, reveling in the wonderful wildness about us, were soon to be mingled with the hard work of making a farm. I was first put to burning brush in clearing land for the plough. Those magnificent brush fires with great white hearts and red flames, the first big, wild outdoor fires I had ever seen, were wonderful sights for young eyes. Again and again, when they were burning fiercest so that we could hardly approach near enough to throw on another branch, father put them to awfully practical use as warning lessons, comparing their heat with that of hell, and the branches with bad boys. "Now, John," he would say,—"now, John, just think what an awful thing it would be to be thrown into that fire:—and then think of hell-fire, that is so many times hotter. Into that fire all bad boys, with sinners of every sort who disobey God, will be cast as we are casting branches into this brush fire, and although suffering so much, their sufferings will never, never end, because neither the fire nor the sinners can die." But those terrible fire lessons quickly faded away in the blithe wilderness air; for no fire can be hotter than the heavenly fire of faith and hope that burns in every healthy boy's heart.

From *The Story of My Boyhood and Youth* by John Muir, 1913. Reprint (Madison: University of Wisconsin Press, 1965).

✳ HAMLIN GARLAND, 1860-1940 ✳

Raised on farms in Wisconsin and Iowa, Hamlin Garland focused much of his early fiction writing on the hardships faced by midwestern farmers, including his family, and actively promoted economic reforms that he believed would help them prevail against powerful, moneyed interests. In an influential speech at the 1893 Chicago World's Fair, Garland also called for a new American literature written by and about those living in the rural center of the nation, railing against an educational system that privileged the teaching of European and eastern American writers. He eventually moved to the coasts himself, however, and by the time he passed away in Los Angeles, had authored over forty books and earned many literary honors, including the 1922 Pulitzer Prize. While living in New York, he completed his acclaimed Middle Border series, four autobiographical books that sought to record pioneer life for posterity, while also criticizing the idealized pioneer myths that led to so many struggles for farmers and their families. In this excerpt from *A Son of the Middle Border* (1917), Garland shares his conflicted feelings about plowing the wild prairies near his boyhood home in Mitchell County, Iowa—a landscape he now wishes he could share with his children.

The Homestead on the Knoll

Spring came to us that year with such sudden beauty, such sweet significance after our long and depressing winter, that it seemed a release from prison, and when at the close of a warm day in March we heard, pulsing down through the golden haze of sunset, the mellow *boom, boom,*

boom of the prairie cock our hearts quickened, for this, we were told, was the certain sign of spring.

Day by day the call of this gay herald of spring was taken up by others until at last the whole horizon was ringing with a sunrise symphony of exultant song. "*Boom, boom, boom!*" called the roosters; "*cutta, cutta, wha-whoop-squaw, squawk!*" answered the hens as they fluttered and danced on the ridges—and mingled with their jocund hymn we heard at last the slender, wistful piping of the prairie lark.

With the coming of spring my duties as a teamster returned. My father put me in charge of a harrow, and with old Doll and Queen—quiet and faithful span—I drove upon the field which I had plowed the previous October, there to plod to and fro behind my drag, while in the sky above my head and around me on the mellowing soil the life of the season thickened.

Aided by my team I was able to study at close range the prairie roosters as they assembled for their parade. They had regular "stamping grounds" on certain ridges, where the soil was beaten smooth by the pressure of their restless feet. I often passed within a few yards of them.—I can see them now, the cocks leaping and strutting, with trailing wings and down-thrust heads, displaying their bulbous orange-colored neck ornaments while the hens flutter and squawk in silly delight. All the charm and mystery of that prairie world comes back to me, and I ache with an illogical desire to recover it and hold it, and preserve it in some form for my children.—It seems an injustice that they should miss it, and yet it is probable that they are getting an equal joy of life, an equal exaltation from the opening flowers of the single lilac bush in our city back-yard or from an occasional visit to the lake in Central Park.

Dragging is even more wearisome than plowing, in some respects, for you have no handles to assist you and your heels sinking deep into the soft loam bring such unwonted strain upon the tendons of your legs that you can scarcely limp home to supper, and it seems that you cannot possibly go on another day,—but you do—at least I did.

There was something relentless as the weather in the way my soldier father ruled his sons, and yet he was neither hard-hearted nor unsympathetic. The fact is easily explained. His own boyhood had been task-filled and he saw nothing unnatural in the regular employment of his children. Having had little play-time himself, he considered that we were having a very comfortable boyhood. Furthermore the country was new and labor scarce. Every hand and foot must count under such conditions.

There are certain ameliorations to child-labor on a farm. Air and sunshine and food are plentiful. I never lacked for meat or clothing, and mingled with my records of toil are exquisite memories of the joy I took in following the changes in the landscape, in the notes of birds, and in the play of small animals on the sunny soil.

There were no pigeons on the prairie but enormous flocks of ducks came sweeping northward, alighting at sunset to feed in the fields of stubble. They came in countless myriads and often when they settled to earth they covered acres of meadow like some prodigious cataract from the sky. When alarmed they rose with a sound like the rumbling of thunder.

At times the lines of their cloud-like flocks were so unending that those in the front rank were lost in the northern sky, while those in the rear were but dim bands beneath the southern sun.—I tried many times to shoot some of them, but never succeeded, so wary were they. Brant and geese in formal flocks followed and to watch these noble birds pushing their arrowy lines straight into the north always gave me special joy. On fine days they flew high—so high they were but faint lines against the shining clouds.

I learned to imitate their cries, and often caused the leaders to turn, to waver in their course as I uttered my resounding call.

The sand-hill crane came last of all, loitering north in lonely easeful flight. Often of a warm day, I heard his sovereign cry falling from the azure dome, so high, so far his form could not be seen, so close to the sun that my eyes could not detect his solitary, majestic circling sweep. He came after the geese. He was the herald of summer. His brazen, reverberating call will forever remain associated in my mind with mellow, pulsating earth, springing grass and cloudless glorious May-time skies.

As my team moved to and fro over the field, ground sparrows rose in countless thousands, flinging themselves against the sky like grains of wheat from out a sower's hand, and their chatter fell upon me like the voices of fairy sprites, invisible and multitudinous. Long swift narrow flocks of a bird we called "the prairie-pigeon" swooped over the swells on sounding wing, winding so close to the ground, they seemed at times like slender air-borne serpents,—and always the brown lark whistled as if to cheer my lonely task.

Back and forth across the wide field I drove, while the sun crawled slowly up the sky. It was tedious work and I was always hungry by nine, and famished at ten. Thereafter the sun appeared to stand still. My chest caved in and my knees trembled with weakness, but when at last the white flag fluttering from a chamber window summoned to the mid-day meal, I started with strength miraculously renewed and called, "*Dinner!*" to the hired hand. Unhitching my team, with eager haste I climbed upon old Queen, and rode at ease toward the barn.

Oh, it was good to enter the kitchen, odorous with fresh biscuit and hot coffee! We all ate like dragons, devouring potatoes and salt pork without end, till mother mildly remarked, "Boys, boys! Don't 'founder' yourselves!"

From such a meal I withdrew torpid as a gorged snake, but luckily I had half an hour in which to get my courage back,—and besides, there was always the stirring power of father's clarion call. His energy appeared super-human to me. I was in awe of him. He kept track of everything, seemed hardly to sleep and never complained of weariness. Long before the nooning was up, (or so it seemed to me) he began to shout: "Time's up, boys. Grab a root!"

And so, lame, stiff and sore, with the sinews of my legs shortened, so that my knees were bent like an old man's, I hobbled away to the barn and took charge of my team. Once in the field, I felt better. A subtle change, a mellower charm came over the afternoon earth. The ground was warmer, the sky more genial, the wind more amiable, and before I had finished my second "round" my joints were moderately pliable and my sinews relaxed.

Nevertheless the temptation to sit on the corner of the harrow and dream the moments away was very great, and sometimes as I laid my tired body down on the tawny, sunlit grass at the edge of the field, and gazed up at the beautiful clouds sailing by, I wished for leisure to explore their purple valleys.—The wind whispered in the tall weeds, and sighed in the hazel bushes. The dried blades touching one another in the passing winds, spoke to me, and the gophers, glad of escape from their dark, underground prisons, chirped a cheery greeting. Such respites were strangely sweet.

So day by day, as I walked my monotonous round upon the ever mellowing soil, the prairie spring unrolled its beauties before me. I saw the last goose pass on to the north, and watched the green grass creeping up the sunny slopes. I answered the splendid challenge of the loitering crane, and studied the ground sparrow building her grassy nest. The prairie hens began to seek seclusion in the swales, and the pocket gopher, busily mining the sod, threw up his purple-brown mounds of cool fresh earth. Larks, bluebirds and king-birds followed the robins, and at last the full tide of May covered the world with luscious green.

Harriet and Frank returned to school but I was too valuable to be spared. The unbroken land of our new farm demanded the plow and no sooner was the planting on our rented place finished than my father began the work of fencing and breaking the sod of the homestead which lay a mile to the south, glowing like a garden under the summer sun. One day late in May my uncle David (who had taken a farm not far away), drove over with four horses hitched to a big breaking plow and together with my father set to work overturning the primeval sward whereon we were to be "lords of the soil."

I confess that as I saw the tender plants and shining flowers bow beneath the remorseless beam, civilization seemed a sad business, and yet there was something epic, something large-gestured and splendid in the "breaking" season. Smooth, glossy, almost unwrinkled the thick ribbon of jet-black sod rose upon the share and rolled away from the mold-board's glistening curve to tuck itself upside down into the furrow behind the horse's heels, and the picture which my uncle made, gave me pleasure in spite of the sad changes he was making.

The land was not all clear prairie and every ounce of David's great strength was required to guide that eighteen-inch plow as it went ripping and snarling through the matted roots of the hazel thickets, and sometimes my father came and sat on the beam in order to hold the coulter to its work, while the giant driver braced himself to the shock and the four horses strained desperately at their traces. These contests had the quality of a wrestling match but the men always won. My own job was to rake and burn the brush which my father mowed with a heavy scythe.—Later we dug post-holes and built fences but each day was spent on the new land.

Around us, on the swells, gray gophers whistled, and the nesting plover quaveringly called. Blackbirds clucked in the furrow and squat badgers watched with jealous eye the plow's inexorable progress toward their dens. The weather was perfect June. Fleecy clouds sailed like snowy galleons from west to east, the wind was strong but kind, and we worked in a glow of satisfied ownership.

Many rattlesnakes ("massasaugas" Mr. Button called them), inhabited the moist spots and father and I killed several as we cleared the ground. Prairie wolves lurked in the groves and swales, but as foot by foot and rod by rod, the steady steel rolled the grass and the hazel brush under, all of these wild things died or hurried away, never to return. Some part of this tragedy I was able even then to understand and regret.

At last the wide "quarter section" lay upturned, black to the sun and the garden that had bloomed and fruited for millions of years, waiting for man, lay torn and ravaged. The tender plants, the sweet flowers, the fragrant fruits, the busy insects, all the swarming lives which had been native here for untold centuries were utterly destroyed. It was sad and yet it was not all loss, even to my thinking, for I realized that over this desolation the green wheat would wave and the corn silks shed their pollen. It was not precisely the romantic valley of our song, but it was a rich and promiseful plot and my father seemed entirely content.

Meanwhile, on a little rise of ground near the road, neighbor Gammons and John Bowers were building our next home. It did not in the least resemble the foundation of an everlasting family seat, but it deeply excited us all. It was of pine and had the usual three rooms below and a long garret above and as it stood on a plain, bare to the winds, my father took the precaution of lining it with brick to hold it down. It was as good as most of the dwellings round about us but it stood naked on the sod, devoid of grace as a dry goods box. Its walls were rough plaster, its floor of white pine, its furniture poor, scanty and worn. There was a little picture on the face of the clock, a chromo on the wall, and a printed portrait of General Grant—nothing more. It was home by reason of my mother's brave and cheery presence, and the prattle of Jessie's clear voice filled it with music. Dear child,—with her it was always spring!

From *A Son of the Middle Border* by Hamlin Garland, 1917. Reprint (New York: Penguin, 1995).

The Homestead on the Knoll

✳ ZITKALA-ŠA, 1876-1938 ✳

Zitkala-Ša [es-SAY] (Red Bird), Dakota, was born on the Yankton Reservation in South Dakota, where she participated in the traditional life of her tribe and played in the prairies—"as free as the wind that blew my hair" (8). At age eight, she left with Quaker missionaries for White's Manual Labor Institute in Wabash, Indiana, where she and other Indian children, while being taught to read and write and play music, experienced a difficult process of assimilation that involved cutting their hair and other ways of separating them from their heritage. During the following years, she became an accomplished violinist, writer, political activist, and advocate for women's and Native American rights, while working to bridge the gulf between white society and the traditions of tribal cultures. Toward that end, she authored numerous works of fiction, poetry, and nonfiction, including *American Indian Stories* (1921), which was intended for young readers. This essay, a revised version of a 1902 piece she published in the *Atlantic Monthly* titled "Why I Am a Pagan," speaks to the spiritual influences of the prairie environment.

The Great Spirit

Whe n the spirit swells my breast I love to roam leisurely among the green hills; or sometimes, sitting on the brink of the murmuring Missouri, I marvel at the great blue overhead. With half-closed eyes I watch the huge cloud shadows in their noiseless play upon the high bluffs opposite me, while into my ear ripple the sweet, soft cadences of the river's song. Folded hands lie in my lap, for the time forgot. My heart and I lie small upon the earth like a grain of throbbing sand. Drifting clouds and tinkling waters, together with the warmth of a genial summer day, bespeak with eloquence the loving Mystery round about us. During the idle while I sat upon the sunny river brink, I grew somewhat, though my response be not so clearly manifest as in the green grass fringing the edge of the high bluff back of me.

At length retracing the uncertain footpath scaling the precipitous embankment, I seek the level lands where grow the wild prairie flowers. And they, the lovely little folk, soothe my soul with their perfumed breath.

Their quaint round faces of varied hue convince the heart which leaps with glad surprise that they, too, are living symbols of omnipotent thought. With a child's eager eye I drink in the myriad star shapes wrought in luxuriant color upon the green. Beautiful is the spiritual essence they embody.

I leave them nodding in the breeze, but take along with me their impress upon my heart. I pause to rest me upon a rock embedded on the side of a foothill facing the low river bottom. Here the Stone-Boy, of whom the American aborigine tells, frolics about, shooting his baby arrows and shouting aloud with glee at the tiny shafts of lightning that flash from the flying arrow-beaks. What an ideal warrior he became, baffling the siege of the pests of all the land till he triumphed over their united attack. And here he lay, — Inyan our great-great-grandfather, older than the hill he rested on, older than the race of men who love to tell of his wonderful career.

Interwoven with the thread of this Indian legend of the rock, I fain would trace a subtle knowledge of the native folk which enabled them to recognize a kinship to any and all parts of this vast universe. By the leading of an ancient trail I move toward the Indian village.

With the strong, happy sense that both great and small are so surely enfolded in His magnitude that, without a miss, each has his allotted individual ground of opportunities, I am buoyant with good nature.

Yellow Breast, swaying upon the slender stem of a wild sunflower, warbles a sweet assurance of this as I pass near by. Breaking off the clear crystal song, he turns his wee head from side to side eyeing me wisely as slowly I plod with moccasined feet. Then again he yields himself to his song of joy. Flit, flit hither and yon, he fills the summer sky with his swift, sweet melody. And truly does it seem his vigorous freedom lies more in his little spirit than in his wing.

With these thoughts I reach the log cabin whither I am strongly drawn by the tie of a child to an aged mother. Out bounds my four-footed friend to meet me, frisking about my path with unmistakable delight. Chän is a black shaggy dog, "a thoroughbred little mongrel" of whom I am very fond. Chän seems to understand many words in Sioux, and will go to her mat even when I whisper the word, though generally I think she is guided by the tone of the voice. Often she tries to imitate the sliding inflection and long-drawn-out voice to the amusement of our guests, but her articulation is quite beyond my ear. In both my hands I hold her shaggy head and gaze into her large brown eyes. At once the dilated pupils contract into tiny black dots, as if the roguish spirit within would evade my questioning.

Finally resuming the chair at my desk I feel in keen sympathy with my fellow-creatures, for I seem to see clearly again that all are akin. The racial lines, which once were bitterly real, now serve nothing more than marking out a living mosaic of human beings. And even here men of the same color are like the ivory keys of one instrument where each resembles all the rest, yet varies from them in pitch and quality of voice. And those creatures who are for a time mere echoes of another's note are not unlike the fable of the thin sick man whose distorted shadow, dressed like a real creature, came to the old master to make him follow as a shadow. Thus with a compassion for all echoes in human guise, I greet the solemn-faced "native preacher" whom I find awaiting me. I listen with respect for God's creature, though he mouth most strangely the jangling phrases of a bigoted creed.

As our tribe is one large family, where every person is related to all the others, he addressed me: —

"Cousin, I came from the morning church service to talk with you."

"Yes?" I said interrogatively, as he paused for some word from me.

Shifting uneasily about in the straight-backed chair he sat upon, he began: "Every holy day (Sunday) I look about our little God's house, and not seeing you there, I am disappointed. This is why I come today. Cousin, as I watch you from afar, I see no unbecoming behavior and hear only good reports of you, which all the more burns me with the wish that you were a church member. Cousin, I was taught long years ago by kind missionaries to read the holy book. These godly men taught me also the folly of our old beliefs.

"There is one God who gives reward or punishment to the race of dead men. In the upper region the Christian dead are gathered in unceasing song and prayer. In the deep pit below, the sinful ones dance in torturing flames.

"Think upon these things, my cousin, and choose now to avoid the after-doom of hell fire!" Then followed a long silence in which he clasped tighter and unclasped again his interlocked fingers.

Like instantaneous lightning flashes came pictures of my own mother's making, for she, too, is now a follower of the new superstition.

"Knocking out the chinking of our log cabin, some evil hand thrust in a burning taper of braided dry grass, but failed of his intent, for the fire died out and the half-burned brand fell inward to the floor. Directly above it, on a shelf, lay the holy book. This is what we found after our return from a several days' visit. Surely some great power is hid in the sacred book!"

Brushing away from my eyes many like pictures, I offered midday meal to the converted Indian sitting wordless and with downcast face. No sooner had he risen from the table with "Cousin, I have relished it," than the church bell rang.

Thither he hurried forth with his afternoon sermon. I watched him as he hastened along, his eyes bent fast upon the dusty road till he disappeared at the end of a quarter of a mile.

The little incident recalled to mind the copy of a missionary paper brought to my notice a few days ago, in which a "Christian" pugilist commented upon a recent article of mine, grossly perverting the spirit of my pen. Still I would not forget that the pale-faced missionary and the hoodooed aborigine are both God's creatures, though small indeed their own conceptions of Infinite Love. A wee child toddling in a wonder world, I prefer to their dogma my excursions into the natural gardens where the voice of the Great Spirit is heard in the twittering of birds, the rippling of mighty waters, and the sweet breathing of flowers.

Here, in a fleeting quiet, I am awakened by the fluttering robe of the Great Spirit. To my innermost consciousness the phenomenal universe is a royal mantle, vibrating with His divine breath. Caught in its flowing fringes are the spangles and oscillating brilliants of sun, moon, and stars.

From *American Indian Stories* by Zitkala-Ša, 1921. Reprint (Lincoln: University of Nebraska Press, 1985).

Born in Valley, Nebraska, Melvin R. Gilmore became a pioneer in enthno-botany, blending what some considered to be disparate fields of study. In addition to his botanical studies at the University of Nebraska, he also studied for five years among the Omahas, and later among the Arikaras, Oglalas, Tetons, Poncas, Santees, Winnebagos, and Pawnees, the last adopting him as an honorary member of their tribe. He wrote numerous scholarly articles on indigenous plant and animal lore, and held several academic and museum positions, including curator of ethnology at the University of Michigan. His influential book, *Prairie Smoke* (first published in 1921), was intended for a popular audience, and employed a dramatic, mythic, and often romantic style in the hopes of inspiring new appreciation for prairie ecosystems and tribal cultures. The following excerpt from the expanded 1929 edition includes the traditional stories and almost biographical characteristics of four plants native to the tallgrass prairie: prairie rose, sunflower, spiderwort, and pasqueflower.

The Plant Tribes

The Prairie Rose

The prairie was gray and drab, no beautiful flowers brightened it, it had only dull greenish-gray herbs and grasses, and Mother Earth's heart was sad because her robe was lacking in beauty and brightness. Then the Holy Earth, our mother, sighed and said: "Ah, my robe is not beautiful, it is somber and dull. I wish it might be bright and beautiful with flowers and splendid with color. I have many beautiful, sweet and dainty flowers in my heart. I wish to have them upon my robe. I wish to have upon my robe flowers blue like the clear sky in fair weather. I wish also to have flowers white like the pure snow of winter and like the high white cloudlets of a quiet summer day. I wish also to have brilliant yellow flowers like the splendor of the sun at noon of a summer day. And I wish to have delicate pink flowers like the color of the dawn light of a joyous day in springtime. I would also have flowers red like the clouds at evening when the sun is going down below the western edge of the world. All these beautiful flowers are in my heart, but I am sad when I look upon my old dull gray and brown robe."

Then a sweet little pink flower said, "Do not grieve, mother. I will go up upon your robe and beautify it." So the little pink flower came up from the heart of Mother Earth to be upon the sad prairie of her mother's robe.

Now, when the Wind Demon saw the pink flower there, he said, "Indeed she is pretty, but I will not have her trespassing in my playground." So the Wind Demon rushed at her, shouting and roaring, and blew out her life, but her spirit returned to the heart of Mother Earth.

And when the other flowers ventured, one after another, to come out upon the prairie, which was Mother Earth's robe, the Wind Demon destroyed them also, and their spirits returned to the heart of Holy Mother Earth.

At last Prairie Rose offered to go and brighten the appearance of Mother Earth's robe, the prairie. Mother Earth said fondly, "Yes, dear, sweet child, I will let you go. You are so lovely and your breath is so sweet, it may be that the Wind Demon will be charmed by you, and that he will let you remain on his ground." And Prairie Rose said, "Yes, dear mother, I will go, for I desire that my mother's robe shall be beautiful. But if the Wind Demon should blow out my life, my spirit shall return home to the heart of my mother."

So Prairie Rose made the toilsome journey up through the dark ground and came out upon the sad gray prairie. And as she was going, Mother Earth said in her heart, "Oh, I hope the Wind Demon will allow her to live, for I wish my robe to be beautiful!"

Now, when the Wind Demon saw Prairie Rose, he rushed at her, shouting, and said, "Indeed, though she is pretty, I shall not allow her to be upon my ground. I will blow out her life." So he came on, roaring and drawing his breath in strong gusts. Just then he caught the fragrance of the breath of Prairie Rose. "Ah," he said, "how sweet her breath is! Why, I do not have it in my heart to blow out the life of such a beautiful little maiden whose breath is so sweet! I love her. She shall stay here with me. And I must make my voice gentle and sing a melodious song, for I wish not to frighten her with my awful noise."

So he became quiet and breathed gentle breezes which passed over the prairie grasses whispering and humming little songs of gladness.

Then the other flowers also came up through the dark ground and out upon the dull gray prairie and made it bright and joyous with their presence. And the wind came to love all the flowers and the grasses.

And so the robe of our Mother Earth became beautiful because of the loveliness and the sweet breath of the Prairie Rose.

Sometimes the Wind forgets his gentle songs and becomes loud and boisterous, but he does not harm a person whose robe is ornamented with the color of Prairie Rose.

The Sunflower

Once on a time, long ago, a company of Dakota men were going upon a war expedition. And now, as they were within the country of the enemy, they were proceeding very cautiously. One morning very early, they heard what seemed to be the sound of someone singing in a tremulous voice, coming from the direction toward which they were marching. They stopped and stood still to listen.

As they stood thus listening, it seemed to them that the singer, whoever he might be, must be a clown, for he was singing a clown song. There was not light enough to see the singer. But they waited, silently and anxiously peering ahead in the direction from which came the sound of the singing. At the first glimmer of the dawn light they were able to make out the appearance of a man walking with an awkward shuffling gait. His robe was ragged and his leggings drooped down slouchingly in wrinkles about his ankles as he walked. He had circles about his eyes painted a bright yellow, and he was singing a clown song in a husky, wheezy voice.

So they stood wondering at the clown who was coming toward them. He was coming toward the rising sun, and as the daylight grew brighter they were astonished to see the man suddenly changed to a sunflower.

And ever since that time, it is said, the sunflower faces toward the sun.

The Spiderwort

The spiderwort is a beautiful native prairie flower which is known under numerous popular names. It is called spiderwort, spider lily, ink flower, king's crown and various other names. It has been proposed to add to the list another name, "flower-of-romance." This name is proposed from the circumstance of a bit of pleasing sentiment connected with this flower in the folklore of the Dakota nation of Indians.

It is a charmingly beautiful and delicate flower, deep blue in color, with a tender-bodied plant of graceful lines. There is no more appealingly beautiful flower on the western prairies than this one when it is sparkling with dewdrops in the first beams of the rising sun. There is about it a suggestion of purity, freshness and daintiness.

When a young man of the Dakota nation is in love and walking alone on the prairie finds this flower blooming, he stops and sings to it a song in which he personifies it with the qualities of his sweetheart's personality as they are called to his mind by the appearance of the flower before him, its characteristics figuratively suggesting the characteristics of her whose image he carries romantically in his mind and heart. In his mind the beauties of the flower and the charms of the girl are mutually transmuted and flow together into one image.

The words of his song, translated from the Dakota language into the English, are something like this:

> Tiny, gladsome flower,
> So winsome and modest,
> Thou art dainty and sweet.
> For love of thee I'd die.

The Song of the Pasque Flower

The pasque flower has a very extensive range upon the northern prairies, reaching from about 43 degrees north latitude to the Great Slave Lake, above 60 degrees north latitude. It is the earliest flower to put forth its blossoms in the springtime, often appearing before all the snow is gone. Its bluish, lavender-colored flowers gladden the bare brown hillsides with great profusion of bloom, an earnest of returning life. For this reason it has a strong hold upon the affections of all the native tribes throughout all its extended range. The plant is closely related to the anemone, which is sometimes called the windflower.

The people of the Dakota nation have a number of pretty little folk stories concerning the pasque flower. One story is that in the long ago, whenever any of the people happened to pass by where these flowers were blooming, the flowers tried to show the friendliness which they felt for human beings by nodding their heads in the chilly spring wind, showing their smiling faces and saying, "Good morning! Good morning!" But the people passed them unheeding. The flowers became abashed at this indifference, and so nowadays, still feeling friendly towards the people in spite of such rebuffs, they bashfully turn their heads to one side as they nod and call their kindly greetings in their sweet low voice.

There is another pretty conceit connected with the pasque flower. Indians generally are keenly observant of all things in nature, and reverent towards them. They feel reverence for all living creatures, whether plant or animal. They have songs and stories about many of the species of plants and animals with which they are acquainted, such a song being the expression of the life or soul of the species to which it pertains. The song of the pasque flower, translated out of the Dakota language into English runs something like this:

> I wish to encourage the children of other flower nations
> Which are now appearing over all the land;
> So, while they waken from sleep and rise from the bosom
> Of Mother Earth, I stand here, old and gray-headed.

The saying, "I wish to encourage the children of other flower nations," refers to the very early prevernal blossoming of this plant and its consequent ripening while the other flower species (nations) are just peeping through the ground. The entire plant is hairy, and when mature its seed head is plumose and white, similar to the clematis head, suggesting the head of a very old man with long white hair. This explains the allusion in "I stand here, old and gray-headed."

When in springtime an old man of the Dakota nation first finds one of these flowers, it reminds him of his childhood, when he wandered over the hills at play as free from sorrow and care as the birds and the flowers. He sits down near the flower, upon the lap of Mother Earth, and takes out his pipe and fills it with tobacco. Then he reverently holds the pipe towards the earth, then towards the sky, then in turn towards the four quarters of the horizon. After this act of silent invocation and thanksgiving, he smokes. Tobacco was sacred and was used ceremonially as an incense. The pipe was therefore a sort of censer, and was accordingly treated with respect and reverence. In smoking, Indians did not seize the pipestem in the teeth. Such an act would be sacrilegious. The mouthpiece of the pipestem was gently presented to the lips and the breath was drawn through. By this inspiration the smoker united the mystery of the tobacco, the mystery of fire and the mystery of the breath of life.

While the old man sits by the flower and smokes, he meditates upon all the changing scenes of his lifetime—his joys and sorrows, his youthful hopes, his accomplishments, his disappointments, and the guidance of the Unseen Powers accorded to him thus far upon the journey of life—and he is encouraged to believe that he will be guided to the end of life's journey, "beyond the fourth hill" of life. As he has been guided over the hill of childhood, the hill of youth, and the hill of manhood's prime, so he will also be guided over the last hill, the hill of old age.

After finishing his pipe, he empties the ashes reverently upon the ground near the pasque flower which he has been contemplating. Then he rises and plucks the flower prayerfully and carries it carefully home to show to his grandchildren, singing, as he goes, the song of the pasque flower, which he learned as a child and which he now teaches to his grandchildren, commending to them the example of the flower in its courage and endurance and faithfulness.

From *Prairie Smoke* by Melvin R. Gilmore, 1929. Reprint (St. Paul: Minnesota Historical Society Press, 1987).

Meridel Le Sueur was born in Murray, Iowa, but spent much of her child-hood moving around the Midwest, including Kansas and Minnesota, where she was nurtured "in the weathers of three fertile and giant prairie women" (Le Sueur, "The Ancient People and the Newly Come," *Ripening: Selected Work* [1990], 42)—her mother, her maternal grandmother, and their Mandan friend, Zona. These women, along with her stepfather (once the socialist mayor of Minot, North Dakota), sensitized her to Native American perspectives, class issues, and regional environmental history. During the Depression she published numerous pieces of reportage about the struggles of the impoverished and oppressed, and was blacklisted during the McCarthy era, limiting her access to mainstream publication. This essay, which describes the impact of the Dust Bowl on farm families and the land, was included in her innovative history of Minnesota, *North Star Country* (1945), which focused on the lives of "ordinary" citizens, allowing them to speak in their own voices. An earlier version was published in 1934, in the *American Mercury*, which is why it is placed here in the chronology.

Drought

Because the ground is chapt, for there was no rain in the earth, the plowmen were ashamed and covered their heads. Yes, the hind calved in the field and forsook it, because there was no grass. And the wild asses did stand in the high places, they snuffed up the wind like dragons; their eyes did fail, because there was no grass. —Jeremiah 14:6

Instead of fraternity you will get isolation; instead of inalienable peasant alot-ment the land will be drawn into commerce; instead of a blow at the grabbing speculators, the basis for capitalist development will be expanded. But . . . it is historically good, for it will frightfully accelerate social development and bring much nearer new and higher forms. —Lenin on Karl Marx's Answer to Krieg as to the Opening of Free Western Lands in America

That year, in the early days of the depression, after the bank holiday— what has come to be known as the "crash"—came the drought. I drove through the country trying not to look at the ribs of the horses and the cows, but you got so you couldn't see anything but ribs, like beached hulks on the prairie, the bones rising out of the skin. You began to see the thin farmer under his rags and his wife lean as his cows.

This was not something sudden. It had been happening a long time and came sooner than Hill had predicted. In the spring, after the terrible winter, added to mortgages and low prices, there was no rain. The village where I lived did not exchange money for two years; they bartered and exchanged what produce they had.

I drove to Dakota in the bus through the hot, stifling country, with dark clouds of dust moving steadily eastward into Illinois and Ohio; Dakota walking east. Everyone talked about horses, cattle, seed, land, death, hunger. The bus went steadily through the bleak country, and they pointed out the window at land now owned by the insurance companies, and we saw it splitting open like rotting fruit after years of decay and erosion, exposing the gashed core; it was the fifth year of the depression.

We stop at a little town and some drivers get on, space-drunk young men, their skin pocked, burnt from the dust and wind, driving from Dakota east, driving trucks back and forth. They like to drive swift and mad over the country, a girl in every city. "We went three hundred miles yesterday," one of them says, slouching down, sleeping on the go—huge half-man, half-boy; mid-American faces lolling.

There is something about the American earth that is curiously loved. Even like this, with this dark doom of ruin over it, everyone sees it as it originally was: lovely, green, eldorado, with clear streams and broad pastures beside the still rivers.

The country looks so lonely and silent; I realize with a start that the ground has not been plowed or planted. There was no seed. A farmer answers me before I have spoken, "You can't get you no seed. Last year the drought ruined everything and this year they ain't no seed. My farm plumb blew away. The land is sure ruined this year, and mine for a sight longer. It blew right out from under me, clean as a whistle. The fields packed against the barn so you couldn't see the top, the thistles and tumbleweed caught the dust at the fences so you could walk right over them on solid sand. That there land won't grow a Russian thistle."

"What's going to happen?" another says. "People from my home town, pioneers, mind you, gone to Alaska now. Pioneering some more. It beats the Dutch."

"Holy mackerel!" says the grieving men. "Whole villages ruined! What will they do there?"

And the man talk goes on:

This is the bread-basket of the world going up in dust.

Yes, sir, we were starved, stalled and stranded. Dried out three years runnin'. My wife and child jest dried up and blowed off.

Did ye hear this one? A Dakota farmer was waitin' at the hospital for his first baby. The nurse announced an eight-pound baby to the man waiting with him and to the Dakota farmer she announced a three-pound one and to the surprise of the other he seemed very happy. "Man," said the father, "three pounds is a smidgen of a baby. Don't look to me that baby's got a chance. Why are you so happy?" "I'm a farmer from North Dakota." "What's that got to do with it?" "Why, everything. In North Dakota we're darn glad to get our seed back!"

Got to keep movin'. Jest lookin' fer a home. We didn't stop to shut the door.

Dust got so thick the prairie dogs dug their holes right into the air!

Yes, sir, the government bet me a hundred and fifty acres against my belly that I couldn't stick it out. I won.

That land now jest fittin' to hold the world together.

We seen many a time, many a season. We come a far trek and we'll go farther yet. We'll ride through more seasons and buck more journeys yet.

And the folksay echoes back from the years:

My whole place dried up and blowed away.

I didn't sell out, I give out.

Tractors and the wind is our enemy. Should never have plowed up the buffalo grass.

An Indian turned back the freshly plowed prairie sod and said to the new settler who was showing him how to plow, "Wrong side up!"

Another Indian about to be hanged said, "The land you now take by force is the flesh of my fathers."

The conversation continues in a beer joint when the bus stops for lunch. A man in a tattered coat and split shoes joins us in hopes of a free drink, and the young man has been tippling from a bottle in his pocket and makes a speech. The Constitution is a good one, he says; there's a difference between the Constitution and the government, you can bet your bottom dollar on that. According to the Constitution there is no such thing as sedition and the Constitution permits life, happiness, and shall we say the pursuit of freedom—yes, we will say it. "Your mother told you you were going to be President. That's the bunk. Who in hell *wants* to be? All I want is to find some water to go fishing in. Why didn't she tell you you were going to live a good life, that's all—just that stuff, a good life? That's what the Constitution says for a man: life, liberty, and the pursuit of happiness."

The man in the tattered clothes waves his hands surprisingly. "But what about private capital," he says, and you could have knocked us over with a straw. "Yes, sir, what about private capital? Suppose," he says, "I want to put up a skyscraper, or build a railroad, what about it?"

THE BUS FLED through the country and the sun was obscured by the moving dust. The women put wet cloths over the children's noses. And you could see the kind of terror that had grown in everyone. The winter wheat had died; the peas had gone in and fallen down as if mowed by a hot scythe. The corn never came up. The onion seed blew away. And it was too much, at last. A high wind is an awful thing; it wears you down, it nags at you day after day, it sounds like an invisible army, it fills you with terror as something invisible does. It was like the flu terror. No one went outdoors. People shut up their houses as if from some horrible invasion, some massacre on the streets. The radio took to announcing that there was no danger in the dust, which was not true; many children died of it, of dust pneumonia and other diseases that were never named. When you looked out the window you saw the black cloud of dust going over, you saw the fields whiten and die and the crops creep back into the ground.

At first the farmers kept on plowing, first with two horses, then you had to use four to rip the earth open, and when you did a fume of dust went up like smoke and a wind from hell whipped the seed out.

Every day the pastures got worse. The grass was dry as straw and the cattle lost their flesh quickly. You had to look for a green spot for them every morning. Children were kept out of school to herd the cattle around near streams and creeks. Some farmers cut down the trees so the cattle could have the poor wind-bitten leaves. Some farmers have turned their cows into what was left of their winter wheat, which is thin as a young man's first beard.

Then on Decoration Day all hope went—the wind started again, blowing hot as a blast from hell, and the young corn withered as if under machine-gun fire; in two hours the trees looked as if they had been beaten.

We pass mules hauling a light wagon full of barrels, and with a start I realize that they are hauling water. The wells are dry. The hills are bare and we see no cattle. We pass a freight taking John Deere machinery back to Minneapolis. It is being repossessed. We begin to pass old cars loaded to the gills with household furniture and the back seat full of waving children. They are going west again.

When late afternoon came down not a soul was in sight: the houses closed up tight, the blinds drawn, the windows and doors closed. There seemed to be menace in the air. It was frightening—you could hear the fields crack and dry, and the only movement in the down-driving heat was the dead writhing of the dry blighted tumbleweeds.

There was something terrifying about this visible sign of disaster. It went into your nostrils so that you couldn't breathe: the smell of hunger. It made you count your own ribs with terror. You don't starve in America. Everything looks good. There is something around the corner. Everyone has a chance. Is that all over now?

The dust now becomes so thick, the driver must drive very slowly. It grinds against the windshield. We drive as if going to a funeral; the corpse is the very earth. The houses are closed and stand in the haze hardly visible, unpainted, like the hollow pupa when life has gone. But you know that everywhere in those barricaded houses are eyes drawn back to the burning windows, looking out at next winter's food slowly burning.

The whole countryside bears not only its present famine but its coming hunger. No vegetables now and worst of all, no milk. It is monstrous with this double doom! Every house is alike in suffering; hundreds of thousands of such houses from state to state.

An awful thing happened. The sun went down behind the hot rim of the horizon, and the men and women began to come out of the houses, the children lean and fleet as rats, the tired, lean farm women looking to see what was left. The men ran into their fields, ran back for water, and began to water what was left of their gardens with buckets and cups, running, pouring the puny drops of water on the baked earth as if every minute might count now. The children ran behind the cows, urging them to eat the harsh, dry grass. It looked like an evacuated countryside with the people running out after the enemy had passed. Not a word seemed to be spoken. In intense silence they hurried down the rows with buckets and cups, watering the wilted corn plants, a Gargantuan and terrible and hopeless labor. One man came out stubbornly with a horse and plow and began stirring up the deadly dust. Even the children ran with cups of water, all dogged, mad, without a word. There was a terrible madness in it, like things that are done after unimaginable violence.

A farmer gets on and says he's going anywhere. He's got to get away from the sound of his cows crying for food. He says that the farmer across from him yesterday shot his twenty-two head of cattle and then shot himself.

When I shut my eyes, the flesh burns the eyeballs and all I can see is the sign visible now of starvation and famine, ribs, the bones showing through the skin, rising over the horizon.

IN THE SMALL TOWN, Main Street is crowded and the farm woman and I look in the window of the grocery store. "A few nubbins of corn," she says, "is all that come up. Nobody in our parts could use the tractors, no gasoline, we got out the old mules. We'll be goin' back to the hand plow and the sickle. We lost our mule then and sold the hogs to feed the chickens, and ate the chickens ourselves. Now our place is bare as a hickory nut. I don't think you want to come out. Listen, I haven't got—Oh, it would make me feel bad."

"Never mind," I said, and she looked at me with her bleached eyes, with a touch of strange awkwardness, and bowed quaintly and we got in the old surrey driven by the only son left at home now. The old nag walks through Main Street and the men lounge far back in the shadow of eaves, the street looks stripped as if buzzards had been at it. We move out into the prairie, through the sulphurous haze; the ruined earth slopes into the sun and, strangely, a young cycle of moon shows white in the scorched sky.

We drop into the smoky pit of prairie and have to cover our faces against the blowing sand which will cut into your skin. The bald prairie, without a blade of grass, is mounded unfamiliarly, some houses almost buried, the marks of fences sticking up. The grasshoppers are ticking like clocks.

We drive in on a road marked out since their departure by shifting sands and one side of the two-room shack is banked with sand. The mother scrapes open the door and I go in with her while the son puts away the horse. It is early but she has to light a lamp to see, and the shivering light falls through the yellow dust onto the chairs, and over a bare table covered an inch with dust, which she quickly wipes off. There is a picture on the wall of a strong man with black mustaches; perhaps the man, her husband, who plowed this land. The mother moves around the room, her shy eyes on me, her body taking on the grace of a hostess, "I tell Joe not to hang the towel on the line—the grasshopper'll eat it clean out. They are mighty powerful hungry this year, like everything."

We move closer to each other, spreading the scrubbed wooden table. We become tenderly acquainted in the room as she begins to get excited, in the way of lonely people, telling about her husband—birthing her children right in there, she says, right on that bed without hide or hair of a doc. She is like a drunken person, recounting now her lonely pain, bearing children that are gone and will never return. "There was a time," she said, "when I was afraid to look out the window down past the lower forty—afraid they'd be a-comin' to take somethin' away from us. But that's one relief now; we ain't got nothin' now and nobody is a-comin'!"

We ate dinner, the boy shy at first; then he too is hungry for talk and tells about working on the road, about how he wants to work—he's real strong, about the Knutsons, and he gets up to show me their light eerily spread in the dust down the coolie—they are leaving tomorrow for the west, the whole kit and kaboodle of them. They're getting off that farm where they been all their lives.

"Yes," the mother said, "where all their children was borned and raised. It's shameful. It's a terrible thing."

After supper we can still see the eerie light and imagine the Knutsons moving in their sad tasks. The mother can't keep away from the window and I know she is thinking that they could be put off, too. The boy goes outside and I know he's looking up the hill. He comes in, cracking his knuckles, "They shouldn't do it. They oughtn't get out like that without kicking up a row. What about their money? What about the years of work they got in there? What about it all? You don't hear nothin' about that. The mortgage ain't nothing to have in that land to what they got there, their whole endurin' lives."

"Be quiet," the mother says. "God'll punish you. We are law-abidin' and that's the law."

We go into the lean-to. I am to sleep with the mother. Joe on the floor where he always sleeps. She tells me shyly she hasn't sheets, but her blanket's clean. She blows out the light modestly and we undress in the dark. I feel good to be near her. We lie side by side in the old bed and I know I am sleeping on the side of her dead husband. Accidentally I touch her dry hand and she clutches me, "Do you think they will take it away from us?" I am startled by the fierce vitality of her hand, like a strong bird grasping my fingers. I grasp her hand warmly and she is shaking—not crying, just shaking like a bird captured and straining.

In the morning the sun came up naked as a hot plow in the sky, and there truly seemed as much of the earth in the sky as below it. We start out early for the Knutsons'. Their old Ford is packed and an old rickety trailer sags beneath their household goods. Mrs. Knutson has a baby and is with child, and there must be eight children in the back seat. We stop and we all get out and stand in the dust looking at each other. Nobody knows what to say now. They haven't lived all their lives for this moment. We drive down the hill again. "They have a lot of children," I say inanely, and the mother says bitterly, "That's one crop that never fails—never a crop failure of babies."

The sun is already hot. Before us, standing in the dust, a clot of men are gathered at the gate of a farmhouse; the gate is half buried in dust; the dust sits gray on the clothes of the men, on their slouching hats. The sheriff and three men are standing on the stoop. A man stands in the center of the group of farmers, speaking. As we drive up he calls out, "Join us, brother." Joe stops the horse. "We can't back down," the man is saying. "We can't be afraid. Let him post the notice of auction; we'll be here. Now is the time to do it. You got rights. We got to begin to go forward. Everything is dry as a contribution box, you can see that. You're willin' to pay but you *can't* pay. We got no wheat, no hogs, no nothin'. We got time to think about it, to figure it out—"

Just then the Knutsons come down the hill and drive by without stopping, the children peering back around the sides, Mrs. Knutson not looking back, not waving; and we watched the old gas buggy go down the road and pass out of sight in the dust.

"They's no use going off now," Joe says, "for the West, for Oregon, for Alaska—They's no use goin' off now—"

From *North Star Country* by Meridel Le Sueur, 1945. Reprint (Lincoln: University of Nebraska Press, 1984).

Osage author John Joseph Mathews was born on tribal territory in Pawhuska, Oklahoma, served in the military during World War I, and was educated at the University of Oklahoma and Oxford University in England, before returning to live in his homeland. During the 1930s, Mathews served on the Osage Tribal Council, worked to restore tribal self-government, and helped to establish the Osage Tribal Museum in Pawhuska. A recipient of a Guggenheim Fellowship, Mathews authored several books with Native American themes, including the best-selling novel, *Wah'kon-tah: The Osage and the White Man's Road* (1932). *Talking to the Moon* (1945) recounts his observations of the natural world during his years living in the savanna-like Blackjack country of northeastern Oklahoma, and is organized according to the different Osage "moons" or months of the year. In this chapter, he recounts the agricultural activities associated with the Planting Moon, describes the mating dance of the prairie chicken, and considers the changes in his own life and relationship to the land.

※ ※ ※

Planting Moon

The planting moon corresponds with the month of April and is the shortest name in the Osage calendar. They say simply, "Wah-Pee," and that includes all the rather involved ceremonies of planting and growth: the collection and care of wild lily roots, the gathering of wild onions, and the preparation of the symbolic robes for the children.

All this is woman's work, the female children of the earth planning with Mother Earth and the Moon Woman for the coming of the fruits of the earth; preparing for the nuptials of ageless earth and Grandfather the Sun and all his male manifestations. Certain ceremonies were performed and songs were sung, and often, as they sang their planting songs, the April rains would slant from the sky with a solitary cloud as their source, while the sun, still present, made the raindrops sparkle with life.

Old Ee-Nah-Apee told me one time in her slow, broken English: "My son, you asked about this here Wah-Pee, what we call Plantin' Moon. Some time we call it woman's moon—that's what we say; woman's moon, but we just say that 'cause it's Plantin' Moon. Some time we call Just-Doing-That Moon, crazy woman moon too, but it ain't that.

"L-o-o-o-ong time ago we used to have them bags, I guess you call it, made out of grass. We put corn in these here bags, and we put them on our back; we tie them too with buckskin strings on our back. We go there with long pole in our hand too. We stand there at that place where we gonna plant that corn. We stand there with that pole in right hand, and we look at Grandfather the Sun. We have cleaned all them weeds and stuff off from that place where we gonna plant this here corn; we make little opah, hehn? What you call it hills, hehn?

"Purty soon womens go to them little—hills, I guess, and they make hole with that pole on south side of that there hill. They used to say Grandfather sure would see them holes in them hills on south side, that-a-way. We put corn in them hills, in them little holes; and when we have all of 'em with corn in it, we put our feets on it. We stand on them little hills and make drum against the earth with them poles and sing purty song."

She would then sing the song for me, and, though it fascinated me, I am not sure that I remember it very well, so I have gone elsewhere for a free translation of it:

> I have made a footprint, a sacred one.
> I have made a footprint, through it the blades push upward.
> I have made a footprint, through it the blades radiate.
> I have made a footprint, over it the blades float in the wind.
> I have made a footprint, over it the ears lean toward one another.
> I have made a footprint, over it I pluck the ears.
> I have made a footprint, over it I bend the stalks to pluck the ears.
> I have made a footprint, over it the tassels lie gray.

I have made a footprint, smoke rises from my lodge.
I have made a footprint, there is cheer in my lodge.
I have made a footprint, I live in the light of day.

When she had finished, she would look at me with a new light in her eyes and then laugh as though she were embarrassed by recalling such primitive things that the tribe was now attempting to put away forever. "That was a l-o-o-o-ong time ago," she would say in a patronizing manner, but even so she had lived in those days, and she had great happiness in recalling them.

"Which foot, mother," I once asked her, "did you use to cover the corn?"

"Like always for Chesho it is lef foot, and for Hunkah right one."

She referred to the grand division of the tribe represented by Chesho, the Sky People, and Hunkah, the Earth People, the Sky People being the peace division and the Earth People being the war division.

We do no planting on the ranch, except in a small way. We sometimes plant small fields to kaffir for barnyard feeding and lespedeza in thrown-out fields that are slow in coming back to bluestem. I often have kaffir planted in little half-acre clearings for the prairie chickens and quail. Naturally it is never harvested but left for their harvesting.

I do little planting, then, but that planting song of the Osage runs in my head when the Planting Moon comes and the earth bursts into life and fantastic promises. Rain squalls come often and hang from the sky like some diaphanous drapery while the sun shines.

The Planting Moon could be called the moon of promise, for with it comes hope eternal, and the character of my soil on the sandstone ridge is forgotten. The ground seems rich and fruitful in my yard as I dig with the sharpshooter spade. I dig more holes than I can ever fill with plants because the earth invites digging, and the feel of it and the odor of it inspire dreams. I plan a garden in the savage blackjacks each spring. Sitting in the wheelbarrow with my tools lying about, I dream of the garden beautiful as the wrens sing ecstatically about me, and that garden is a real ambition as I berate myself for the years wasted.

The wrens, however, as they flit nervously here and there exploring every upturned vessel and every cranny in the eaves, realize the dreams about which they sing in a nest and family, but my dreams fade with the song.

One year they built a nest in the pocket of my leather jacket which I had hung on the fence as I delved happily. I had to break up their plans for a home, but I did so on the theory that inasmuch as they are incessant nest-builders anyway, this nest in my pocket was only an overflow expression and had nothing to do with the harsh business of life; an ornamental expression very much like the manuscript which lay neglected on my desk.

[...]

The new hope that comes to me now after ten years in the blackjacks, I express by my deep appreciation of the far horizon and the bubbling life and throbbing earth. The fact that I accept the miracle tranquilly, and stay close to the little sandstone house when the ragged clouds play their mad game, is never put up against my mad desire for action of ten years ago; my desire to struggle playfully against the madness of the elements. The hope that comes with the Planting Moon is all absorbing, ignoring man's limitations and scoffing at experience quite often. Senility puts forth its last tassels, and life overflowing with emotion exuberantly expresses the force within; even to overflowering in artificial ornamentation when maturity is reached.

The biological changes within myself during the last ten years have dimmed the glory of moonlight, dulled my sensitiveness to the waving of a grass blade or the movement of leaf shadows perhaps. They have dulled my high pleasure in driving snow, lightning display, and in driving or riding madly across the prairie to keep in sight of the wolf hounds, but the keen emotion which I feel in watching every spring the dance of the prairie chicken never dims.

Every morning before sunrise during the Planting Moon, when the rain is not falling, I drive the station wagon out onto the high prairie and back it close to a dancing ground, placing it between the dancers and the sun, and wait for the birds to gather. I stop before reaching the ground to let the endgate down and raise the rear window. In this way when the car is backed up to the dancing place, you can have complete freedom to take notes or use a camera, if you do not make any noise or carelessly stick a hand or arm out of one of the windows.

The prairie chicken has been slow in adjusting himself to man and man's mechanism. During the oil exploitation in my country he was almost extinct. A few clung on to their range in the middle of the great ranches and lived out the great frenzy of oil development. He has only just now come back to a nice balance with his food supply, but he has not yet learned that a car may carry hunters with guns, and instead of freezing effectively in the grass of the prairie that so well protects him, he will only squat and keep his head above the grass to see what is going on. He knows that a man on horseback may be dangerous and that a man on foot usually is, but his failure to see danger in the cars that travel over the prairie may prove to be quite important to his future existence.

Before the white man came to the Osage these grouse populated the range to the point of saturation and on April mornings filled the prairie with their booming as they danced on the high hills.

Naturalists sometimes call the nuptial activities of birds "antics," but the nuptial ceremonies of the prairie chicken are much more than antics. They are true dances which have fascinated most of my visitors to The Blackjacks, people who are not particularly interested in wild life. These dances have something of the same early-morning fervency of an Easter sunrise service on a mountain, with the advantage, of course, of not being cluttered with a breakfastless mob.

On the high prairie hills where the prairie grass is short, and the early-spring grass or late-winter grass covers the highest points of the hill with emerald, the dances begin just at sunrise. To these spots the cock prairie chicken comes every year on early April mornings. He flies there just before sunup and starts his dance slowly and tentatively, like a participant in the Osage social dances. When he seems to be sufficiently warmed by the emotion that the mating urge inspires, he begins to dance with great fervor and sends his sonorous booming over the heavy currents of the spring prairie. One seldom sees the hens, but they seem to be sitting about in the tall, winter bluestem which surrounds the dance grounds, demurely watching the glorious male dancing for their benefit.

Each cock has a beat, and he will dance up and down this beat which is a straight line, inflating the saffron sacs under the long pinnate feathers on each side of his neck until they change color and resemble small oranges. As he inflates these sacs, he makes a sound which is something like "ooh den doo den do-o-o-o-o-o" and which has a ventriloquisitic effect on the heavy morning air. His wings trail and his tail shapes itself into a fan and stands erect, while the pinnate feathers on each side of his neck stick up above his head like horns. His head is lowered as the sacs on his neck fill and glow in the pale light of morning. He takes a few short, mincing steps with his head down, his wings atrail, and his tail erect, then stops and prances in one spot, lowers his head and shakes it slightly as the climax comes. The resulting rolling, sonorous booming can be heard for a great distance, and he may repeat this several times before going back along his beat to repeat the procedure at the other end of it.

There may be four or five, or there may be twenty or forty, cocks dancing on the same dance ground, and the result is a rolling, booming sound that is the expression of full-blooded life, and this sound, with the sad whistling of the upland plover and the irresponsible song of the meadowlark, makes up the theme song of the early-morning April prairie.

Sometimes when the cocks have brought their wings back to normal position and have lowered their tails, and the pinnate feathers again cover the saffron sacs, now deflated, they will cackle like a domestic chicken and fly straight into the air to the height of two or three feet out of the sheer joy of living. If the ventriloquial sound of their booming has misled a searcher, this flying into the air with inane exuberance will certainly attract his attention.

After they have danced for several mornings, the fights occur. One cock will come too close to the beat of another, or they may suddenly decide to run at each other. At first there is bluff, then they lower their heads and, with beaks close to the ground and pointed at each other, they peck tentatively, then back away. But suddenly they will fly into the air; much higher than two game cocks, but in much the same manner, except that they depend more on their beaks than on their heels, which are undeveloped. If a slight prairie breeze springs up before the dancing is over, barred feathers from the fighting cocks will float over the grasses.

Later, the hissing of the breezes through the dead grasses of the last season will bring to their consciousness the danger that is ever present, and they seem gradually to become sober and more alert. The sun, barely peeping over the prairie's rim, sets up the little breezes and restless currents, and, as the dancers become more calm and alert, the least little noise will frighten them, and they will be off, alternately flapping and sailing far across the hills.

I have been much interested in the impressions which these nuptial dances have on people of divers interests, and for that reason I take casual Blackjack visitors out on the prairie during the season of this activity. All of them express surprise that birds and animals should behave in such "a human manner," but not one of them has failed to be impressed as we talk about the experience over coffee under the Blackjack. Like all sex-inspired emotions and actions, the dance has a fundamental appeal to man and inspires the same interest in some of my visitors, who are usually bored by nature, as does the eternal drama of sex in literature, on the screen, and in music.

I have watched this dance every spring for years, and, as in the case of the Osage dances, I have never grown tired of it. Every spring, along with the renewal of the dance, comes a renewal of my intense interest; as fresh as the urge that inspires it in a greening world filled with renewed hope.

One of the men who had photographed the last heath hen cock on Martha's Vineyard explained to me how the last heath cock in all the world had come up to the old dancing ground, to strut instinctively before hens that were gone from the earth; the last dramatic gesture of a species, with only the eyes of cameras hidden in the bush to record the farewell.

I passed the first Planting Moon in the blackjacks, expressing in action the wonder of being alive. I planted roses, wisteria, ivy, lilacs, honeysuckle, and the indigenous trumpet vine. I was thrilled with my solitude and with the madness of that particular Planting Moon. I splashed the station wagon through the east pasture over the road I had just made, when the winds whipped the rain across the prairie. I enjoyed the drive from town over the dirt county road and through the slush and mud of my private road, when the sheet-lightning played over the prairie and there was some doubt whether I could get through.

I was proud of harmony with the life about me. I became a part of it as I had wished. The fox squirrels were the first to ignore my presence. They lay along the limbs of the yard trees in catlike poses or played over the roof. They even became accustomed to the two Irish setters and worried them into a frenzy.

The coyotes trotted nonchalantly over the ridge at dusk or stood off and yelped at the house when the first lamp was lighted and gave out on the dark trees. On rainy mornings I often saw them sneaking down the canyon from the wrath of Tony of the ranch house at the head of the canyon, but we lived in harmony.

I saw a bobcat's track in the ashes where I put the garbage, just outside the west gate, and the prairie chickens came in flocks of forty and fifty to feed on the acorns during winter months, sitting in the trees around the house. Once when the setters were in town, a skunk walked into the house through the open door. I heard the dry clicking of his claws on the cement floor and looked up from my book. We looked at each other for some time. He had caught my movement in looking up from my reading. If I had had hackles, they might have risen, but instead the back of my neck tingled where racial hackles once grew. The fear of discomfort in men and animals is really great. I kept my immobility consciously and he kept his instinctively; the fear, however, was on my side, as there was none in his innocent self-assurance.

Assured that I was harmless, he went into the kitchen, and I could hear him tearing the paper from the garbage box, the bacon wrapping. He stayed so long that I continued reading for some time, then I heard his claws ticking against the cement, and he passed on out into the yard. I got up and closed the door.

Naturally the setters didn't like this harmony of life on the ridge. They cursed the coyotes from the safety of the woven-wire fence and worried when the prairie chickens came whishing in great flocks to settle in the trees. They slavered as they watched the squirrels jump from tree to tree, eternally hoping for a misstep.

Thus for over a year I lived as a part of the balance. There was no shouting and no firing of guns around the little house. But a strange thing happened; I, under the influence of the following Planting Moon, broke the truce. I brought pheasants, chickens, and guineas to the ridge.

Man is under the same natural urges as the other species of the earth. He mates, he fears, he struggles to survive, and he expresses himself in song and in play; he goes even further and acknowledges the progression of life through his dream of God. But why can't he be satisfied without responsibility? Perhaps it is the primal urge to protect his mate and young diverted into other channels when mate and young are lacking or not in primal need of his protection.

In any case, with all my plans to become a part of the balance of nature on the ridges, I brought conflict, after the period of a year. Perhaps my position was unnatural, living as I did, not from the ridge, but feeding myself artificially from cans brought from town and food from the ranch. I was not a part of the economic struggle of the ridge which results in the balance, and therefore I was really an anomaly, as far as my own survival was concerned. After bringing pheasants, guineas, and chickens to the ridge, I had to fight for the survival of my charges against my predacious neighbors, which was probably a more natural state and in the end more satisfying than the "friends and neighbor" idea. I became important to my predacious neighbors; the presence of my charges whetted their desires and sharpened their cunning. We learned more about each other; we found ourselves in struggle now and pitted our wits against each other and saw no more of each other in repose. We had greater mutual respect, and I became a part of the struggle and remained a part of the balance through my strength to protect my flocks. Thus, I achieved a greater harmony with my environment and found that there is no place for dreams in natural progression, and it seems to me that I had realized for the first time that with responsibility come enemies.

From *Talking to the Moon: Wildlife Adventures on the Plains and Prairies of Osage Country* by John Joseph Mathews, 1945. Reprint (Norman: University of Oklahoma Press, 1981).

Considered one of the most influential voices in environmental ethics and the science of wildlife management, Aldo Leopold was born in Burlington, Iowa, where he spent much of his youth exploring natural areas along the Mississippi River. A graduate of the Yale Forest School, he assumed U. S. Forest Service posts in the Southwest before being hired as a professor of game management in the Agricultural Economics Department at the University of Wisconsin–Madison, a few hours from where he grew up. In 1935, he purchased eighty degraded acres along the Wisconsin River outside of Baraboo, where he and his wife and children spent many days living in the renovated chicken coop known as "The Shack" and working to restore the land. That experience became the heart of his posthumously published book, *A Sand County Almanac: And Sketches Here and There* (1949), in which he proposed his famous land ethic: "A thing is right when it tends to preserve the integrity, stability, and beauty of the biotic community. It is wrong when it tends otherwise" (262). By this standard, our treatment of the tallgrass prairies has been a massive failure, as he concludes here while observing the fate of a compass plant in a roadside cemetery.

Prairie Birthday

During every week from April to September there are, on the average, ten wild plants coming into first bloom. In June as many as a dozen species may burst their buds on a single day. No man can heed all of these anniversaries; no man can ignore all of them. He who steps unseeing on May dandelions may be hauled up short by August ragweed pollen; he who ignores the ruddy haze of April elms may skid his car on the fallen corollas of June catalpas. Tell me of what plant-birthday a man takes notice, and I shall tell you a good deal about his vocation, his hobbies, his hay fever, and the general level of his ecological education.

EVERY JULY I WATCH eagerly a certain country graveyard that I pass in driving to and from my farm. It is time for a prairie birthday, and in one corner of this graveyard lives a surviving celebrant of that once important event.

It is an ordinary graveyard, bordered by the usual spruces, and studded with the usual pink granite or white marble headstones, each with the usual Sunday bouquet of red or pink geraniums. It is extraordinary only in being triangular instead of square, and in harboring, within the sharp angle of its fence, a pinpoint remnant of the native prairie on which the graveyard was established in the 1840's. Heretofore unreachable by scythe or mower, this yard-square relic of original Wisconsin gives birth, each July, to a man-high stalk of compass plant or cutleaf Silphium, spangled with saucer-sized yellow blooms resembling sunflowers. It is the sole remnant of this plant along this highway, and perhaps the sole remnant in the western half of our county. What a thousand acres of Silphiums looked like when they tickled the bellies of the buffalo is a question never again to be answered, and perhaps not even asked.

This year I found the Silphium in first bloom on 24 July, a week later than usual; during the last six years the average date was 15 July.

When I passed the graveyard again on 3 August, the fence had been removed by a road crew, and the Silphium cut. It is easy now to predict the future; for a few years my Silphium will try in vain to rise above the mowing machine, and then it will die. With it will die the prairie epoch.

The Highway Department says that 100,000 cars pass yearly over this route during the three summer months when the Silphium is in bloom. In them must ride at least 100,000 people who have 'taken' what is called history, and perhaps 25,000 who have 'taken' what is called botany. Yet I doubt whether a dozen have seen the Silphium, and of these hardly one will notice its demise. If I were to tell a preacher of the adjoining church that the road crew has been burning history books in his cemetery, under the guise of mowing weeds, he would be amazed and uncomprehending. How could a weed be a book?

This is one little episode in the funeral of the native flora, which in turn is one episode in the funeral of the floras of the world. Mechanized man, oblivious of floras, is proud of his progress in cleaning up the landscape on which, willy-nilly, he must live out his days. It might be wise to prohibit at once all teaching of real botany and real history, lest some future citizen suffer qualms about the floristic price of his good life.

THUS IT COMES to pass that farm neighborhoods are good in proportion to the poverty of their floras. My own farm was selected for its lack of goodness and its lack of highway; indeed my whole neighborhood lies in a backwash of the River Progress. My road is the original wagon track of the pioneers, innocent of grades or gravel, brushings or bulldozers. My neighbors bring a sigh to the County Agent. Their fencerows go unshaven for years on end. Their marshes are neither dyked nor drained. As between going fishing and going forward, they are prone to prefer fishing. Thus on week ends my floristic standard of living is that of the backwoods, while on week days I subsist as best I can on the flora of the university farms, the university campus, and the adjoining suburbs. For a decade I have kept, for pastime, a record of the wild plant species in first bloom on these two diverse areas:

Species First Blooming in	Suburb and Campus	Backward Farm
April	14	26
May	29	59
June	43	70
July	25	56
August	9	14
September	0	1
Total visual diet	120	226

It is apparent that the backward farmer's eye is nearly twice as well fed as the eye of the university student or businessman. Of course neither sees his flora as yet, so we are confronted by the two alternatives already mentioned:

either insure the continued blindness of the populace, or examine the question whether we cannot have both progress and plants.

The shrinkage in the flora is due to a combination of clean-farming, woodlot grazing, and good roads. Each of these necessary changes of course requires a larger reduction in the acreage available for wild plants, but none of them requires, or benefits by, the erasure of species from whole farms, townships, or counties. There are idle spots on every farm, and every highway is bordered by an idle strip as long as it is; keep cow, plow, and mower out of these idle spots, and the full native flora, plus dozens of interesting stowaways from foreign parts, could be part of the normal environment of every citizen.

The outstanding conservator of the prairie flora, ironically enough, knows little and cares less about such frivolities: it is the railroad with its fenced right-of-way. Many of these railroad fences were erected before the prairie had been plowed. Within these linear reservations, oblivious of cinders, soot, and annual clean-up fires, the prairie flora still splashes its calendar of colors, from pink shooting-star in May to blue aster in October. I have long wished to confront some hard-boiled railway president with the physical evidence of his soft-heartedness. I have not done so because I haven't met one.

The railroads of course use flame-throwers and chemical sprays to clear the track of weeds, but the cost of such necessary clearance is still too high to extend it much beyond the actual rails. Perhaps further improvements are in the offing.

The erasure of a human subspecies is largely painless—to us—if we know little enough about it. A dead Chinaman is of little import to us whose awareness of things Chinese is bounded by an occasional dish of chow mein. We grieve only for what we know. The erasure of Silphium from western Dane County is no cause for grief if one knows it only as a name in a botany book.

Silphium first became a personality to me when I tried to dig one up to move to my farm. It was like digging an oak sapling. After half an hour of hot grimy labor the root was still enlarging, like a great vertical sweet-potato. As far as I know, that Silphium root went clear through to bedrock. I got no Silphium, but I learned by what elaborate underground stratagems it contrives to weather the prairie drouths.

I next planted Silphium seeds, which are large, meaty, and taste like sunflower seeds. They came up promptly, but after five years of waiting the seedlings are still juvenile, and have not yet borne a flower-stalk. Perhaps it takes a decade for a Silphium to reach flowering age; how old, then, was my pet plant in the cemetery? It may have been older than the oldest tombstone, which is dated 1850. Perhaps it watched the fugitive Black Hawk retreat from the Madison lakes to the Wisconsin River; it stood on the route of that famous march. Certainly it saw the successive funerals of the local pioneers as they retired, one by one, to their repose beneath the bluestem.

I once saw a power shovel, while digging a roadside ditch, sever the 'sweet-potato' root of a Silphium plant. The root soon sprouted new leaves, and eventually it again produced a flower stalk. This explains why this plant, which never invades new ground, is nevertheless sometimes seen on recently graded roadsides. Once established, it apparently withstands almost any kind of mutilation except continued grazing, mowing, or plowing.

Why does Silphium disappear from grazed areas? I once saw a farmer turn his cows into a virgin prairie meadow previously used only sporadically for mowing wild hay. The cows cropped the Silphium to the ground before any other plant was visibly eaten at all. One can imagine that the buffalo once had the same preference for Silphium, but he brooked no fences to confine his nibblings all summer long to one meadow. In short, the buffalo's pasturing was discontinuous, and therefore tolerable to Silphium.

It is a kind providence that has withheld a sense of history from the thousands of species of plants and animals that have exterminated each other to build the present world. The same kind providence now withholds it from us. Few grieved when the last buffalo left Wisconsin, and few will grieve when the last Silphium follows him to the lush prairies of the never-never land.

From *A Sand County Almanac: With Essays on Conservation from Round River* by Aldo Leopold, 1949. Reprint (New York: Ballantine Books, 1984).

At age twenty-four, Josephine W. Johnson won the 1935 Pulitzer Prize for her first novel, *Now in November*, a story about the impact of that decade's drought and economic depression on farm families. It was a life she knew well from her childhood in Kirkwood, Missouri. After marrying Grant G. Cannon, editor of the *Farm Quarterly*, they moved from the St. Louis area to Iowa City (where she taught at the University of Iowa) before settling in Hamilton County, Ohio. In 1956, they purchased thirty-seven acres of land outside Cincinnati, where they raised their children and later established a nature preserve—what she deemed an "inland island" within suburbia. Johnson became a nationally respected nature writer and environmental activist, especially after the publication of *The Inland Island* (1969), which was praised by Edward Abbey in the *New York Times* and favorably compared to *Walden*. In this widely read book, Johnson offered month-by-month observations of the wildlife and natural cycles on their land—which included woods, wetlands, and remnant prairie—interspersed with reflections on her personal life and, as in this excerpt, the tumultuous world of the late 1960s, and the relationship between children and nature.

July

Skies, fields, ponds, a mass of redwing blackbirds. Saw a hawk laboring over the fields, covered by a cloud of blackbirds. His head in a gnat-ball of birds. Where did all these flying things come from? These blackbirds flocking suddenly. Noisy and restless. Wave after wave over the pond. The trees full as monkeys in a jungle. How sweet we would have thought that domed nest in another spring. But not this time.

Armies of bulrushes are taking over the pond. Moving out toward the center, pushed forward by the willows. Tried to cut some of the willows with a corn knife. The old branches clanged like steel. The new ones gashed open like white-throated snakes. I'm a lousy cutter. The big frogs are sluggish on chilly days. Almost stepped on them. The purplish-black dragonflies are huge. Their wings have a fierce metallic sound. Gathered cattails for our little friend, the stocky, very small little boy who comes down to visit. "I am a *professional* swimmer," he said. "I caught a fish once that was *probably* a shark." He was proud of his new sandals, very leathery and strappy. Once he came alone very late in the evening. "I came to see you," he said, "because there was nobody else to see. Nobody on my block was allowed out." He spent a long time arranging the cattails to stick out like rockets behind his bike. He looked at the flowerpots on our cistern top—our instant patio. "You would have a nice backyard," he said, "if you had a swimming pool in it." He punctuated all his sentences with the name of God, like periods and commas. His family is large. Extended by his confusion of cousins with sisters and brothers. He did not smile often, but when the smile came it was very sweet and broad. He was a square little package of brightness and life, and we felt bad when he said he was going away to live in Kentucky. No more professional swimmer on the doorstep. No more advice.

TODAY IS WET, damp, soggy and swollen. My fingers are swollen. The air is swollen and warm. The butterfly wings are swollen. The wool rug is big with water. It is rising off the floor. The furniture is getting enormous, fat and undulating with water. Everything sticks. Is up tight. My brain is a big sponge. It rises against the skull. It will pour through the fontanel.

The far hills are covered with blue mist. The blackbirds sound like a rush of rain. They are pests. I hate them and all their speckled young. A breeze tries to move through the soggy air. The leaves let down the gathered rain.

Sun comes through the sog. Makes everything ten times worse. Thank God the house is clean. The wet and dirty clothes fought down into the cellar (where they soak up damp and grow enormous, ectoplasmic forms— cloth dough). My stomach swells and invades my throat. My feet are getting bigger and bigger. Overflowing the sticky leather shoes. Even my hollow heart is heavy. Brain, heart, feet, all sponge. I weigh a thousand pounds this morning.

The grass loves this world swamp, this massive aerial soup. You can see it grow before your eyes. The pine trees reach the house, flop down exhausted on the roof. Their great, green, bristly arms embrace us. The rain from every needle rolls down the roof.

The sky is grey. It is mashing the swallows low. Something is pressing the grief button. Tears are appropriate. Toads swell with fear and rage. They swell with rain. Their soft, fat throats vibrate. Their gold-rimmed eyes have no expression. They have horns on their toes.

Down under the ground the moles are *expanding*.

ALL THE WARM honey things, musky things flowering, the dusty, musky milkweed balls, the honey-sweet buckwheat vines remind me of summer nights riding past fields of flowering buckwheat in Columbia, Missouri.

Aunt Laura died on the fifth of July, at three in the morning. Ninety-three years old this month. The last of the Franklin sisters. It's all over now. Everybody's suffering is over now. There was no joy for any of them at the end.

This is one of the years of the Vietnam war, this is one of the years of the Negro revolution. This is one of the years of the Israel-Egyptian war.

A move has been started to educate the poor to file bankruptcy claims. A splendid idea. Start new again. Break out of the whole rotten coffin.

Two white butterflies go by the window. I must get out. Out! It's a panic. I can't stand the walls. You haven't a character vigorous enough to panic. You are just sitting there. You lie a lot.

I feel tired and heavy and without grace. Without grace of God or flesh, or mind.

I shall file bankruptcy of another sort.

IN THE WALNUT GROVE, whose leaves are gold in the evening light, the blue bellflowers are beginning to bloom, tall and delicate spires along the path. And the small blues, the butterflies, open their mini-wings that alternate blue and grey like dusty jewels. A bit of fudge hops in the path, a toad traveling. The path is a delight. Broad and covered with small leaves, second growth of thimbleweed and clover, close to the ground. Each leaf like a flower, so that one walks on a delicate carpet, winding through the grove under the thin trees bordered with bellflowers, and emerges into the sunlight of the pasture.

Blue is rare in nature. Sky and sea have most of it. True blue is hard to find on earth. Most of what's called blue runs off into the purple. But there's the bluebird, a piece of sky. The indigo bunting of unearthly dark vibration. No color like it. No mineral, no jewel. There's the bluejay, whose big bold blue is muted by the black leading which holds his panes of sapphire glass in place. There are the dark-blue seeds of wild grape, almost black. The big round balls of moonseed vines, dark blue and shiny. There are the hanging blue balls of the Solomon's seal in autumn, dark as wild grape. Blue seeds of sassafras trees held in scarlet cups. Once I saw a cerulean warbler. So high up, so far away, that only once the blue shone through the opening of green leaves. But I'm positive of it. There are the blue eyes of chalky blue-green dragonflies. And in summer mornings, drifts of blue along the road from chicory flowers. But here we start creeping into the lavender, the true blue fading; the wild hyacinths aren't really blue, we start calling violets blue, pale wild pansies rising to purple, rich, shining purple in the wild larkspur flowers. And the whole field of purple broadens out and carries one away. There's no end to this purple. Color is a strange thing. I always loved purple as a child. Purple velvet. Purple pansies. Now all its variations disturb me. With sunlight coming through it, or in the dark woods, the larkspur is beautiful. But only there. (Larkspur is poison, those fat lively seeds are deadly. Staphisagroine. Death in the fine-cut leaves.)

In the pasture, only a name now for this open piece of land beyond the walnut grove, the wild bergamot fills the air with its warm mint smell. It is lavender with no pretense of blue. We had red bergamot once, the domestic relative of this wild pasture. Bee balm, enormous, like Fourth of July explosions. A peculiar red, neither crimson nor luminous. A wood red. The dazzling eyes of blackberries are there. Polished and abundant. Thorny and sharp and acid. Gathered a cup for Grant and was raked by those tiny little thorns, bitten from head to foot by a world of insects, in five minutes.

JOSEPHINE W. JOHNSON

The teasels are high and crowned with purple flowers on their brushes, surrounded by long green horns, like Texas steers. They grow higher than my head, and hum with bees.

I am feverish and irritable with chiggers. Toads have chiggers, too. A fresh approach is needed to the chigger. Where to begin? The chigger is one one-hundredth of an inch long. His poison spreads into a lump a thousand times bigger. It rises into a red loaf of flesh. But he, the chigger, is bright yellow and blind. He runs swiftly, apparently sniffing the air for flesh, or, hopefully, to encounter edible objects. A large mass moving through the grass is a great flesh-fall for the chigger. He likes the base of a hair follicle and, perching upside down, buries his beaky face and hangs on.

It is said that he drops off in the autumn, lies dormant in the soil—no, not dormant, but molting. Two or three weeks are spent as "quiescent nymphs," then they molt again and become adults. Still they sit there anyway, adults, quiescent, until spring. Then—and here the record becomes confused and cloudy—ten months are spent in adult stage. Adults do not attack man. Adults lay eggs. Eggs hatch, and what hatches normally lives under the overlapping scales of snakes, or on land turtles. Or on rabbits. Why, then, all this scurrying rapidly about to find man? What's the matter with the cool scale of a snake? The slow ride on a turtle? The long, germy hair of a wild rabbit? That expression, "normally lives," needs explanation.

Metcalf and Flint write of the chigger: "If one anticipates a visit to the domain of these small tormentors, almost complete freedom from attack can be obtained by treating the clothing with dimethyl phthalate, dibutyl phthalate, or benzyl benzoate."

The whole midwestern world is his domain.

[...]

Patience pays off. Silence and stillness bring reward. (Nothing moral about it, but it's the only way you'll ever see anything but green vegetables and bugs.) At the far end of the pool something slipped into the water. Something bigger than a frog in broad daylight moving around among the rocks. The raccoon was fishing.

Where are all your animals? the little children cried, running, shouting, through the woods. Twenty little children, panting, shouting, screaming, throwing rocks into the streams. Birds, frogs, even bugs went rushing to the hills. How sad that they thought they might see what only hours of silence, days of watching, ever bring to sight. They liked the water. They saw a bird or two. They found green walnuts. "They smell like lemons," they said. They were afraid of snakes they never saw. One little girl put her tiny dark hand in mine. She did not want to walk in the woods alone, but she wanted to be there. "We could get a little bag, if you had one, and come back later all alone," the little girl whispered, "and pick up the walnuts all by ourselves." Children should have greenness. There should be parks of greenness everywhere. This is a mad world of roads and concrete blocks and people going somewhere else because they can't stand to be where they are, because they've ruined where they are, to get somewhere else which is ruined, to come back to where they were. There should be parks for people, and woods for animals, while there is still room on earth for both, and time. Put the parks back in the cities where the people are. A park every block. That's not too much. That is as important as eating and drinking, for the body and the soul. What kind of life is it to live on the edge of a concrete street, four lanes, six lanes wide? What kind of a neighborhood is a neighborhood of streets and cars and cars and streets? Not all the plumbing or the new paint in the world can make that into a neighborhood, or into a life.

The raccoon was fishing. He fished as though he did not really want to get wet. Sometimes he hopped with his tail high, as his hands pursued his food under the water. He ate whatever he found right there in the stream. He rested his chin on a rock and felt underneath it, brought up a dripping crawfish and ate it right there in the stream; green water dripped from his fur. He came downstream, zigzagging along the margin. He ate several times, then slowly, dreamily, climbed the bank. He saw me, or saw a silent alien thing, and paused and stared. I did not move. He stared, I stared. He moved away and climbed a tree. There he stopped in a crotch and dried and cleaned his fur, pausing sometimes to look down at me with a wondering, idle peer. He seemed sleepy, and after awhile he climbed higher and disappeared.

The sun is high and hot. A single yellow jewelweed blossom, large and pale yellow, not orange like the jewelweeds in the pasture and the draws, dangles above the stream. A hummingbird zooms and sizzles through the flowers. The propeller wash of his wings shakes the whole plant from stem to stern. Mulleins along the stream have reached their peak. Spires plastered with pale-yellow blossoms, pale-green and hairy leaves flopped out. Gold coral fungus grows out of old stumps. One could go about gathering an enormous yellow bouquet out of the heart of July. Fill the arms with orange lilies, the ubiquitous gold; sweep up the sunflowers, the gold honeysuckle, the great stalky suns of compass plants and prairie dock; find clumps of black-eyed Susans rimmed with gold, and wild coreopsis, ragwort and squawweed through the high, dry woods, yellow hearts of daisies and bedstraw in the fields.

Gold bugs, yellow butterflies, orange lilies. What to do with this great fragrant glow? Well, hold on to it. You'll want it in the days to come.

From *The Inland Island* by Josephine W. Johnson, 1969. Reprint (New York: Simon and Schuster, 1971).

Born and raised in Emmetsburg, Iowa, Winifred Mayne Van Etten worked her way through undergraduate school at Cornell College in Mount Vernon (IA), where her interests in writing and literature were first kindled. After earning an MA at Columbia University, she was hired as a professor of English at Cornell, but was let go in 1933 when she got married, as administrators felt it was unethical to employ women whose husbands already drew salaries—a common practice during the Depression. She was eventually rehired, but in the interim wrote her only novel, *I Am the Fox*, which in 1936 won the prestigious Atlantic Monthly Press and Little, Brown and Company Award. In a radio essay at the time, she talked of the conflict modern women felt between family and career, a struggle that left them feeling "maimed, deprived of the full and complete expression of [themselves]." In this 1978 essay—one of her few publications since the 1930s—she links the fate of the prairie in her home state to the struggle against conformity and the search for peace during and after both world wars, in her life and those of others.

Three Worlds

Someplace in the family archives there is a large mounted photograph. It shows my father standing in the midst of his field of corn. He was a tall man, six feet two or more, and he was stretching as far as he could above his head to hang his derby hat on the highest ear of corn he could reach. There were higher ears, but this one was the best he could do. We lived on the edge of town and just beyond the town limits he had seven acres of land that he treated as a garden. No weed dared intrude. In the days when forty or forty-five bushels an acre was usual he in some years produced one hundred. As a child I assumed that a hundred bushels was normal. Sometimes I was allowed to help with the plowing. I drove the team; my father walked behind guiding the plow. When I reached the point at which, coming to the end of a row, I could turn the team, reverse the direction, and start back the other way without tearing up any corn hills, I regarded myself as an expert.

These were the days of the tall corn in Ioway.

Years later my husband and I picked up a hitchhiker from Connecticut. We noticed him staring at the crops along the highway. Finally, he spoke. "I don't think it's so very tall," he said. Obviously the only thing he had ever heard about Ioway (at least he did not confuse it with Ohio or Idaho) was that it is the place where the tall corn grows.

We tried to explain to him that tallness was no longer an object. No more did farmers cut their tallest stalks and take them to town where rival mast-high specimens stood in a row in front of the bank. The corn picker had put an end to the tall corn. Uniformity was now the goal, and Ioway was the place where the tall corn grew—once—but now conformed to the demands of machines. Uniformity—that was the criterion of excellence. In Iowa. Ioway was gone.

The First World / *The Prairie*

Ioway was my first world, the world of my very early childhood. No buffalo roamed there, but it was still pretty close to true prairie. We took long excursions across it. Rover trotted along in front, tail waving as debonairly as though he had not already covered many miles. Trinket stepped it off lightly, five people and a surrey apparently nothing to her flying slender legs. We had been on the way since before dawn on our annual trip to visit an uncle who lived fifty miles north of us near the Minnesota border. We had shivered at first in the morning chill, but now it was getting warm and there were many miles still to go. Fifty miles in a day for one horse pulling a carriage and five people was a long journey, but the trip could be made no other way. There was no railroad, hardly anything that could even be called a road, only a sometimes barely discernible trace through the tall prairie grass, and as we got farther north, skirting the edges of a big slough, there was water standing along or on what passed for a road. Now and then a fox eyed us suspiciously, and once an animal my father said was a wolf sent delicious chills through us as we recalled those stories of Russians tossing first one passenger then another from their vehicle to delay the ravening wolves.

But this wolf, if it was a wolf, seemed of a mind to tend to its own business or else it was of a peaceful disposition. Everything was peace: a soft June morning, banks of wild roses, the tall grasses, the reeds and cattails. The bobwhites whistled, the meadowlarks sang. Even my father sang. He couldn't carry a tune, but there were two lines of one ditty that he droned over and over in his monotone as we rode along. I never heard him sing more than the two lines and never except on the occasion of one of these family safaris:

> Oh, green was the grass on the road, on the way
> And bright was the dew on the blossoms of May.

Sometimes we all sang, mostly hymns, since those were the songs we all knew. We did the Crusaders' Hymn over and over. It seemed to fit this lovely, lonely world through which we were passing.

> Fair is the meadow
> Fairer still the woodlands
> Robed in the blooming garb of spring.
> Jesus is fairer, Jesus is purer
> Who makes the woeful heart to sing.

I was never a religious child in any conventional way, but somehow I sensed a connection between the woodlands, the meadow, and that purity we were taught to call Jesus at Sunday school. The son of God and man was a concept beyond me, the Holy Ghost complete befuddlement, but fairest Lord Jesus, the fair meadow, the fair woodlands, these gave me an intuition of oneness that, without owning the capacity to think about it, I felt included me.

DARK WOULD BE falling by the time we reached my uncle's farm, all of us hungry, all so dead tired that even the mattresses stuffed with corn husks felt and smelt good and whispered and rustled soothingly. In the morning there would be prayers in the neat little living room with its center table bearing a ten-pound Bible with Doré illustrations, a tall kerosene lamp, its double globes hand painted with flowers, perhaps a big rosy shell on the lower shelf. Of course there was the cottage organ. If the decor was what every housewife felt it should be, there would be on the organ shelves some small ornaments; a china boot; a fancy box or two; and crowning glory of home decorations, a tall vase of peacock feathers. Such proofs of culture could only be surpassed by a stereopticon set by means of which we acquainted ourselves with far climes and countries while gaudy Japanese wind chimes tinkled in the open window. It never occurred to us that we might actually sometime see these places ourselves.

Our world was bounded by what amounted to a slightly modified frontier to the north of us and to the south the little town from which we had come. The town, a sort of rough excrescence on the smoothness of the prairie, had mud streets and electric light only in the stores except for a few affluent residences and those who had their own Delco systems. But it had the county courthouse and the jail, and once in a while an automobile was to be seen, usually driven by some young blade viewed with as much disapproval as one of Hell's Angels would be a half century later. Such a monster, if my mother was driving, would set Trinket prancing on her hind legs until some gallant from the sidewalk came out and held her head until the horror had passed.

The town had eight churches: six Protestant, two Catholic, plus a Catholic grade school and the quarters for the nuns and priests. God's geese (the nuns), so-called by one of our family friends always admired for the rightness of her views, were seldom seen on the streets, but we were conscious of their presence just as we were of some sinister existence in the priest's house. The town also had at one time according to my elders nine saloons. One story said that an Irish emigrant came out in the early days, set a jug of whiskey down on the prairie, and the town grew up around it and fell at once to fighting as though Cadmus had sown the dragon's teeth there and not in Greece. For it was a divided town, half Irish Catholic, half Protestant, a mixture of Scandinavians, Germans, Scots, and English, the latter two often referred to as "remittance men" because their money was remitted to them from the old country.

But for my family the most important element in the town was the existence of three weekly papers, two Catholic and Democrat (they were considered one and the same thing), one Protestant and Republican of which my father was the owner and editor. No issue was too small for a Battle of the Boyne and my father was always in the midst of the fray. He used to try to prevent our seeing the other two papers, but somehow we always found out what he was being accused of that week. Sometimes he was said to have been found lying drunk in the gutter (he who was a teetotaler) surrounded by dead soldiers. At other times, in fact at all times, he was up to some political skullduggery designed to keep Catholics and Democrats out of their rights. Nearly every week there was recorded some new villainy on his part, and my view of my native town was colored for years by these bitter battles. For bitter they were. What goes on in Ulster today seems more comprehensible to me because of the tenacious feuding imported to this peaceful prairie from the Old World. Even the dogs were Catholic or Protestant. On a Saturday my brother would take Rover, cross the tracks to the south side, and there take on all comers, whether curs or Catholics, in epic battle. Few of us, however, were really personal enemies.

One of my most abhorred duties as a young girl was "to meet the trains." There were trains in those days, two of them, the Milwaukee and the Rock Island, two each way, each day. When everyone else in the office was too busy, I had to do this chore. "Meeting the train" meant walking to the depot, going up to each person who bought a ticket and inquiring who he was, where he was going, and why. An ally and helper, a great icebreaker, was Rover's successor, Old Shep. He was a stray that had learned to be a station beggar until he adopted us. But he still met the trains, spotted everyone who appeared to be carrying a lunch and, though a big dog, sat up on his hind legs before his victims and tried to look underfed. He usually got something and so did I—many a handout I would not otherwise have had the courage to ask for.

Another ally, truly unexpected, was the train meeter of the rival paper, a middle-aged man, who out of pure pity for my shyness often handed over gratis some of his own garnerings. After that I was never able really to believe that all Democrats and Catholics, one and the same thing of course, were children of mischief. Still, any exception to the rule was rare and change came to Ioway very slowly. After I had become a schoolteacher, I overheard at a club meeting one middle-aged woman whisper to another behind her hand in the tone of one relating a shocking scandal about another person present, "You know, she's a Democrat."

On the other hand, one of my Ulsterite father's best friends was a priest, who for years on their wedding anniversary, sent my parents a gift of silver, a spoon, a cold meat fork, a ladle. I do not remember him, but his name was that of a friend in our house for years.

Later I came to see that probably my father hadn't minded as much as the rest of the family did. For he was Irish too, an Orangeman, and though he wanted to be sure he was on the right, the moral, side of a controversy, the fact remained that he loved a fight for its own sake. It was not so nice for his daughters. In spite of an occasional detente there were many times when we hated to go out in public or to be seen on the streets. We yearned for the day when we could go to college and become teachers, somewhere, anywhere except at home. It never occurred to us or to anyone else that we could do anything except teach. What else could a nice girl do? She filled in the time between high school or college and marriage by teaching—for pin money, for funds for filling a hope chest, or simply for something to do. One difference for us was no hope chests. In our family it was not expected that we would marry, for only one man in the world was fit to marry, our mother had married him, and it followed logically that we would do nothing so morally equivocal as to marry at all. The teaching itself with a few notable exceptions was not really very important to anyone, the students, the teacher herself, the community. About all that was required of her was "to keep order in the classroom."

I remember vividly some of the women who toured the state campaigning for women's suffrage. Strange creatures, they seemed to me. Normally a woman's public appearances were confined to reading in mincing, ladylike tones a "paper" to her social club. But these women stood up and talked straight out as though they didn't even realize they were women. Since my parents were strong believers in votes for women, one or two of these campaigners were our house guests, and we heard many anecdotes of their adventures that we would not have heard in a public meeting. They laughed about them, but they were bruised; the obscene remarks made to them on the streets; the filthy places in which they sometimes had to stay, collapsing country hotels patronized chiefly by bedbugs, hamlets with no hotel at all where they stayed in some private home and got revolting food to eat. One told of having to swallow for breakfast eggs boiled (it was washday) in the same boiler in which the family's clothes were simmering; the stern disapproval of most of the men and the timid reluctance of most of the women to say what they really felt. How proud I was when my mother stood up at a meeting in the county courthouse and spoke for five minutes on the issue. Of course, she was a university graduate and had been a teacher. What could you expect from such ruined females?

The day my mother went to the polls and cast her first ballot was a prideful occasion. She voted, no protest, exactly as my father "suggested" to her. And when I cast my first vote as a college senior, my father saw to it that I received an absentee ballot and a list of "suggestions" about various candidates. It never occurred to me or my father that there was anything out of the way about this manner of exercising one's citizenship. After all, what did I know about politics? I had been taking courses in history and government, but all that was theory. The thing was to know the candidates. This sort of thing occurred long after my fundamental attitudes had been fixed in childhood.

For us, young females, the world was a place governed by taboos. The stern Calvinist morality we inherited from one side amalgamated with a late Victorian gentility in a way that produced some pretty potent forbiddings. The taboo was usually related to sex, of course. Anything a girl did that in any way resembled what a boy or man did was taboo. Looking back, I can understand some of these forbiddings; others seem totally incomprehensible.

Taboo: A girl must not part her hair on the side. Boys wore theirs that way.

Taboo: Everybody knew about the bad end to which whistling girls came. In some quarters, it was no joke.

Taboo: Legs didn't even get mentioned. Before leaving for church on a Sunday morning, each of us was stood up in front of the east door through which the summer sun sent stabbings of light. If there was the slightest sign of a shadow, back we went to put on another petticoat. I have worn as many as five. I used to entertain myself in church by imagining all those buttons simultaneously bursting off and all those petticoats descending in starched white heaps to the floor. They were fun to iron, too, with "sadirons" (no object was ever more accurately named) heated on a coal-burning range. Shirts for three men. Petticoats for four women—each petticoat with a flounce, each flounce with a ruffle, each ruffle with embroidered or lace trimming consisting of two or three rows of "insertion" and a final lace frill.

Summer or winter females wore hats in the house of God; St. Paul said they had to. My hair was tied with a ribbon at the top of my head. The stiff crown of the hat pressed hard on the ribbon knot, held by an elastic under the chin, well chewed in an effort to loosen it a little. The result was that I spent nearly every Sunday afternoon lying in the hammock nauseated with headache and making occasional dashes for the porch railing.

Taboo: Card playing. Whenever cards were mentioned my mother saw the Devil, horns, hooves, and tail complete.

Taboo: Dancing. There were whispered tales of dissolute girls who checked their corsets in the cloakrooms at the public dance halls in order to "enjoy as much sensual pleasure from the dancing as possible."

Taboo: Looking into the open doors of a blacksmith shop as you went by.

Taboo: Walking on any street where there was a saloon.

Taboo: Walking by the horse doctor's office or the livery stable.

Taboo: Scraping your feet on the sidewalk. I got mud on my shoes one day, and when I tried to scrape it off on the sidewalk the elders with me fairly snarled reproof.

Taboo: Walking close to a business place in basement quarters. You were to take to the outer edge of the sidewalk lest some lascivious male might look up and be filled with glee to discover that you had legs.

Taboo: Owning a dog. No nice girl had a dog of her own. The family dog was all right, but as a piece of personal property the dog was taboo, especially if it was a female. The owner might as well advertise her profession.

The sheer irrationality of most of these notions not surprisingly led before too many years had passed to the so-called revolt of youth.

Actually, there was little for any young person to do, even those not reared in so austere a regime as ours. There was a lake, weedy, muddy, and scummy, but it had several small islands, one, our favorite, about the size of a haystack, good for picnicking. Later the lower end of the lake was dredged and became a favorite playground for many. Naturally no fastidious person would go into the water. After all, who knew what was going on down there under the surface? It couldn't be for nothing that those prickly wool swimsuits sold in the stores carried the label, "Urine rots wool." We could go in far up the lake, the undredged part, where some of our farming friends had houses close to shore. But the first time I tried it I came out covered with leeches. I nearly screamed myself into spasms while my country friend matter-of-factly plucked them off one by one, leaving me streaming a pale mixture of blood and water. Eventually the town made a park, its trees still baby-sized, and put in a tennis court and some swings. The park was little used for a while. Tennis, like that other foreign game, golf, or pasture pool as we called it, was for effete foreigners like the remittance men. We had the true bumpkin mentality. Whatever was outside our limited experience was bound to be either wicked or ludicrous.

The high school had a football team, but athletics were certainly not overemphasized in the schools. I never saw a gymnasium until I went to college. Once in a while we were made to stand between the rows of classroom desks and wave our arms and bend our knees and feel silly. This was called calisthenics and happened only occasionally. What we did get was a core of required courses and a very few electives: four years of English, four of history or civics, four of mathematics, and a language, Latin or German (until the First World War when French patriotically replaced German). I had a remarkable Latin teacher as a result of whose influence I studied Latin for four years and went on to another four in college. I had another reason to be grateful to her. Girls were required to take domestic science (no one called it home economics) in the seventh, eighth, and ninth grades. I loathed it so much that in the seventh and eighth grades I never would have passed if my teacher, who roomed next door to us, had not taken pity on me and finished my aprons and nighties herself. Now as I entered high school I swallowed my fear of the formidable Latin teacher-principal and begged off the domestic science requirement. There was sympathy in her soul someplace, for she let me take courses in physiology and "physical geography" instead. Perhaps she, like me, was a feminist in the cradle. It did not take me long to observe that taboos applied to girls more frequently and more firmly than to boys. The son was able to escape. His sisters could not. We were in effect incarcerated. Therefore I hated being a girl.

Before I had ever seen or heard of a penis I envied boys. Freud, I knew when I read him years later, was wrong altogether about the envy of the female for the male. I envied males because their very existence conditioned and controlled mine. Femaleness was a condition of innate inferiority. When I once confided to our family doctor that I, too, would like to become a doctor when I grew up, he laughed his booming laugh. "Ha, ha. A hen Medic!" Clearly, an obviously absurd idea. I gave it up on the spot.

What it came down to was this. A female was both inferior and for some unknown reason obscene. The best thing a girl could do was to pretend she didn't exist. Her best strategy was silence and a poker face.

Even girls reared in a more lenient pattern than ours had no greater freedom of choice. Teach. Be a stenographer. Be a nurse. Even nursing was suspect. I heard a mother telling her daughter just what kind of duties she as a nurse would be expected to perform for male patients. "And nine times out of ten the man for whom she does these things will insult her." A sexual insult then was something only slightly less appalling than rape is now.

My first theatrical experience was a heady trip to the local nickelodeon. What was offered there (fare five cents) was ordinarily deemed trash but this time *The Odyssey* and *The Last Days of Pompeii* were to be shown. These two were considered educational. They proved at least unforgettable; the smoke belching from the upside down ice cream cone of Vesuvius, Scylla thrusting out head one and snatching a Greek, then head two and nabbing another, then at precise mechanical intervals head three and the rest. I saw her in dreams for years.

Thirty years later when audiovisual aids were the modish thing in educational circles, my English department, feeling the obligation to be up-to-date, ordered a catalogue, examined it, found there a film of *The Odyssey* (which our classes were at the time reading), sent for the film, showed it. Lo, it was the precise film I had seen as a child at the nickelodeon. I collapsed in mirth and so did all the students. Onward and upward with the arts. The majesty of *The Odyssey* must have stayed with that group forever.

But there were better things in that town.

The town did have aspirations to culture. The women's clubs worked hard and intelligently on worthwhile studies and projects. The men had a weekly debating club characterized by furious differences of opinion. There was an adequate opera house. Maude Adams played there. Schumann-

Heink sang. We heard, oh wonder, a major symphony orchestra. There was a lyceum course and, inevitably in a self-respecting town, a chautauqua where we swallowed huge gobbets of culture and were rewarded at the end with a play (always a farce or a melodrama) or a "humorous lecturer." It was William Jennings Bryan and "A Cross of Gold" to Strickland Gilliland and "Bibbety, Bob."

The town had two or three highly trained musicians. They shook their heads over the introduction of the phonograph, "canned," machine-made music. They were seers. The opera house soon became a motion picture theater, and no more road companies visited the town. Mothers need worry no longer about their sons hanging around stage doors. Instead they could wonder what they were up to in those Tin Lizzies.

What does growing up mean? I thought at one time that it meant that one had arrived at an immutable view of life and the world. The grown-ups I knew never seemed to change. They had all the answers, the same yesterday, today, and forever. But such a definition of growing up did not apply to a child of the twentieth century. I grew up all over again in each of my three worlds.

That first world of my early childhood was bucolic, idyllic, and, I believed at the time, everlasting. We knew, naturally, that there had been wars and torturers in the past. But those things were relics of ancient evil. They could never happen again. It was still to us, just emerging from childhood, a world of peace.

The Second World / *Gopher Prairie*

That child's world abruptly ended with the outbreak of World War I. It almost destroyed those innocent, provincial conceptions of my childhood. But not quite. We still had our faith in our American rightness. Even if the President was a Democrat, we were making the world safe for democracy. Our kind of world would prevail. We were putting a stop forever to that evil anachronism, war.

The war had its compensations, too. Though remote, it was exciting, a kind of melodrama itself with a satanic villain to hate, and dozens of nice songs, sweet or jolly, and grown-up girls marrying soldiers and weeping on the railroad platform as they left.

I remember the end of the war. In the middle of the night a truck manned by loud-voiced citizens with megaphones toured the streets shouting, "Germany has signed the armistice terms. Germany has signed the armistice terms." My father put his head out the window and roared. Then he sped to town to join the dancing, shouting celebration in the streets. We stayed at home. Girls, you know. I still relive my white rage as I lay that night in bed and listened to the revelry afar. Bed in my opinion was no place to spend a night like this one. The next night there was a proper, organized celebration with many speeches and all moral things of that kind. To that we went. And learned again that justice, goodness, and Americanism had triumphed. Our faith had been vindicated. Before too long, however, there would begin an erosion of that faith. Some not too much older than we were already thinking of the world as a place where war was inevitable, and each new war sadder and uglier than the last.

I was, in years, no child by now. Though still in high school, I was aware of fierce controversies and undercurrents of feuds on an international scale that seemed a sad magnification of the feuds and little wars of my home town. Inside ourselves, little had changed. Therefore war. But external changes came furiously. We found them delightful. The car was the great emancipator of youth. Those brothels on wheels drove our elders crazy though they were as mad about them as their offspring who, if they considered themselves of the intelligentsia, left the Prairie behind with all possible speed to live in Greenwich Village. Paris would have been better but next best was the Village where Ioway origins, Iowa itself, could be forgotten or concealed, and the frenzied world of Fitzgerald was the model, with his "gold-hatted, high bouncing lovers."

But those of us who attended small, midwestern colleges scarcely knew we were living in the Jazz Age. Nor did we know that we belonged to the Lost Generation.

Of the writers of the period, Sinclair Lewis gave us the hardest jolt. There wasn't much resemblance between the way he presented Gopher Prairie and the Prairie we thought we knew. That anyone should describe it as a place where "dullness was deified, dullness was God" was at first unthinkable. But the thinking ones thought and many ended by repudiating their childhood, the American past, and America itself. They had grown to a new stage. They were in revolt against war, against the "back to normalcy" of Harding, proud product of the Middle West. Even those of us who stayed in the Middle West didn't want to go back. We didn't want normalcy, which we now interpreted as Babbittry, and viewed the ordinary American as "Boobus Americanus." For Mencken was our guru. He revealed normalcy to us as an abomination. We went about carrying copies of the *American Mercury* to prove how sophisticated we were though far from Paris where all the sad young men held forth, and Hemingway's Parisian expatriates and roving alcoholics were in reality as alien to us as Fitzgerald's frenetic youth. Actually, Fitzgerald was not as alien as we thought him to be if we had read him right. He blinded us by his glitter. We did not discern that he was writing of the death of the American Dream, that prairie land of innocence from which he too had originally come.

But Boobus Americanus we knew. He lived next door. He was Babbitt, Elmer Gantry, the Rotarian, the booster, the flapper, the campus sheik with his coonskin coat and unbuckled galoshes. Above all, he was a Methodist as were most of us. To Mencken there was something very very funny about a Methodist. It was startling to learn that we were boobs. Perhaps we were so ready to accept Mencken because we already were feeling that this was not necessarily the best of all possible worlds. In 1931 the tomb of the Unknown Soldier was completed and inscribed.

Here rests in
Honored glory
An American Soldier
Known but to God.

But the glory and the honor had somehow faded. The war had then been fought for nothing. Prohibition increased lawlessness, the presence of women at the polls failed to produce the purification of politics expected of it. The government was corrupt. You could take your choice: since the dream was gone, you might as well have fun — bounce, bounce, bounce. Or settle down, make a living, make if possible more than a living, get rich. Be a Babbitt.

But that didn't work out either. Our elders had been telling us all along:

"Come boom, come bust." And "bust" came. The period between the wars was almost two different eras, boom and bust queerly combined.

In midwestern colleges there had never been much boom. Nor had the mores changed much. Maybe the boys found a jazz age off campus someplace. On campus girls still had to be in at eight o'clock. The doors were locked. One minute's tardiness drew penalties. When it was discovered that fire escapes could provide exit from or to things other than fires, the screens on the escapes were nailed shut. "Better to burn here than hereafter," quipped one faculty member in a committee set up to survey safety conditions on campus.

Many students lived in conditions of incredible poverty. When the banks closed, we thought the end of the world had come. And we could see another war in the making. Instead of an anachronism war was turning into a chronic condition. Four freedoms tried hard to take the place of making the world safe for democracy, but it was a sad war hated even by those who conceded it had to be fought. And if war were a chronic disease, it must be that it proceeded from something in ourselves. Like the creeping horrors I now knew existed under the fair meadow where a praying mantis could teach men lessons in ingenious cruelty, it must be that there was some horror in ourselves. Boobus Americanus was admirable compared to what we now saw ourselves to be.

The night World War II came to an end I happened to be spending in a berth in a sleeping car. I did not sleep. At every station stop, I raised the blind a few inches and peeked out at the sorry little knots of trying-to-celebrate citizens. Before I left home there had been an impromptu gathering to take note of the end of war. It couldn't be called a celebration. It was more like a prayer meeting. I did not go. I felt it just as well not to call God's attention to what we had been up to on this planet if by some good fortune his attention had been fixed on some other part of the universe at the time. The bomb put an end forever to any good I had hoped might come from the war. Not for one moment did I believe those who were saying that this was the ultimate weapon, so hideous that war could never be used again lest we prove T. S. Eliot wrong and end the world with a bang, not a whimper. Sooner or later the bomb would be used. Whenever had men invented some fascinating, lethal toy and refrained from using it? I said a word I wouldn't even have heard in my childhood. The bomb would be used just because we had it.

This was my last growing up. I no longer expected any sort of change that involved a change in persons. I couldn't foresee all the uses of nuclear energy in addition to the bomb. All I saw was Hiroshima.

Now it was not Lewis, not Fitzgerald, not Mencken who came to mind. It was Hemingway. I had thought I hated him for seeing what he thought he saw in bullfights. Now I believed he was right. A bullfight was simply something in us coming out.

He was right, too, about our loneliness. We were not merely a lost generation but a lost humanity. A few "clean, well-lighted places" were to be found where chronic loneliness became more bearable. But loneliness was palpable. Every celebrator on the station platforms looked lonely. In every crowd each looked separate, alone.

I recalled the celebrating at the end of the first war. Then we were "we" or had thought so. But this time was different. There was fear and dread and something else. Hemingway had named it. *Nada.* "Some lived in it and never felt it but he [Hemingway's Spanish waiter] knew it was all *nada y pues nada y nada y pues nada y nada.*" Hail nothing. All the lives, all the money, all the destruction. And all for nothing. Hail nothing, the nothing that would always be. A sleeping car is a fine place for not sleeping. "After all," Hemingway's waiter said to himself as he hunted for a bodega, open all night, to take the place of the clean, well-lighted cafe which insisted on closing up, "it is probably only insomnia. Many must have it."

The Third World / *To Make a Prairie*

My third world is still in the making. It is a world of marvel, of scientific miracles. It still has its moments of glory. When America put a man on the moon, I wanted to live forever in order to see what those mighty machines and mighty men of science would discover as they started gadding about the universe. "One giant leap for mankind." Except for one thing. Something science could not provide. We are all insomniacs. And worse than insomnia is nada. Old people complain of loneliness. Young ones say, "I can't find myself. I just can't, you know, get my head together, you know. What's the use of it all, you know?" Hail nothing. Such feelings set many to rummaging about in old religions or new cults. But perhaps for the old the best that could be hoped was a clean, well-lighted place that stayed open all night and an always available game of bingo. And for the young a commune, a gang, a new monasticism, a gulping materialism.

The earth too suffered. Polluted air, polluted water, whole species of animals disappearing forever. In their place emptiness. Ioway was gone, Iowa going, the black earth disappearing under concrete, a whole farm swallowed by every cloverleaf on a new four-lane highway.

Now every bit of the prairie that had seemed so endless to Trinket, Rover, and three children in the back seat of a surrey was something to cherish. Old bits of surviving prairie were hunted out, restored, and new ones made.

Perhaps it was recoverable after all—that feeling I had had in childhood that I, the meadow, the woodlands, were all one. But from now on, I knew, we would have to make our own prairie.

> Prairie
> To make a prairie it
> takes a clover and one bee,
> One clover and a bee
> And revery.
> The revery alone will do
> If bees are few.
> —Emily Dickinson

Insomnia must give way to revery. But we must understand Emily. The revery alone will do only if it restores what I had thought as a child that the prairie was: a fair meadow of peace inside and outside ourselves. To have peace within oneself—that would be to be grown up. To have it outside ourselves—there may be a chance if the world can learn the meaning of revery of Emily's sort. If it can, the revery alone will do. But the revery we must have.

From *Growing Up in Iowa: Reminiscences of 14 Iowa Authors* edited by Clarence A. Andrews (Ames: Iowa State University Press, 1978).

Considered a founder of the modern prairie restoration movement—and one of its greatest literary voices—John Madson was born and raised in Ames, Iowa, served with the U. S. Army Air Force in Europe during World War II, and earned a degree in wildlife biology at Iowa State University in 1951. His writing career started with the Iowa Conservation Commission (now the Iowa Department of Natural Resources) as editor of *Iowa Conservationist* magazine, and he later worked for the *Des Moines Register*, before moving to Godfrey, Illinois, near St. Louis. An avid sportsman, his essays and articles about a variety of landscapes were published in such magazines as *Smithsonian*, *Audubon*, *National Wildlife*, and *Field and Stream*. His writing often included a lively blend of science, history, memoir, storytelling, humor, and close observation of the natural world. Through landmark works such as *Where the Sky Began* (1982) and the following essay (first published in *Audubon* in 1972 and reprinted in his 1979 collection, *Out Home*), he introduced a national audience to the magnificent beauty and natural history of the tallgrass prairies, as well as the urgent need for preservation and restoration.

The Running Country

Lieutenant Colonel Stephen Kearny's orders were clear: proceed northwest into the Iowa wilderness with a unit of cavalry and survey the Des Moines River for possible fort sites.

They left Keokuk in June 1835 with a couple of transport wagons and a small herd of beef cattle, riding along the crest of a sun-drenched prairie ridge between the Des Moines and Skunk rivers, bound upstream into unknown country.

It was a beautiful time, as only an early prairie summer can be beautiful. The ridges glowed with flowers, and when the wind parted the new grass ahead, Kearny saw wild strawberries "that made the whole track red for miles" and stained the horses' hooves and fetlocks.

Slowed by the wagons and cattle, the party was averaging fifteen miles each day. As it turned out, that was about the same rate that the strawberries were ripening; the soldiers and the berry ripening were traveling north together. As if that weren't lucky enough, one of the cows freshened and began giving milk, and the troop dined on fresh strawberries and cream all the way to the headwaters of the Des Moines.

First-class foraging, and it sure beat beans. But then, this strange new country beat almost anything.

It was virgin tallgrass prairie, and Kearny and his men rode stirrup-deep through young bluestem grass and flowers, the hooves muffled in a loamy wealth that had been accruing annual interest for twenty thousand years. Later that summer, when they skirted ridges and crossed flats, their horses would vanish in a sea of Indian grass and big bluestem so tall that it could be tied in knots across the pommel of a cavalry saddle. It was a land belonging to grass, flowers, and sun, a new sort of land that was open to the sky, and trees and shadows shrank from it. For a long time, so did people.

THE FIRST SIGNS of prairie began back in Ohio as little natural clearings in the great eastern hardwood forest. They were strange clearings filled with strange grasses, and they saved a lot of axework. These eastern outriders of the prairie were quickly filled with fields and people, and vanished before men had a chance to really know them.

Scattered openings continued across northern Indiana, becoming larger, although the land was mostly forest. And then suddenly, twenty miles west of the Wabash River, the world opened up. A man would walk up out of the forested floodplain, step through a screen of sumac and wild plum, and stand blinking in a land that blazed with light and space. He was at the eastern edge of the Grand Prairie of Illinois; from there, north to Lake Michigan and west to the Mississippi, the prairies opened and broadened, sometimes spanning fifty miles without a tree or any other object to break the fabric of the grassland.

From the Wabash, this tallgrass prairie ran to the Missouri River and beyond, covering the western parts of Missouri and Minnesota and almost all of Iowa, extending into the eastern Dakotas, deep into Nebraska, and down into Kansas, Oklahoma, and Texas. It was called *tallgrass prairie* because it was a region dominated by huge grasses—Indian grass, the cordgrasses, and big bluestem, which might grow twelve feet high. It was a special region, labeled clearly and precisely with special plants. At its western boundary, out around the 100th meridian, the tallgrass prairie merged with shorter mixed grasses and midgrasses, which merged in turn with the short grasses of the Great Plains.

True prairie was not a matter of location, but of composition. The lie of the land had nothing to do with whether it was prairie or not; if it was tallgrass prairie it included the tallgrass communities. Some prairie was flat, much of it was rolling, and some was broken and rocky. But it needed tallgrasses if it was to qualify as true prairie—the most easterly of the great American grassland societies that sprawled between the Rockies and the eastern forests.

IT WAS HERE that the forested East ended, and the West really began. It stunned the pioneers coming from their Ohio and Kentucky forests, and one old journal effused: "The verdure and flowers are beautiful, and the absences of shade and the consequent profusion of light, produces a gaiety which animates every beholder."

Open as it was, it was not treeless. My home country in Iowa was a series of named prairies, such as the Ross Prairie or the Posegate Prairie, or whatever, and these stretches of shaggy grass were more or less fenced with groves and timbered valleys. Such prairies were said to resemble lakes, with boundary timber as shorelines indented with deep vistas like bays and inlets, throwing out long points that were capes and headlands.

Years ago, when Americans were less homogenized and outlandish accents still drew attention, a man in Maine asked me where I hailed from. I told him I was from Iowa. He shuddered, and calc'lated that his forests and hills must seem very beautiful to me. I replied that it was sure different from anything back in Story County, all right, and that seemed to please him.

The truth of it was, my home country had about all the hills and trees that we needed.

Our prairie country had a marked pitch and roll to it, like an ocean quieting after a bad storm. There is prairie country that's about as flat as land can be, but most prairie has a fine roll and break, with the land billowing off to the skyline and some timber down in the folds.

Many of the original trees were the same as back East: elms, hard maples, silver maples, shagbark hickories. But in Iowa, the beech tree's place was taken by basswood; I never saw a beech until I was thirty years old.

The prairie forests varied, depending on where you found them. Along smaller streams were wild plum, box-elder, wild cherry, soft maple, elm, and wild grape. If the floodplain was flat and wide, there were walnuts, hackberries, and great cottonwoods. These floodplain forests were densely undergrown, but this wasn't true of forest higher up on the ridges. There the trees were encroaching on prairie domain, and had to recognize the sovereignty of King Grass. These upland forests, what there were of them, were open. The groves of oaks and maples, huge-boled and ancient, looked like royal parks. The ground beneath them was free of undergrowth and carpeted with shade-tolerant grasses and flowers. Those old upland groves are gone today, mostly, but now and then you'll see a sentinel oak at the edge of a high pasture. After a hard rain, the closely grazed ground on such a ridge may glitter with flint flakes—the tribes knew good summer camps when they saw them.

Primeval prairie woods were found on sand or clay ridges, rocky outcrops, or the floodplains of streams. All the richest parts of the original prairie country were in grass; the forest existed at the sufferance of grass, and only on places that grass did not choose to occupy.

I've never thought of such country as monotonous, although my Maine friend would have felt so, I'm sure. But I've had four generations to get the forest out of my blood. By 1800, some of my people had begun to peer warily out of the Ohio forests; by 1825 they had gotten as far west as the eastern edge of the Grand Prairie. Bit by bit, they crept out of the trees and weaned themselves away from the Wooden Country. In 1853 my great-grandad grasped the nettle (and a new bride) and left the old states, heading west into the prairie frontier of central Iowa.

The strain has bred true; I like to return to trees, and sit and walk and hunt under them, but I could never live under them if they kept out all the sky.

Those royal groves must have been something, though.

WHY ALL THAT grass?

Why, suddenly, twenty miles west of the Wabash, did the land begin running out of trees?

Some early settlers thought the land was just too poor to grow trees—but it didn't take them long to find that wasn't true. They finally decided that prairie was caused and maintained by the fall and spring infernos that swept through the grasslands, leaving leagues of blackened ash and carbonizing any tree seedling that had the temerity to invade prairie.

There was a lot to this. Certain islands in prairie lakes had fine groves of ash trees where fire could never reach. Yet the fire theory had gaps in it, and some early ecologists began to suspect that fire was an effect of prairie rather than a cause.

Mostly, it was a matter of rainfall. Eastern forests are in humid climate while grasslands are in drier country. The plains and prairies lie in the rain shadow of the Rocky Mountains—a lofty barricade to the moisture-laden winds from the Pacific. Prevailing winds carry enough rain to grow grass, but not unlimited forests. Then, too, prairie maintains itself well. Even at the eastern edge of the grassland, where rainfall was sufficient to support either tallgrasses or trees, the dividing line was abrupt. Tallgrass prairie is a closed community that rarely admits aliens; tree seedlings can seldom live in prairie sod with its intense competition, crowding, and ground-level shading.

Years ago, a pioneer ecologist named Bohumil Shimek began to suspect that the grasslands weren't just caused by low rainfall and maintained by fire, but resulted from evaporation caused by incessant exposure to wind, low relative humidity, and frequent high temperatures.

One of the greatest of these, Professor Shimek felt, was wind. Wind breaks twigs and leaves, and drives dust and sand against delicate tissues, abrading and tearing them. The effect of constant wind on tree leaves is also physiological. Shaking a plant increases its rate of transpiration—in a climate where a tree can't afford to transpire too much water vapor. Of course, this is checked by the closing of the leaves' stomata. But that also checks the processes of respiration and assimilation. If physical shaking continues long and violently, the plant can be weakened and even killed.

It follows, then, that trees most exposed in their spring and summer leaf seasons to hot, constant winds would be in the greatest danger.

This is apparent in Iowa, where spring and summer winds often blow from the southwest. Western Iowa streams that drain into the Missouri flow southwest, directly into these winds. Eastern Iowa streams flowing into the Mississippi run southeast—at right angles to the prevailing winds and with maximum protection. Iowa streams flowing southwest may have scanty, brushy timber if they have trees at all, and some flow for miles through open prairie. The Maple, Little Sioux, Boyer, and Nishnabotna are examples. In eastern Iowa, the southeast-flowing streams—the Cedar, Wapsipinicon, Des Moines, Iowa, and Skunk—have dense floodplain forests.

In western Illinois we can see the effects of wind and sun on the hills above the Mississippi. Certain forested headlands and bluffs above the river are capped with little tallgrass prairies. You won't find these on east-facing hillsides, looking away from the river. The hill prairies face west, at right angles to the full blaze of an August afternoon, looking out over a broad floodplain where the wind has a long fetch. Of all our Illinois landscapes, none are more exposed to intense sun and wind than these west-facing river bluffs and hills. They are ecological niches of native prairie.

The steep ground that creates these prairies may also protect them. Some of the limestone river bluffs near my home are nearly two hundred feet high, towering over the Great River Road and the Mississippi in white, buttressed walls. The edges of these cliffs are dangerous pastures, as some farmers have found. Cows, unlike cats, can't land on their feet nor spare eight lives. Just behind these river bluffs may be rough, wooded valleys that discourage any farming. Guarded from front and behind, the little prairies along the brinks of these river cliffs have survived.

Our family often goes up there for Sunday lunches, climbing gargantuan limestone stairs to the clifftop. With forest at our backs and the broad Mississippi out in front, we bask in original bluestem high above the beaten highway of a newer, noisier world.

OLD-TIME PRAIRIE was a grandmother's quilt of color and form that shifted constantly as the wind breathed life into the grasses. Willa Cather remembered Nebraska when "there was so much motion to it; the whole country seemed, somehow, to be running." One of Harvey Dunn's finest paintings of prairie life was of a Dakota girl pumping water, her skirts blowing, an embodiment of the old prairie adage: "There's just nothin' prettier than a girl pumping water in the wind!"

The play of wind on tallgrasses, with the land running beneath a towering sky, is something we may not know again, for we will not see such vistas of grass again.

The tallest of the prairie grasses — big bluestem and cordgrass or slough grass — never reached their highest growth in the richest soil, but in lower, marshier land. There, where substrata of clay lay near the surface, the big bluestem grew to twelve feet. Cows could be lost in it, and might be found by a mounted man only when he stood in the saddle or rode up a nearby ridge and watched for the cows moving through the deep grass. The size of prairie grass was proportional to moisture. The dark green slough grass, called "rip-gut" because of its saw-edged leaves, grew in dense stands on low flats. Pioneers avoided low prairie swales that were marked by this "black grass." A traveler had to pick his way carefully over the ridges in spring, for the low places were impassable. When incumbent frontier politicians made soaring reelection promises, they often swore that they had "waded sloughs" in the interests of their constituents, for no work was harder than that.

Another lowland grass was Indian grass — tall, coarse, and up to eight feet high, usually found in more southerly regions of tallgrass prairie. All of these are fine livestock feeds, and make excellent hay if cut before their stems are too tough and fibrous.

The higher and better-drained the prairie became, the finer and shorter the grasses. True prairie uplands are dominated by little bluestem, with rich stands of Junegrass, side-oats grama, needlegrass, and prairie dropseed.

The tallgrasses need moisture and plenty of it, and were the dominant grasses in the prairies east of the Mississippi. Farther west, as rainfall diminished, tallgrasses retreated to lower parts of the prairie, and the well-drained uplands were covered with shorter midgrasses. Still farther west, with even less rainfall, those midgrasses were replaced by the shortgrasses of the true plains: western wheatgrass, buffalo grass, and blue grama.

Colors and texture of tallgrass prairies varied with the season and elevation. In early spring, the bleak prairie hillsides might be brownish gray from the weathered ash of the fall fires. Then, often well before Easter, the prairie pasqueflowers would appear. Of all early wildflowers these are the bravest, not blooming in sheltered woods, but out in the big open on glacial moraines where the wind cuts to the bone in late March.

Then, one bright morning, the south-facing slopes would look as if patches of spring sky had fallen on them, and you knew that the bird's-foot violets were in bloom. There was white, woolly, prairie cat's-foot coming on, and the first green blush of new grass on the slopes.

The hilltops were splashed with early spring flowers: false dandelion, cream-colored paintbrush, and mats of groundplum vetch. The new grass would be spangled with tiny purple, blue, and white grass flowers, and perhaps yellow upland buttercups and yellow lousewort.

When the bluestem grasses began to appear in mid-April it was a signal for the spring flowers to hurry, for they were small plants that were easily overpowered by the growing grasses.

Prairie and meadow violets appeared, with vetch and false indigo. Along the streams and low places were marsh marigold, yellow stargrass, and purple heart-leaved violets. The prairie pinks came into bloom and enameled a landscape of young grass with pink, white, and purple. With them came the puccoons, splashes of rich orange in the greens and pinks.

By early June, most of the spring flowers were gone. The flowers were taller now. Daisies began to appear, and larkspur and purple coneflower. There was a foot-high prairie lily with a red bloom, and with these lilies came clouds of prairie roses. About the only thing that "them politicians down at Des Moines" ever did that pleased our Grandma Tut was to make the prairie rose Iowa's state flower.

By summer there were myriads of blossoms, all holding their own with the lofty grasses. Wild indigo, with its heavy, creamy blooms, stood tall. White larkspur stood above many grasses; so did oxeye daisy, many sunflowers, goldenrod, and compassplant—the set of its oaklike leaves marking the prairie meridian. Leadplant, with its silver-gray leaves and purplish flowers, was everywhere on the upland prairie. From August on through autumn, wild asters bloomed white, lavender, and purple. Deep in the ripening grasses, almost hidden, were the fringed gentians and bottle gentians.

This was the season when the prairie flamed with blazing-star or gayfeather, a tall purple spike of blooms whose root bulbs were fed to Indian ponies to increase speed and endurance.

Some of the finest floral displays were on low ground, hidden around the prairie marshes. Pioneers recall four types of wild orchids there. The smallest, and possibly the rarest, was yellow. There was a white orchid with purple mottling, and a larger yellow orchid. The largest and finest of the wild orchids in Iowa grew two feet high far back in the marshes; on each stalk there were two or more white and purple blooms with subdued mottling, and floral pouches "big enough to hide a hummingbird."

The prevailing colors of prairie flowers were white, purple, and yellow, although some phloxes came in tints of red, and there were red lilies and the orange-red butterfly weed with its masses of tiny hourglass-shaped flowers. The first of the year's prairie flowers were the smallest and most delicate; as the seasons went on, flowers became taller, gaudier, and stronger-stemmed to compete with the rising tide of grasses that deepened and coarsened with the summer.

In color and form, those grasses had the last word. By Indian summer they had ripened, rich and stately, each clan with its own colors, and those colors shifting and changing with the wind. By September the grasses had lost their greens and had deepened into tones of gold and bronze. There were tawny stands of ripe Indian grass, patches of airy switchgrass heavy with seed, and the wine-red fields of bluestem. The colors did not last; as winter came on the grasses bleached and faded as the prairie world retrenched, sending its vigor underground, to await spring and the time of pasqueflowers.

THIS TALLGRASS PRAIRIE swarmed with wildlife.

It wasn't uniform, featureless range. There was a variety of habitats: ridgetops with short midgrasses, hillsides and flats with deep grasses, upland groves, heavily timbered floodplains, and the endless sumac and plum woods borders.

Out on the open ground there were bison, although probably never as many as on the shortgrass plains farther west. There were bear, ruffed grouse and turkey in the forests, and deer and elk in and around the forest edges.

The prairie then was strewn with small lakes, potholes, and marshes, and veined with tiny creeks that ran over beds of bright glacial pebbles, hidden beneath the arching grasses. The openness of these prairies, and their frequent wetness, drew countless shorebirds and waterfowl. The prairies teemed with ducks, geese, pelicans, sandhill cranes, whooping cranes, bitterns, egrets, and whistling swans. Shorebirds came in vast spring clouds; in March and April, golden plovers began arriving from Argentina on their way to the Arctic, and they would always stop on the prairie. These plovers (settlers called them "prairie pigeons") hunted burned hillsides for grubs and insect eggs, and the dark slopes were spangled with moving gold and silver as the great flocks of plovers fed.

Swans and geese arrived in squadrons to graze on the new bluestem shoots, and with them came huge, loose wedges of Eskimo curlews, long-billed curlews, and upland plovers.

Even without trees, the prairie drew songbirds. The bobolink was one of the old-timers, and so was the kingbird. There was our western meadowlark, singer laureate of the tallgrass prairie, almost identical with the eastern meadowlark but singing a longer and more tuneful song. Lacking trees, the meadowlarks and dickcissels and song sparrows made do, and sang from tall flower stalks; many of the grasslanders, such as longspurs, horned larks, and bobolinks, simply sang on the wing.

Some prairie women sorrowed at leaving their eastern songbirds behind them in the forests, but this began to change as homestead windbreaks of soft maples, box-elders, and willows began to mature. The "timber birds" did some pioneering of their own, and joined the settlers out on the prairie. Many prairie-born children could easily remember the first robins that they ever saw—an event marked not only by a brand-new bird, but by the delight of their parents.

Such are the accounts that come to us, recalling the old prairie as a place rich in form, life, and color—a trans-Mississippi Eden that was fresh and new. (Many of these writers returned East to do their euphuizing.)

But there were those who hated prairie, and spoke of it as "a fearful place."

Washington Irving wrote: "To one unaccustomed to it, there is something inexpressibly lonely in the solitude of a prairie. The loneliness of a forest seems nothing to it. There the view is shut in by the trees and the imagination is left to picture some livelier scene beyond. But here we have the immense extent of landscape without a sign of human existence . . ."

The prairie had many faces, and men saw them in different ways. But whatever the prairie's variables, there were some constants. Big, it was, and overwhelming. Lonely, it was.

The second wave of pioneers, passing through tiny settlements on the prairie frontier, were greeted with pathetic eagerness by the original settlers, who implored them to stop and stay. Some of this was a practical need for neighbors; much was loneliness. But the greater part may have been a profound insecurity—of being part of something wholly new, with no ancestral precedents in Europe or New England. In a forest a man is partly hidden; he has carved a niche with sheltering hills and walls of trees. There is intimacy in a forest. But not on a prairie; there are no walls on a prairie. The prairie man and woman, and their puny fields, were exposed to a vast and pitiless sky. There was no snugness, no security, nothing to shield the family; a man was alone and naked, bared to an openness of terrifying intensity and magnitude.

It must have been hardest for women, for they have the deepest and most tender roots in tradition. The prairie men left many things in the East and good riddance to most of them, the stump farms and taxes and fields that were more rock than soil. But a woman left the church were she had been married; she left a stone-and-frame home with a rock maple by the kitchen window, and familiar birdsong, and followed her man into a heathen wilderness of cyclones, blizzards, flaming horizons, and an everlasting wind that would turn her skin to leather and end her youth.

Even as the people changed the prairie, it changed them. That prairie and the vastness beyond it were uniquely American. The men and women who settled there were also unique, or became so, and they may have been the first total, genuine Americans. They were new people in a new land—and such people and land had not been known before.

IT WAS, AS THE people finally found, a land of incredible richness and fecundity.

At first, they deduced that if the land was too poor to grow trees it was probably too poor to grow much in the way of crops—overlooking the point that their crops were actually grasses, and that the prairie was the world's finest grower of grasses.

That was one reason the first of the prairie pioneers settled in bottomland timber. It was the logical thing to do. Forests had building materials and fuel, clear streams and abundant game. Then, too, these settlers were accustomed to clearing fields from forest—that was simply the way men farmed. They failed to see that trees grew on the prairie's poorest soils, and that forest soils were a thin veneer of fertility that was a delusion, for it would be quickly eroded away when there were no longer trees to stabilize it. Finally, the early settler didn't know how to cope with open prairie, and he was terrified of the prairie fires that did not enter the bottomland forests.

So, for a long time, home base remained in the timber, and familiar forest was laboriously cleared in preference to breaking virgin prairie sod. In Marshall County, Iowa, as late as 1867, prairie was selling for $3 to $10 per acre while timbered lands sold for $30 to $50. This was the first stage in prairie pioneering, with families living in forested clearings as they had always done, farming some small fields, grazing some livestock, and hunting and trapping.

But in spite of the prairie's strangeness, the pioneer farmers soon found that it was where the action was. They left the trees and began to learn what the land really offered.

Building was one of the first and greatest problems. Forests were often miles distant, and there were no railroads to bring building materials, so the settlers had to invent. In the western parts of the longgrass prairie they built with sod, cutting thick turfs a foot wide and two feet long. A course would be laid two sods wide; the next course would be at right angles to the first. This sealed all joints, and the wall of a sod house would be from two to three feet thick, warm in winter and cool in summer. The roofs, made of pole framing, would be covered with thin turfs and perhaps thatched with slough grass. This same cordgrass served as fuel; tightly twisted into faggots, it would burn hotly as long as ten minutes. In one hour a prairie boy or girl could twist a day's supply of fuel. Years later, their grandchildren would gather corncobs from the hog lot for the kitchen range. In one form or another, grass cooked and heated.

Some "soddies" were grand, two-storied, shingle-roofed houses, and a sod home with touches of Victorian architecture was about as wild a concession to function and fashion as we'll ever know. But as lumber became available at reasonable prices in the 1870s and 1880s, and freighters and railroads began bringing in lumber and glass, sod houses began to vanish.

The second great problem was sodbreaking. The native sod was a solid mass of locked, intertwined rootlets that were studded with the massive rootstocks of compassplant, redroot, and other forbs.

There were commercial plowmen who would break your claim for $12.25 per acre with a "breaking plow"—a huge, wheeled device drawn by five-yoke of oxen that cut a furrow two feet wide. The moldboard of the old bar share-breaking plow was seven feet long.

A breaking plow could also be made from the wheels of the settler's wagon, a handmade axle, and a long beam that could be raised or lowered to regulate the depth of the furrow. Still, there was the problem of the moldboard. Eastern plows had great, heavy, cast-iron moldboards that were designed for the rocky fields of New England. They were almost useless on the prairie, for they would not "scour" and rid themselves of clinging loam.

A great breakthrough came in 1837 when John Deere of Illinois invented his new prairie plow. It was a walking-plow that could be drawn through prairie sod with a three-horse team—a handsome implement, light and artistic, with a bright, shining, tempered steel moldboard. It was one of those great pieces of original art that combine utility with beauty, like the Lancaster rifle or the gracefully curved helve of our American ax.

With such a plow, the farmer could break the dense prairie sod with amazing efficiency. As the furrow was cut, there was a constant popping sound, like a tiny volley of pistol shots, caused by the breaking of tough roots and spurs. This incessant cracking and popping had a slight ring to it, amplified by the tempered steel of the polished moldboard.

It was best to break new ground between early May and mid-July. The sod was plowed as shallow as possible; the thinner it was cut, the better it would rot. A settler might strike an ax into the sod and drop seed into the hole, and while the overturned sod was decomposing and mellowing it was also nourishing the growing corn. Or, you could drop corn into every fourth row while plowing, and this would be covered by the plowing of the fifth row. Neither the yield nor the quality of corn was high that first year, but it was sufficient, and a man might make fifteen to twenty-five bushels per acre of "sod corn."

It might be several years before the grassroots had completely rotted and become rich, smooth loam. But it was all worth it. A man could spend half his lifetime back in the Wooden Country clearing land and just getting ready to farm. On the prairie, one man and his team, working for only two months, could have an "eighty" broken and planted with sod corn or flax.

It was a hard and lonely life for the prairie families that came first, but there were neighbors on the way. During one month in 1854, over 1,700 covered wagons passed near Peoria, Illinois, bound for Iowa. The real pioneering era of Iowa's frontier—the sod shanty and semisubsistence living—lasted no more than a decade.

But during the years of primitive abundance, there were strange and primitive dangers.

UNLIKE TREES and shrubs, the tops of annual grasses die each autumn. Chlorophyll fades, and life retreats into the roots, rhizomes, and seeds of the grass. The dead materials left above may return to the earth. Or, in a sense, to the sun.

Tall prairie grasses, especially in dry autumns, became tinder of almost explosive quality. Prairies have always burned from fires set by lightning, and sometimes by man. Indians were known to fire the prairies to drive game or defend themselves from enemies. Settlers commonly caused fires that they blamed on the Indians. But whatever the cause, one thing was sure: each autumn and spring the fires would come.

By day there might be a strange haze in the air, and a smokiness flowing down the watersheds. By night, a reddish-orange glow was reflected from behind the horizons. If there was no wind, the haze and the glow might last for days before the settler actually saw flames, and he had time to plow fire lanes around his buildings and stack-yards, and burn the grass within the plowing. The windless prairie fire would advance deliberately, marching across the grassland while tufts of bluestem vanished in puffs of flame, the white smoke drifting into blue sky. Coyotes and foxes exploited this, hunting before the flames to catch dispossessed mice and rabbits.

But if there was wind, there was a fire blizzard—one of the greatest horrors of prairie life.

It came with walls of flame thirty feet high and a deep devouring roar, and black smoke instead of white, and the sun darkened and animals went mad. The glow of these great prairie fires could be seen for forty miles, and showers of ash and flake would be carried that far ahead by the wind. Single prairie fires were known to have burned over two hundred square miles, and one fire traveled over twenty-two miles "as fast as a horse could run."

Within five miles of where I was born, a family of five Ohio emigrants died on a lovely October day in 1860 when they were trapped in a ravine by prairie fire.

In 1873, in Saline County, Nebraska, a fire blizzard roared across the grasslands toward a prairie schoolhouse. A mother who lived in a nearby soddy ran over to the school for her own children and some young relatives. The teacher begged her not to take them, but the hysterical mother would not listen. The ten children and the woman tried to outrun the fire, but lost their race. All eleven died in the flames. The teacher and the other pupils were safe on a nearby plowed field, and the school never did burn.

Old Sitting Bull himself, in the dry, flaming autumn of 1885, warned some Dakota schoolchildren that they could never run away from a prairie fire. "Go to bare ground," he counseled them, "or onto sand, gravel, or plowing. Or set a backfire. Go to a place with no grass. But do not run."

Entire towns were destroyed by some of these prairie fires; in Leola, South Dakota, all but twelve of the town's hundred buildings were burned in 1889 by a prairie fire that traveled forty miles in four hours.

Prairie fires were feared by almost every pioneer, and even the small boys who usually found them exciting had to admit there were drawbacks. Herbert Quick, the Iowa writer, told of the sharp grass stubs on burned-over prairie that pricked a schoolboy's bare feet and caused festering sores. Of course, this could have been solved by simply wearing shoes to school, but that's a stupid solution when it's spring and you're ten years old and the prairies are greening up. Those prairie boys found a friend in need in the pocket gopher. The big gophers threw up mounds of soft, fine, cool earth, often in long lines. The ultimate luxury was to walk all the way to a country school without stepping off a soft gopher mound. It never worked that way, of course. The gopher mounds usually wandered off in the wrong direction and a boy ended up walking farther to school through grass stubs than if he'd taken a direct route in the first place.

That problem, like all others caused by prairie fires, vanished with the bluestem. There came a time when the autumn horizons no longer glowed red at night, for the prairie was gone.

WE SPENT OUR tallgrass prairie with a prodigal hand, and it probably had to be that way, for these are the richest farm soils in the world. There were certain wilderness things that were fated to be spent almost to the vanishing point: bison in shortgrass plains, lobos and grizzlies in settled cattle country—and the vistas of true prairie.

But spending, is one thing; bankruptcy is another.

To squander the last stands of true prairie would wipe out a valuable index to original quality. It is important that our agronomists, botanists, zoologists, and soil physicists have reference points to the original plants and soils of our most valuable ecosystem. We may someday have to rebuild those soils, or try to. Native prairie is a baseline from which creative research can depart, and return for reference. We'd never dream of melting down the platinum meter in Paris and converting it to jewelry; it is the master rule, an original measurement upon which so much engineering and science are based. And so, in an even greater sense, is native prairie.

Just as important is the maintenance of certain tangible links with the old time. To destroy the last of the native prairie would be as criminally stupid as burning history books, for prairie is a chronicle of human courage, endurance, and victory, as well as a finished natural system.

Deciding sometime tomorrow afternoon that maybe we should have a few more prairies around, and then abandoning a few cornfields to that end, just won't get it. Original tallgrass prairie is the endpoint of twenty-five million years of evolution; it cannot be restored overnight, if indeed it can be restored at all. If the job of prairie restoration were left to nature alone, and if there were adequate sources of seed, nature might be able to convert cornfields to "native" prairie in two hundred to three hundred years. Imitation prairies have been built, but with perhaps thirty plant species instead of the original two hundred or more, each occupying a special niche in a special way. Prairie in its full form cannot be reconstructed by man—Eric Hoffer notwithstanding.

However, tallgrass prairie is the most difficult of all native America to conserve. This is because it is the world's most valuable farm soil—and it must be conserved in quantity if it is to mean what it should. There are still many "splinter prairies" in Midwestern states, some of several hundred acres. A man could stand in a grove of virgin white pine of the same size and feel that he was in primeval forest. Not so with prairie. To the average man, a scrap of native prairie is just a shaggy weedpatch between cornfields. Prairie must have sweep and perspective to look like prairie. It is more than just native grasses and forbs; it is native sky, and native horizons that stretch the eye and the mind. To be prairie, really good prairie, it must embrace the horizons. That is the ideal, and the only places where you will still find it are in parts of Nebraska's Sand Hills and in the Flint Hills of eastern Kansas.

The Flint Hills prairie has survived because beds of cherty limestone lie so close to the surface that the land can't be plowed. It is heavily grazed, and has been for a hundred years, but it is still prairie, rolling in long waves from the Nebraska line down into the Osage Hills of Oklahoma. It's country worth seeing.

One morning early last April, Kansas biologists Bob McWhorter and Bob Henderson hauled me out of bed at 3:00 A.M. for a prairie chicken count in the northern Flint Hills.

Before dawn we were on the ancient dancing grounds of the prairie chicken, with a traditional stage setting. The southern horizon was flame-torn with spring prairie fires that reddened the sky and gave the strange impression of sunset at dawn. Somewhere behind us several coyotes were swapping hunting yarns and settling down for the day, and even before it was light enough to see, we could hear the haunting, hollow booming of prairie chickens from several directions, some nearby and others dim with distance. With first light we could see them on the prairie ridge before us, over thirty of them in the closest flock, the sun glowing orangely on the inflated air sacs of the dancing, posturing males. As we watched, a phalanx of upland plovers swept past. Just over the ridgetop to the north, not twenty yards above the fire-blackened prairie that was beginning to blush green with new bluestem, a column of Canada geese moved out for breakfast. Another prairie day had begun, much the same as spring days have begun on Kansas prairies since the Miocene.

OF THE RELIC PRAIRIES that I know, none is as poignant as the tiny scrap that I found years ago in the center of an intensely farmed Iowa section.

It was a small, lost graveyard, all that remains of the little settlement of Bloomington, wiped out by diphtheria over a hundred years ago. About a dozen weathered stone markers leaned and lay in a patch of original blue-stem. Among the graves were those of a young mother and her children, and when I found the place in late summer their graves were set about with a few tall, magenta torches of blazing-star, stateliest of all prairie flowers. It was part of an original time and place, and it held fitting memorials. There were the flowers of gayfeather to lift the spirits of lonely, beauty-starved women. There was bluestem for the men, for their haycutting and prairie chicken hunting; for the children, there was compassplant, with its wonderful chewing gum, and wild strawberrries hidden in the grass.

That patch of tallgrass prairie was a more enduring memorial than the stones that stood there, and infinitely more appropriate. Today, our memorials reflect our values, and we will probably be interred in manicured "memory gardens," our graves decked with plastic blossoms that are imitations of imitations. That, too, may be appropriate.

MY FEELING FOR tallgrass prairie is like that of a modern man who has fallen in love with the face in a faded tintype. Only the frame is still real; the rest is illusion and dream. So it is with the original prairie. The beautiful face of it had faded before I was born, before I had a chance to touch and feel it, and all that I have known of the prairie is the setting and the mood—a broad sky of pure and intense light, with a sort of loftiness to the days, and the young prairie-born winds running past me from open horizons.

A strong place puts a mark on all that lives there, and the mark may outlast the place itself. Prairie people are like their western meadowlarks, seeming to be the same as their eastern relatives, but with a different song. It was the prairie that changed all that. It gave them a new song, and a new reason for singing.

From *Out Home* by John Madson, 1979. Reprint (Iowa City: University of Iowa Press, 2008).

JOHN MADSON

The author of nine books of poetry and prose, Bill Holm spent his youth on a farm near Minneota, Minnesota, before eventually attending Gustavus Adolphus College in St. Peter and then graduate school at the University of Kansas in Lawrence. Although Holm traveled the world, living for extended periods in China and Iceland, his home residence continued to be Minneota. He taught for twenty-seven years at Southwest Minnesota State University, where he encouraged students to appreciate and write about what was in the world around them, whether boxelder bugs or prairie. In this essay, published in 1987, Holm follows his own advice by exploring the many ways an affinity for the prairie shapes identity and ways of seeing. "There are two eyes in the human head," he writes, "the eye of mystery, and the eye of harsh truth—the hidden and the open—the woods eye and the prairie eye" (42).

Horizontal Grandeur

For years I carried on a not-so-jovial argument with several friends who are north-woods types. They carted me out into the forests of northern Wisconsin or Minnesota, expected me to exclaim enthusiastically on the splendid landscape. "Looks fine," I'd say, "but there's too damn many trees, and they're all alike. If they'd cut down twenty miles or so on either side of the road, the flowers could grow, you could see the sky, and find out what the real scenery is like." Invariably, this provoked groans of disbelief that anyone could be insensitive enough to prefer dry, harsh, treeless prairies. There, a man is the tallest thing for miles around; a few lonesome cottonwoods stand with leaves shivering by a muddy creek; sky is large and readable as a Bible for the blind. The old farmers say you can see weather coming at you, not like woods, where it sneaks up and takes you by surprise.

I was raised in Minnesota, true prairie country. When settlers arrived in the 1870s they found waist-high grass studded with wild flowers; the only trees were wavy lines of cottonwoods and willows along the crooked Yellow Medicine Creek. Farmers immigrated here not for scenery, but for topsoil; 160 flat acres without trees or boulders to break plows and cramp fields was beautiful to them. They left Norway, with its picturesque but small, poor, steep farms; or Iceland, where the beautiful backyard mountains frequently covered hay fields with lava and volcanic ash. Wives, described by Ole Rølvaag in *Giants in the Earth*, were not enamored with the beauty of black topsoil, and frequently went insane from loneliness, finding nowhere to hide on these blizzardy plains. But the beauty of this landscape existed in function, rather than form, not only for immigrant farmers, but for Indians who preceded them.

Blackfeet Indians live on the Rocky Mountains' east edge in northern Montana—next to Glacier National Park. Plains were home for men and buffalo, the source of Blackfeet life; mountains were for feasting and dancing, sacred visions and ceremonies, but home only for spirits and outlaws. It puzzles tourists winding up hairpin turns, looking down three thousand feet into dense forests on the McDonald Valley floor, that Blackfeet never lived there. It did not puzzle the old farmer from Minnesota who, after living and farming on prairies most of his life, vacationed in the Rockies with his children after he retired. When they reached the big stone escarpment sticking up at the prairie's edge, one of his sons asked him how he liked the view. "These are stone," the old man said; "I have stones in the north eighty. These are bigger, and harder to plow around. Let's go home."

When my mother saw the Atlantic Ocean in Virginia, she commented that though saltier, noisier, and probably somewhat larger, it was no wetter or more picturesque than Dead Coon Lake or the Yellow Medicine River and surely a good deal more trouble to cross.

There are two eyes in the human head—the eye of mystery, and the eye of harsh truth—the hidden and the open—the woods eye and the prairie eye. The prairie eye looks for distance, clarity, and light; the woods eye for closeness, complexity, and darkness. The prairie eye looks for usefulness and plainness in art and architecture; the woods eye for the baroque and ornamental. Dark old brownstones on Summit Avenue were created by a woods eye; the square white farmhouse and red barn are prairie eye's work. Sherwood Anderson wrote his stories with a prairie eye, plain and awkward, told in the voice of a man almost embarrassed to be telling them, but bullheadedly persistent to get at the meaning of the events; Faulkner, whose endless complications of motive and language take the reader miles behind the simple facts of an event, sees the world with a woods eye. One eye is not superior to the other, but they are different. To some degree, like male and female, darkness and light, they exist in all human heads, but one or the other seems dominant. The Manicheans were not entirely wrong.

I have a prairie eye. Dense woods or mountain valleys make me nervous. After once visiting Burntside Lake north of Ely for a week, I felt a fierce longing to be out. Driving home in the middle of the night, I stopped the car south of Willmar, when woods finally fell away and plains opened up. It was a clear night, lit by a brilliant moon turning blowing grasses silver. I saw for miles — endless strings of yardlights, stars fallen into the grovetops. Alone, I began singing at the top of my voice. I hope neither neighborhood cows, nor the Kandiyohi County sheriff were disturbed by this unseemly behavior from a grown man. It was simply cataracts removed from the prairie eye with a joyful rush.

Keep two facts in mind if you do not have a prairie eye: magnitude and delicacy. The prairie is endless! After the South Dakota border, it goes west for over a thousand miles, flat, dry, empty, lit by brilliant sunsets and geometric beauty. Prairies, like mountains, stagger the imagination most not in detail, but size. As a mountain is high, a prairie is wide; horizontal grandeur, not vertical. People neglect prairies as scenery because they require time and patience to comprehend. You eye a mountain, even a range, at a glance. The ocean spits and foams at its edge. You see down into the Grand Canyon. But walking the whole prairie might require months. Even in a car at 60 miles an hour it takes three days or more. Like a long symphony by Bruckner or Mahler, prairie unfolds gradually, reveals itself a mile at a time, and only when you finish crossing it do you have any idea of what you've seen. Americans don't like prairies as scenery or for national parks and preserves because they require patience and effort. We want instant gratification in scenic splendor as in most things, and simply will not look at them seriously. Prairies are to Rockies what *Paradise Lost* is to haiku. Milton is cumulative; so are prairies. Bored for days, you are suddenly struck by the magnitude of what has been working on you. It's something like knowing a woman for years before realizing that you are in love with her after all.

If prairie size moves the imagination, delicacy moves the heart. West of Minnesota, the prairies quickly rise several hundred feet and form the Coteau. This land looks more like the high plains of Wyoming. Rougher and stonier than land to the east, many sections have never been plowed. Past Hendricks, along the south and west lake shores, things open up— treeless hills with grazing cattle, gullies with a few trees sliding off toward the lake. Ditches and hillsides are a jumble of flowers, grasses, and this-tles: purple, pink, white, yellow, blue. In deep woods, the eye misses these incredible delicate colors, washed in light and shadow by an oversized sky. In the monochromatic woods, light comes squiggling through onto a black green shadowy forest floor. My eye longs for a rose, even a sow thistle.

A woods man looks at twenty miles of prairie and sees nothing but grass, but a prairie man looks at a square foot and sees a universe; ten or twenty flowers and grasses, heights, heads, colors, shades, configurations, bearded, rough, smooth, simple, elegant. When a cloud passes over the sun, colors shift, like a child's kaleidoscope.

I stop by a roadside west of Hendricks, walk into the ditch, pick a prairie rose. This wild pink rose is far lovelier than hot-house roses wrapped in crinkly paper that teen-agers buy prom dates. The dusty car fills with its smell. I ignore it for a few minutes, go on talking. When I look again, it's dry, as if pressed in an immigrant Bible for a hundred years. These prairie flowers die quickly when you take them out of their own ground. They too are immigrants who can't transplant, and wither fast in their new world.

I didn't always love prairies. On my father's farm I dreamed of traveling, living by the sea and, most of all, close to mountains. As a boy, I lay head on a stone in the cow pasture east of the house, looking up at cloud rows in the west, imagining I saw all the way to the Rockies and that white tips on the clouds were snow on mountaintops or, better yet, white hair on sleep-ing blue elephant spines. Living in a flat landscape drove me to indulge in mountainous metaphor, then later discover that reality lived up to it. When

I finally saw the Rockies years later, they looked like pasture clouds, phantasmagorias solider than stone.

The most astonished travelers do not come from the Swiss Alps, or the California coast. Only William Carlos Williams, who lived in the industrial prairies of New Jersey, would notice the Mexico of *Desert Music*. A southwest poet with a wood's eye would have seen saguaro cactus or medieval parapets. Trust a prairie eye to find beauty and understate it truthfully, no matter how violent the apparent exaggeration. Thoreau, though a woodsman, said it right: "I can never exaggerate enough."

From *Prairie Days* by Bill Holm (San Francisco: Saybrook, 1987). Reprinted in *Inheriting the Land: Contemporary Voices from the Midwest*, edited by Mark Vinz and Thom Tammaro (Minneapolis: University of Minnesota Press, 1993).

Noted travel writer, William Least Heat-Moon (William Lewis Trogdon), was born in Kansas City, earned his doctorate in English at the University of Missouri in Columbia, and continues to live in his home state. Of distant Osage ancestry, he was raised with an appreciation for that heritage and in middle age decided to publicly adopt the native name his father had created for him when he was a boy. In books like the best-selling *Blue Highways: A Journey into America* (1982) and *River-Horse: The Logbook of a Boat across America* (1999), Heat-Moon has sought to honor his dual heritage while exploring the rich historical, cultural, and environmental diversity of the nation. In *PrairyErth* (1991), Heat-Moon chronicles his travels in one place—rural Chase County, Kansas, in the heart of the Flint Hills tallgrass prairie—where he unearths a wealth of historical and geographical information, and in the process affirms: "I was not a man of the sea or coasts or mountains but a fellow of the grasslands" (28). In this chapter, he reflects on the ways the wind (including tornados) has helped create the tallgrass prairies—its plant communities, but also its stories and legends.

Under Old Nell's Skirt

I know a man, a Maya in the Yucatán, who can call up wind: he whistles a clear, haunting, thirteen-note melody set in the Native American pentatonic scale. He whistles, the wind moves, and for some moments the heat of the tropical forest eases. It's a talent there to appreciate. But does he summon the wind, or does he know just the right time to whistle before the wind moves? He says, in effect, that he is on speaking terms with the wind, and by that he means it is a phenomenon, yes, but also a presence, and it has a name, Ik, and it is Ik that brings the seasonal rain to Yucatán. You may call such a notion pantheism or primitivism or mere personification: he wouldn't care, because for him, for the Maya, for all of tribal America, the wind, the life bringer, is something to heed, to esteem: Ik.

In Kansas I've not heard any names for the nearly constant winds, the oldest of things here. When the Kansa Indians were pushed out of the state, they carried with them the last perception of wind as anything other than a faceless force, usually for destruction, the power behind terrible prairie wildfires, the clout in blizzards and droughts, and, most of all, in tornadoes that will take up everything, even fenceposts. But people here know wind well, they often speak of it, yet, despite the several names in other places for local American winds, in this state, whose very name may mean "wind-people," it has no identity but a direction, no epithet but a curse. A local preacher told me: *Giving names to nature is un-christian.* I said that it might help people connect with things and who knows where that might lead, and he said, *To idolatry.* Yet the fact remains: these countians are more activated by weather than religion.

Almost everything I see in this place sooner or later brings me back to the grasses; after all, this is the prairie, a topography that so surprised Anglo culture when it began arriving that it found for this grand-beyond no suitable word in its immense vocabulary, and it resorted to the French of illiterate trappers: prairie. Except in accounts of novice travelers, these grasslands have never been meadows, heaths, moors, downs, wolds. A woman in Boston once said to me, *Prairie is such a lovely word—and for so grim a place.*

More than all other things here, the grasses are the offspring of the wind, the power that helps evaporation equal precipitation to the detriment of trees, the power that breaks off leaves and branches, shakes crowns and rigid trunks to tear roots and disrupt transpiration, respiration, nutrient assimilation. But grasses before the wind bend and straighten and bend and keep their vital parts underground, and, come into season, they release their germ, spikelets, and seeds to the wind, the invisible sea that in this place must carry the code, the directions from the unfaced god, carry the imprint of rootlet and rhizome, blade and sheath, culm and rachis: the wind, the penisless god going and coming everywhere, the intercourse of the grasses, the sprayer of seed across the opened sex risen and waiting for the pattern set loose on the winds today of no name; and so the grasses pull the energy from the wind, the offspring of sunlight, to transmute soil into more grasses that ungulates eat into flesh that men turn into pot roasts and woolen socks.

Now: I am walking a ridge in the southern end of Saffordville quadrangle, and below me in the creek bottom are oaks of several kinds, cottonwood, hackberry, walnut, hickory, sycamore. Slippery elms, once providing a throat emulcent, try to climb the hills by finding rock crevices to shield their seed, and, if one sprouts, it will grow straight for a time, only to lose its inborn shape to the prevailing southerlies so that the windward sides of elms seem eaten off but the lee sides spread north like tresses unloosed in March. If a seedling succeeds on a ridge top, it will spread low as if to squat under the shears of windrush, and everywhere the elm trunks lean to the polestar and make the county appear as if its southern end had been lifted and tilted before the land could dry and set. A windmill must stand straight and turn into the wind to harvest water, but the slippery elm turns away to keep the wind from its wet pulp.

And there is another face to this thing from which life proceeds. Yesterday I walked down a ridge to get out of the November wind while I ate a sandwich, and I came upon a house foundation on a slope bereft of anything but grasses and knee-high plants. It was absolutely exposed, an oddity here, since most of the homes sit in the shelter of wooded vales. This one faced east—or it would have, had it still been there—and the only relief from the prevailing winds that the builder had sought was to set the back of the house to them. There was the foundation, some broken boards, a few rusting things, and, thirty feet away, a storm cellar, its door torn off, and that was all except for a rock road of two ruts. The cave, as people here call tornado cellars, was of rough-cut native stone with an arched roof, wooden shelves, and a packed-earth floor with Mason jar fragments glinting blue in the sunlight; one had been so broken that twin pieces at my feet said:

The shards seemed to be lost voices locked in silica and calling still.

These cellars once kept cool home-canned food (and rat snakes), and, when a tornado struck like a fang from some cloud-beast, they kept families that mocked their own timorousness by calling them *'fraidy holes*, and it did take nerve to go into the dim recesses with their spidered corners and dark, reptilian coils. I stepped down inside and sat on a stone fallen from the wall and ate safely in the doorway, but, even with the sun shafts, there was something dismal and haunted in the shadowed dust of dry rot here and dank of wet rot there. Things lay silent inside, the air quite stilled, and I felt something, I don't know what: something waiting.

Was there a connection between this cave and that house absent but for its foundation? The site, sloping southwest, seemed placed to catch a cyclone in a county in the heart of the notorious Tornado Alley of the Middle West, a belt that can average 250 tornadoes a year, more than anywhere else in the world. A hundred and sixty miles from here, Codell, Kansas, got thumped by a tornado every twentieth of May for three successive years, and five months ago a twister "touched down," mashed down really, a mile north of Saffordville at the small conglomeration of houses and trailers called Toledo; and the newspaper caption for a photograph of that crook'd finger of a funnel cloud was "Holy Toledo!" Years earlier a cyclone wrecked a Friends meetinghouse there, but this time it skipped over the Methodists' church and went for their houses. In Chase County I've found a nonchalance about natural forces born of fatalism: *If it's gonna get me, it'll get me.* In Cottonwood Falls, on a block where a house once sat, the old cave remains, collapsing, yet around it are six house trailers. Riding out a tornado in a mobile home is like stepping into combine blades: trailers can become airborne chambers full of flying knives of aluminum and glass. No: if there is a dread in the county, it is not of dark skies but of the opposite, of clear skies, days and days of clear skies, of a drought nobody escapes, not even the shopkeepers. That any one person will suffer losses from a tornado, however deadly, goes much against the odds, and many residents reach high school before they first see a twister; yet, nobody who lives his full span in the county dies without a tornado story.

Tornado: a Spanish past participle meaning turned, from a verb meaning to turn, alter, transform, repeat, *and* to restore. Meteorologists speak of the reasons why the Midlands of the United States suffer so many tornadoes: a range of high mountains west of a great expanse of sun-heated plains at a much lower altitude, where dry and cold northern air can meet warm and moist southern air from a large body of water to combine with a circulation pattern mixing things up: that is to say, the jet stream from Arctic Canada crosses the Rockies to meet a front from the Gulf of Mexico over the Great Plains in the center of which sits Kansas, where, since 1950, people have sighted seventeen hundred tornadoes. It is a place of such potential celestial violence that the meteorologists at the National Severe Storms Forecast Center in Kansas City, Missouri, are sometimes called the Keepers of the Gates of Hell. Countians who have smelled the fulminous, cyclonic sky up close, who have felt the ground shake and heard the earth itself roar and have taken to a storm cellar that soon filled with a loathsome greenish air, find the image apt. The Keepers of the Gates of Hell have, in recent years, become adept at forecasting tornadoes, and they might even be able to suggest cures for them if only they could study them up close. Years ago a fellow proposed sending scientists into the eye of a tornado in an army tank until he considered the problem of transporting the machine to a funnel that usually lasts only minutes, and someone else suggested flying into a cyclone, whereupon a weather-research pilot said, yes, it was feasible if the aviator would first practice by flying into mountains.

Climatologists speak of thunderstorms pregnant with tornadoes, storm-breeding clouds more than twice the height of Mount Everest; they speak of funicular envelopes and anvil clouds with pendant mammati and of thermal instability of winds in cyclonic vorticity, of rotatory columns of air torquing at velocities up to three hundred miles an hour (although no anemometer in the direct path of a storm has survived), funnels that can move over the ground at the speed of a strolling man or at the rate of a barrel-assing semi on the turnpike; they say the width of the destruction can be the distance between home plate and deep center field and its length the hundred miles between New York City and Philadelphia. A tornado, although more violent than a much longer lasting hurricane, has a life measured in minutes, and weathercasters watch it snuff out as it was born: unnamed.

I know here a grandfather, a man as bald as if a cyclonic wind had taken his scalp—something witnesses claim has happened elsewhere—who calls twisters Old Nell, and he threatens to set crying children outside the back door for her to carry off. People who have seen Old Nell close, up under her skirt, talk about her colors: pastel-pink, black, blue, gray, and a survivor said this: *All at once a big hole opened in the sky with a mass of cherry-red, a yellow tinge in the center,* and another said: *a funnel with beautiful electric-blue light,* and a third person: *It was glowing like it was illuminated from the inside.* The witnesses speak of shapes: a formless black mass, a cone, cylinder, tube, ribbon, pendant, thrashing hose, dangling lariat, writhing snake, elephant trunk. They tell of ponds being vacuumed dry, eyes of geese sucked out, chickens clean-plucked from beak to bum, water pulled straight up out of toilet bowls, a woman's clothes torn off her, a wife killed after being jerked through a car window, a child carried two miles and set down with only scratches, a Cottonwood Falls mother (fearful of wind) cured of chronic headaches when a twister passed harmlessly within a few feet of her house, and, just south of Chase, a woman blown out of her living room window and dropped unhurt sixty feet away and falling unbroken beside her a phonograph record of "Stormy Weather."

London Harness, an eighty-five-year-old man who lives just six miles north of the county line, told me: *I knew a family years ago that was crossing open country here in a horse and wagon. A bad storm come on fast, and the man run to a dug well and said, "I'm going down in here—you do the best you can!" The wife hollered and screamed and run to a ditch and laid down with their two little kids. That funnel dropped right in on them. After the storm passed over, she and the kids went to the well to say, "Come on up, Pappy," but there weren't no water down there, and he weren't down there. If you're in that path, no need of running.*

Yesterday: in the sun the broken words on the Mason jar glinted and, against the foundation, the wind whacked dry grasses and seed pods, *tap-tap-tap, rasp-rasp,* and a yellow light lay over the November slope, and Ma and son: did they one afternoon come out of the cave to see what I see, an unhoused foundation, some twisted fence wire, and a sky turning golden in all innocence?

From *PrairyErth: A Deep Map* by William Least Heat-Moon (Boston: Houghton Mifflin, 1991).

Born near the small town of Peterson in northwestern Iowa, Drake Hokanson moved west when he was a young boy, but spent his summers in Iowa with his grandparents and great-grandparents, who owned an implement dealership and farmed in the area. His first photograph was exhibited at Grinnell College when he was a freshman at the University of Iowa, and his writing and photographs about places and the people who call them home have appeared in numerous books, journals, magazines, and anthologies, as well as public exhibitions across the country. He is the author of three books, including *Lincoln Highway: Main Street across America* (1988) and *Reflecting a Prairie Town: A Year in Peterson* (1994), the latter of which chronicles — in words and photos — an extended return to his homeland, where he once again became part of the rhythms of rural life. In this chapter, Hokanson seeks out what remains of local prairie, identifying communities of plants that, like Peterson itself, have experienced dramatic changes, yet continue to survive.

Habits of the Grass

April 10

At nine-thirty in the morning, I round the corner onto Main Street and am suddenly reminded of busy days of commerce in Peterson's past. There are twenty vehicles parked on Main, with most of them around the Senior Center and Sue's. Downtown looks busier than I've seen it in some time. It is a Wednesday, and there will be a congregate noon meal at the Senior Center, or "Dinner Date" as most call it, and already, at this early hour, card games are in full swing.

At Sue's, a double birthday is in full swing. The celebrants are Jordan Raveling and Wilda Wetherall, and they have pooled their resources to buy the house. People are coming and going, and there is hardly a seat in the place. Chatter is loud and easy. As I come in, Wilda sits with a group of women, a pile of birthday cards in front of her, with more arriving moment by moment. Even though it is his party, Jordan has already left to do field work. It is shaping up to be a good day to be in the field, one of the first of the season.

Sue has baked the cakes for this large crowd, but by the time I arrive, the chocolate is gone; Sue reads my disappointment and points out that the white cake has yogurt in it; maybe she thinks of it as a sort of health cake. With a big piece on a paper plate, I sit down at a table with Carl Brenner and Slim Wetherall, Wilda's husband. As usual, Carl wears a bow tie, blue today; Slim wears a new Garst Seed Company jacket and hat.

Carl offers, "Boy, getting together for coffee like this is really what keeps the town together, you know it? If we didn't have this place, why, I don't know what we'd do."

Slim points out how Linn Grove, seven miles east, has recently lost its coffee shop and how "you don't see many people in town anymore—they come in, go to the bank, and just leave." Slim himself is antsy to leave on this bright spring morning, though Wilda looks like she'll be staying for hours. "The farmers will be hittin' the fields hard today," he says, shifting in his seat. Slim has retired from farming, but the instincts are still strong.

Many farmers feel the pull today. An hour later, north of town, I find the machinery of spring farming on the move; chisel plows, disk harrows, fertilizer applicators, and farm pickups loaded down with bags of seed corn go about their tasks. The soil is drying out, and heat shimmers off the fields as implements dig in. Most farmers today work up the soil or apply fertilizer since it is still too early and the soil too cold for planting, though it won't be long now. For the first time since last fall, the sun again brings noticeable warmth. The high is predicted to be 55 today, with a good chance of rain or snow starting tonight, so farmers are eager to accomplish as much as they can. The forecasters are hedging their bets: if the low pressure system tracks a little south, we'll get snow; if it passes to the north, we'll have rain.

The cool season grasses—brome and bluegrass especially—are greening up quickly. Pastures, hillsides, the cemetery, the ball diamond, and lawns all over town are rich with the spinach-soup green that excites my eyes, so accustomed to the grays and browns of winter. The green stands in sharp contrast to still-drab trees, shrubs, and colorless fields. Many of the trees in farm groves are exhibiting the increased density, that spring fuzziness, that suggests leaves to follow, though their color has yet to change.

Other notices are posted for the return of spring. About eleven, a truck pulls up at Eddie's Hardware to unload the merchandise for his spring sale. Eddie, the driver, and I haul in new rakes, shovels, lawn sprinklers, sprayers, and paint. Last off the truck is a bundle of mailers imprinted with "Eddie's Hardware, Main Street, Peterson, Iowa," mailers that depict sale items amidst riotous spring flowers and Easter-green grass. I notice that Eddie has already set up his racks of spring seeds.

"I sold a bunch already today," he tells me.

Out on the uplands, the western meadowlarks have been back for a couple of weeks. Just now, the red-winged blackbirds are staking their loud claims to cattails around the county, and the turkey vultures are making their first appearances along the north rim of the valley. Warm weather the last two days has brought out a great explosion of Spring Beauties—the first woodland flower—in the woods along the river and an unusual hatch of moths and insects which besmirch my windshield as I drive the rural roads. The smell of warming soil perfumes the air.

Fertilizer arrives in town to enrich the soon-to-be-planted crops. In the afternoon, two semis arrive at the elevator loaded with potash from Rocanville, Saskatchewan. Only the license plates and the logos on the cabs identify these as foreign rigs, but they and the tired drivers have a distinctly foreign feel about them. Lyle Goettsch asks, "How's the weather up your way? Will you get home before the snow?" I find myself wondering what they think of this tiny town in a foreign country.

The elevator and scale office are busy today as fertilizer, seed oats, and livestock feed go out. A smattering of corn comes in, though it tapers off as more and more farmers put aside grain hauling to get into the warming fields.

The river, too, is busy as it pushes snowmelt downstream toward the Missouri, toward the Gulf. It runs higher than any time since last summer. Along the banks, green spikes of new reed canary grass poke through the dense mat of last year's crop.

South of town, in Rod Burgeson's cornfield, the soil waits. Except for a single pass along one edge with a disk to test soil readiness, Rod has done nothing here yet. Today he works on fields to the east as the fertilizer applicator makes the rounds nearby. Here black corncobs lie mixed with gray cottonwood leaves, molding between rows of bleached cornstalks still rooted in the coffee-colored earth. The time will soon come for this place.

On Steele Prairie, two townships west, on this April day, other spring rhythms are beginning to stir in the warming soil, and these signals of the coming season are ancient; like the territoriality of the red-winged blackbird, these rhythms are much older than sales on garden tools or the early greening of brome grass on the face of the Midwest.

On this April 10th, in stark contrast to the rich green of lawns and pastures around the county, Steele Prairie is still altogether brown. The palette runs the range from the silver-brown of a bit of prairie sage to the russet of last year's big bluestem. Mixed in is the khaki of the dormant clumps of little bluestem, the darker brown of great swaths of miscellaneous compacted plants, and the black punctuation of leftover coreopsis stems and coneflower heads waving on stiff dead stalks. On the prairie, early April is not a colorful time.

A breeze stirs the reddish brown stems of the big bluestem. This grand grass dominates the prairie here, just as it once did the entire prairie province. After the compacting effects of the winter's snows, the big blue is about all that remains erect. The four-foot stems lean to the northeast, slightly athwart the wind; I surmise the lean comes less from the wind today than from the first snowstorm of the season back in November; snow and wind from the southwest would push the tall grass into a permanent lean to the northeast. Beneath the tilted grass is the tangled mat of last year's growth: dry stems, leaves, seed heads, and plant material in all stages of decay.

Within Steele Prairie, but still unseen, there is a vernal angst as strong as that of Slim Wetherall and every other farmer in the state. Pulling back a big hunk of the prairie duff, I uncover a tiny forest of green grass spears and new leafy plants pushing upward toward the light. Some are pale, some are bright green. Most will be broad-leafed plants, but some will be grasses, and all are impossible for me to identify at this nascent stage. The ground is warm to the touch.

The month is right and the season is far enough along, but on my walk today I find no pasqueflower, the first spring bloomer on the prairie. No doubt it is here, but the flower is not very large, and its pastel petals blend in easily with the matted confusion.

To the Sioux peoples, the pasqueflower had great significance not only as a harbinger of spring, but as a guidepost of life. As the legend goes, upon seeing the first pasqueflower in the spring, a Sioux man is to fill his pipe, sit close to it, and in silence point the pipe toward the sky, the earth, then the four cardinal directions. He should then smoke the pipe, reflect on his own seasons of life, and consider his triumphs and failures and the unseen creators who guide his life and the world around him. He is then to pluck the bloom and take it to his family, singing the song of the pasqueflower as he goes:

> I wish to encourage the children
> Of other flower nations now appearing
> All over the face of the earth;
> So while they awaken from sleeping
> And come up from the heart of the earth
> I am standing here old and gray-headed.

Although this region is no longer prairie in any botanical sense, the area is still a grassland, even with the comprehensive impact of agriculture across the Midwest and the encroachment of woody plants. Grass grows everywhere that something else hasn't crowded it out. Pavement buries it, woodland shades it, and farm implements battle it every year as they edge close to field fences, but grass persists. Unlike parts of Utah, where bare rock and sagebrush form the predominant texture, or Vermont, where trees of all sizes are the baseline land cover, the Midwest is still deep in grass. In northwest Iowa native and imported grasses vie for every unused sunny place.

The grid lines and fields of the Midwest are appropriately but tiresomely described as a patchwork quilt. But perhaps the cliché has merit in a different sense than the usual. Grasses grow along the edges of most everything, along every stream, in every road ditch; they surround every farmstead and field. If the Midwest is a patchwork quilt, surely the grasses are the stitching that holds each piece to the next, the stitching that anchors the entire landscape. The tangled roots, rhizomes, runners, stems, and leaves of countless individual grasses bind the elements of the land into whole cloth.

The idea that the tendrils of grass hold things together here appeals to me greatly. By percent of ground cover, prairie was 60 percent grass; by weight, grasses accounted for nearly 90 percent of annual plant growth. Prairie grasses have root systems of astonishing dimensions and mass. The roots of prairie cord grass, for example, plunge as much as thirteen feet below the soil surface to find water in dry years. Many grasses send rhizomes, or runners, like relay racers, horizontally just beneath the soil surface to establish new stems and clumps, to colonize every crack, any bare spot of soil. Each rootlet and each runner holds a few crumbs of soil and therefore anchors a piece of the landscape against dislodgement.

J. E. Weaver, a patient Nebraska botanist, knew firsthand how grasses stitched the land together: "One exhaustive study of the binding network of roots showed that a strip of prairie sod 4 inches deep, 8 inches wide and 100 inches long was bound together with a tangled network of roots having a total length of more than 20 miles."

Grasses, perhaps more than any other plant, specialize in simplicity. They produce no showy flowers, develop no extraordinary foliage, and make no unusual demands on their environment. Most grasses, especially prairie natives, concentrate their efforts downward, sending roots to find water. A big bluestem seedling will have developed, after only seven or eight weeks, roots six to eight feet deep. In some grasses, 90 percent of the mass of the plant will be below ground to better gather and conserve both moisture and nutrients. Grasses are well adapted to drought. Supple stems support the leaves and bow easily with the wind, suiting them to the windy plains and prairie. In fact, grasses depend on the wind; they use it for pollen dispersal and therefore have no need for elaborate flowers to attract pollinating insects. Hence, their flowers are small, small enough to require a hand lens for study.

More than being the stitching for the landscape or merely forming a background texture to the land of the Midwest, different species of grass — quite apart from true prairie — have individual personalities. Around Peterson are clumps of native big bluestem, ruddy in the winter light of afternoon; a few sprigs of bottlebrush grass holding their own along a shady roadside through the old Kirchner woods; Indian grass on valley hillsides, Canada wild rye, reed canary grass, poverty grass and cheat grass; side oats grama on dry hills, with its pendant seeds like charms on a bracelet; witch grass and switch grass; prairie three-awn; and lawns full of Kentucky bluegrass and the hated crabgrass. Each has its habitat; each has a story. Some were here to tickle the bellies of the buffalo or even the woolly mammoth; others migrated from Europe just yesterday in seed bags brought by immigrants or as seeds in ship ballast.

Most compelling to me is big bluestem. More than any other plant, big bluestem symbolizes the vegetational history of the tall-grass prairie; the "tall" in tall-grass prairie refers to this grass. Just as it does today on Steele Prairie, big blue dominated the prairie in both size and distribution.

In terms of aggressiveness, big blue is to prairie as bur oaks are to woodland. I once planted a couple of clumps of big bluestem in the garden as an ornamental grass. They were small tufts, about teacup size, bought at a prairie-plant sale. I dug them in one day late in May and didn't expect much that first year. It rained a lot during June and July; the grass stayed green and healthy but didn't grow much at all until midsummer. Then, once it had established a strong root system, the grass went up like a rocket; by the last week of August, it bore no fewer than twenty heavy seed heads, the tallest being over six feet above the soil.

Scattered clumps of big bluestem today dot the valley hillsides around Peterson and announce their native status by turning a reddish brown every fall, and they provide a touch of color all winter against the dull straw tones of their domestic pasture cousins. Though they now stand in pasture, these great clumps remind me, each time I see them, that this was once all prairie.

Roadside grasses tell stories as well. Along most roads, especially along highways and improved county roads where herbicide has long kept native grasses and forbs in check, ditches and embankments have become uniform swaths of smooth brome, imported some years ago as a pasture grass.

Brome is an aggressive grass that can outdo prairie natives unless fire is used to keep it in check. At times an uninterrupted stripe of brome will unroll for mile after mile alongside the car like an unpatterned hall runner. In such homogeneous stands, brome moves before the wind like ripened wheat, etching the eye with flickering lines as its pale stems move against the darker leaves. Individual brome plants are often as uniform as steel fence posts, making the stems, leaves, and seed heads look like pen strokes scribed as hatch marks onto a printing plate, each line exquisitely fine, each line distinct from, but formally identical to, the others.

Most pastures and fallow fields have been seeded to or colonized by smooth brome. It is good livestock forage, having nutritional value and the ability to withstand close grazing. It covers the majority of non-lawn grassy areas in this part of the world.

Brome, like Kentucky bluegrass and other popular pasture grasses, is a cool-season grass, meaning it greens up early in the spring, giving farmers extra weeks of grazing compared to most prairie natives that don't show much color or growth until the weather turns warm. On April 10, any swatches of green I see around the area are certain to be brome, or bluegrass, or cool-season lawn grasses; any place that is green today is not prairie.

Along the back roads, especially the section roads that have warning signs about "Level B maintenance," the story of grass changes. Here the old native grasses appear where they have been safe from road crews and spray trucks. Near Bob White's place is a wet tangle of native, frog-green, prairie cord grass where the ditch dips enough to hold a little water much of the time. The ancestors of this grass grew here in the prairie wet spots years ago; Velma Walrath, at about the same time she was discovering how afraid Gertrude Tolley was of storms, must have played here. The long, sawtooth-edged leaves of prairie cord grass earned it the name "rip gut" among pioneers, a name anyone walking through it would quickly endorse.

Here, too, are occasional clumps of little bluestem, switch grass, Indian grass, and a few great heaps of big bluestem with its graceful, tall stems. In the late summer at the top of each stem of big blue hang three heavy toes of seeds, reminding everyone of its name among the white settlers here: turkey foot. Then there is a thin run of side oats grama grass, a few oats washed from the field, and brome again, stitching in the spaces.

LEAVING THE POST OFFICE one morning with the mail, I found myself wondering what, if any, elements of prairie still hid out in the feral corners of Peterson. Along my route and against the north side of the old Berg Jewelry store grew a few sprigs of grass, and I stopped to investigate. It occurred to me that if there were to be any hints of a lost world, I would find them in the persistent grasses, the grasses that had withstood centuries of drought, fire, hard grazing by bison, and later the assaults of humankind. In town, as in the country, the broadleaf forbs would long ago have been killed off by spray and eager weeding. But could native grasses survive the lawn mower and the sprayer attachment? The season was early, and a close look revealed that most of what grew along the building was in fact crab grass, brome, and common plantain, all aliens. Around Peterson, common plantain, which came from Europe and is not a grass at all, lives up to its name. It is found in sidewalk cracks and any disturbed place. Native Americans called it Whiteman's Foot; it sprung up wherever whites went.

We went looking for clues to prairie, Carol and I, in and around Peterson. This quarter section of land has been streets and yards for a hundred years; before that it had been a cornfield for a few; earlier it had been prairie for millennia. We scoured backyards for a rank holdout of big bluestem hugging a garage against the mower, a stem or two of side oats grama in the compost pile. We walked the back alleys and the perimeter of town looking for any signs. We searched near some old dog kennels; all we found was daisy fleabane, which is at least a native. We poked under farm machinery parked in a town lot; amid the protection of plows and disks we found nothing but bluegrass and quack grass. For several weeks we watched with interest the yard of an abandoned house as it grew up to thick grass and weeds. It was just high enough for the bluegrass to head up and seed out when some good citizen mowed it. A week or two longer and I might have found some old native grasses. But perhaps it was a false hope at that. In town a hundred years of human selection have canceled them out altogether. Our penchant for the fine texture, the tame, our quest for the orderliness of English prototype gardens doomed them.

Around the rough edges of town, the take was slightly better. Bottlebrush grass grows along the road through the woods south of town, and there are some stands of little bluestem on drier hillsides. In the arcs of the river, reed canary grass grows in sweeping, uniform stands. Some strains of this grass are native and have grown here since the glaciers left; others have been imported. I can't tell which this is, but Clint Fraley claims this strain is the aggressive invader.

Unquestionably native are the great bunches of big bluestem and the smaller bunches of Indian grass that surround the town on the hillsides where the trees haven't filled in to shade them out. Like sentries, they grow where they can watch over the town, waving in the wind of a former prairie.

And then we made a discovery, or at least it was a discovery for us. Up behind the schoolhouse, at the east end of the old football field, is what generations of kids have called the sledding hill. I hadn't been there for probably twenty-five years, and all I remembered of it was the rib-crushing dip at the bottom where the steep hill abruptly ended at the level football field. It was late summer, and the grasses everywhere were rank and heavy with seed. As we circled the school and our eyes fell on the vegetation of the hill, I noticed the characteristic russet color of prairie grasses, and spots of silver that suggested leadplant. Closer inspection revealed the sledding hill to be banked deep and undisturbed in fine prairie forbs and homespun grasses.

Here in profusion were Maximilian sunflower, wild garlic, stiff goldenrod, dotted gayfeather, purple prairie clover, prairie coreopsis, and leadplant, a good indicator of undisturbed prairie since livestock particularly like it. Shimmering above and through it all were wild grasses: big bluestem, little bluestem, Indian grass, side oats grama.

Here was a shred of real prairie, a few hundred square feet of what was once several hundred thousand square miles of biome. It lay unseen, as indigenous as a Sioux buffalo hunter and almost as rare, priceless to the eye and heart, and useless to anyone without a sled.

But why here? Why hadn't this scrap of land been put to some use like nearly every other square foot within a hundred miles? The hill is too steep to plow and too small to be worth fencing for livestock; besides, this land has been school property since shortly after the town was platted. Except for sledders and perhaps the browsing of an occasional pony tethered by a schoolchild years ago, this place has been altogether ignored since the bison left.

Carol and I are not the only people who know of the schoolhouse prairie.

One of Peterson's elementary teachers knows all about this place, and has used it for several years as an outdoor classroom. She has brought second-graders out the school door, away from books and videos about important events and places, across the football field, to introduce them to something that is more a part of their past—and present livelihood—than any institution, structure, or idea in the region.

What better place could there be for a discussion about the bison, the Indians, Iowa's three-foot-thick topsoil, complex ecosystems, a lost world? But I was told that one of the kids was stung by a bee on a class trip to the prairie, and, because of the risk, elementary classes don't visit as often anymore.

It's funny what we think is important, what we seek to preserve. Even in so universally agricultural a community as Peterson, we see our significant past as being tied only to human endeavors; our history resides only in log forts, on the rusted share of a breaking plow, behind the plaque announcing the site of the first cabin in the county. The Christian Kirchner home in Peterson is not yet a hundred and fifty years old, and yet it is, and should be, carefully protected by Peterson Heritage and treasured by all who live in Peterson.

But these few square feet of original land cover represent something much deeper and older than mere human activity—of settlers and Indians alike—in the region. It may in fact be as old as the land itself. This ecosystem made not only the Kirchner house possible, but also the entire agricultural economy and social structure of the region, and yet we pay it little mind.

Within sight of this scrap of rare prairie, people tend their irises, weed their vegetable gardens, and fertilize their lawns; up on the sledding hill, each year the sumacs spread their shade a little further, overcoming just a bit more of the prairie, and recently someone looking for an easy source of fill dirt taxied in with a big loader and took out a few cubic feet.

From *Reflecting a Prairie Town: A Year in Peterson* by Drake Hokanson (Iowa City: University of Iowa Press, 1994).

DRAKE HOKANSON

✴ LOUISE ERDRICH, 1954– ✴

Born in Little Falls, Minnesota, Louise Erdrich was raised in southeastern North Dakota, where her parents taught school and her maternal grandfather served as tribal chairman for the Turtle Mountain Chippewas. Although Erdrich left the area to attend Dartmouth College and then Johns Hopkins University, much of her internationally acclaimed writing is set in her homeland, including the novel *Love Medicine* (1984), winner of the National Book Critics Circle Award. Recipient of a Lifetime Achievement Award from the Native Writers' Circle of the Americas, Erdrich currently lives and writes in Minnesota, owns an independent bookstore in Minneapolis, and has co-hosted writing workshops with her sisters on the Turtle Mountain Reservation in North Dakota. In this essay, Erdrich recalls her German American father's love for the prairies near North Dakota's Red River, and expresses her own ongoing familial, even spiritual affinity for the tallgrass and its wildlife.

Big Grass

My father loves the small and receding wild places in the agribusiness moonscape of North Dakota cropland, and so do I. Throughout my childhood, we hunted and gathered in the sloughs, the sandhills, the brushy shelterbelts and unmowed ditches, on the oxbows and along the banks of mudded rivers of the Red River valley. On the west road that now leads to the new Carmelite monastery just outside of Wahpeton, we picked prairie rosehips in fall and dried them for vitamin C–rich teas in the winter. There was always, in the margins of the cornfield just beyond our yard, in the brushy scraps of abandoned pasture, right-of-ways along the railroad tracks, along the river itself, and in the corners and unseeded lots of the town, a lowly assertion of grass.

It was big grass. Original prairie grass—bluestem and Indian grass, side oats grama. The green fringe gave me the comforting assurance that all else planted and tended and set down by humans was somehow temporary. Only grass is eternal. Grass is always waiting in the wings.

Before high-powered rifles and a general dumbing down of hunting attitudes, back when hunters were less well armed, and anxious more than anything to put meat on their tables, my father wore dull green and never blaze orange. He carried a green fiberglass bow with a waxed string, and strapped to his back a quiver of razor-tipped arrows. Predawn on a Saturday in fall he'd take a child or two into the woods near Hankinson, Stack Slough, or the cornfields and box elder and cottonwood scruff along the Wild Rice or Bois de Sioux rivers. Once, on a slim path surrounded by heavy scrub, my father and I heard a distant crack of a rifle shot and soon, crashing toward us, two does and a great gray buck floated. Their bounds carried them so swiftly that they nearly ran us over.

The deer huffed and changed direction midair. They were so close I felt the tang of their panic. My father didn't shoot—perhaps he fumbled for his bow but there wasn't time to aim—more likely, he decided not to kill an animal in front of me. Hunting was an excuse to become intimate with the woods and fields, and on that day, as on most, we came home with bags of wild plums, elmcap mushrooms, more rosehips.

Since my father began visiting the wild places in the Red River valley, he has seen many of them destroyed. Tree cover of the margins of rivers, essential to slow spring runoff and the erosion of topsoil—cut and leveled for planting. Wetlands—drained for planting. Unplowed prairie (five thousand acres in a neighboring Minnesota county)—plowed and planted. From the air, the Great Plains is now a vast earth-toned Mondrian painting, all strict right angles of fields bounded by thin and careful shelterbelts. Only tiny remnants of the tallgrass remain. These pieces in odd cuts and lengths are like the hems of long and sweeping old-fashioned skirts. Taken up, the fabric is torn away, forgotten. And yet, when you come across the

LOUISE ERDRICH

original cloth of grass, it is an unfaded and startling experience. Here is a reminder that before this land was a measured product tended by Steiger tractors with air-cooled cabs and hot-red combines, before this valley was wheat and sugar-beet and sunflower country, before the drill seeders and the windbreaks, the section measures and the homesteads, this was the northern tallgrass prairie.

It was a region mysterious for its apparent simplicity.

Grass and sky were two canvases into which rich details painted and destroyed themselves with joyous intensity. As sunlight erases cloud, so fire ate grass and restored grass in a cycle of unrelenting power. A prairie burned over one year blazes out, redeemed in the absolving mist of green the next. On a warm late-winter day, snow slipping down the sides of soft prairie rises, I can feel the grass underfoot collecting its bashful energy. Big bluestem, female and green sage, snakeweed, blue grama, ground cherry, Indian grass, wild onion, purple coneflower, and purple aster all spring to life on a prairie burned the previous year.

To appreciate grass, you must lie down in grass. It's not much from a distance and it doesn't translate well into most photographs or even paint, unless you count Albrecht Dürer's *Grosses Rasenstuck*, 1503. He painted grass while lying on his stomach, with a wondering eye trained into the seed tassles. Just after the snow has melted each spring, it is good to throw oneself on grass. The stems have packed down all winter, in swirls like a sleeper's hair. The grass sighs and crackles faintly, a weighted mat, releasing fine winter dust.

It is that smell of winter dust I love best, rising from the cracked stalk. Tenacious in its cycle, stubborn in its modest refusal to die, the grass embodies the philosopher's myth of eternal return. *All flesh is grass* is not a depressing conceit to me. To see ourselves within our span as creatures capable of quiet and continual renewal gets me through those times when the writing stinks, I've lost my temper, overloaded on wine chocolates, or am simply lost to myself. Snow melts. Grass springs back. Here we are on a quiet rise, finding the first uncanny shoots of green.

My daughters' hair has a scent as undefinable as grass—made up of mood and weather, of curiosity and water. They part the stiff waves of grass, gaze into the sheltered universe. Just to be, just to exist—that is the talent of grass. Fire will pass over. The growth tips are safe underground. The bluestem's still the scorched bronze of late-summer deer pelts. Formaldehyde ants swarm from a warmed nest of black dirt. The ants seem electrified, driven, ridiculous in tiny self-importance. Watching the ants, we can delight in our lucky indolence. They'll follow one another and climb a stem of grass threaded into their nest to the end, until their weight bows it to the earth. There's a clump of crested wheatgrass, a foreigner, invading. The breast feather of a grouse. A low hunker of dried ground cherries. Sage. Still silver, its leaves specks and spindrels, sage is a generous plant, releasing its penetrating scent of freedom long after it is dried and dead. And here, the first green of the year rises in the female sage, showing at the base in the tiny budded lips.

Horned larks spring across the breeze and there, off the rent ice, the first returning flock of Canada geese search out the open water of a local power plant on the Missouri River. In order to re-create as closely as possible the mixture of forces that groomed the subtle prairie, buffalo are included, at Cross Ranch Preserve, for grazing purposes. Along with fire, buffalo were the keepers of the grass and they are coming back now, perhaps because they always made sense. They are easier to raise than cattle, they calve on their own, and find winter shelter in brush and buffalo-berry gullies.

From my own experience of buffalo—a tiny herd lives in Wahpeton and I saw them growing up and still visit them now—I know that they'll eat most anything that grows on the ground. In captivity, though, they relish the rinds of watermelon. The buffalo waited for and seemed to know my parents, who came by every few days in summer with bicycle baskets full of watermelon rinds. The tongue of a buffalo is long, gray, and muscular, a passionate scoop. While they eat watermelon, the buffalo will consent to have their great boulder foreheads scratched but will occasionally, over nothing at all, or perhaps everything, ram themselves into their wire fences. I have been on the other side of a fence charged by a buffalo and I was stunned to a sudden blank-out at the violence.

One winter, in the middle of a great snow, the buffalo walked up and over their fence and wandered from their pen by the river. They took a route through the town. There were reports of people stepping from their trailers into the presence of shaggy monoliths. The buffalo walked through backyards, around garages, took the main thoroughfares at last into the swept-bare scrim of stubble in the vast fields — into their old range, after all.

Grass sings, grass whispers. Ashes to ashes, dust to grass. But real grass, not the stuff that we trim and poison to an acid green mat, not clipped grass never allowed to go to seed, not this humanly engineered lawn substance as synthetic as a carpet. Every city should have a grass park, devoted to grass, long grass, for city children haven't the sense of grass as anything but scarp on a boulevard. To come into the house with needlegrass sewing new seams in your clothes, the awns sharp and clever, is to understand botanical intelligence. Weaving through the toughest boots, through the densest coat, into skin of sheep, needlegrass will seed itself deep in the eardrums of dogs and badgers. And there are other seeds, sharp and eager, diving through my socks, shorter barbs sewn forever into the laces and tongues of my walking boots.

Grass streams out in August, full grown to a hypnotizing silk. The ground begins to run beside the road in waves of green water. A motorist, distracted, pulls over and begins to weep. Grass is emotional, its message a visual music with rills and pauses so profound it is almost dangerous to watch. Tallgrass in motion is a world of legato. Returning from a powwow my daughter and I are slowed and then stopped by the spectacle and we drive side roads, walk old pasture, until we find real grass turned back silver, moving, running before the wind. Our eyes fill with it and on a swale of grass we sink down, chewing the ends of juicy stems.

Soon, so soon.

Your arms reach, dropping across the strings of an air harp. Before long, you want your lover's body in your hands. You don't mind dying quite so much. You don't fear turning into grass. You almost believe that you could continue, from below, to express in its motion your own mesmeric yearning, and yet find cheerful comfort. For grass is a plant of homey endurance, pure fodder after all.

I would be converted to a religion of grass. *Sleep the winter away and rise headlong each spring. Sink deep roots. Conserve water. Respect and nourish your neighbors and never let trees gain the upper hand.* Such are the tenets and dogmas. As for the practice—*grow lush in order to be devoured or caressed, stiffen in sweet elegance, invent startling seeds*—those also make sense. *Bow beneath the arm of fire. Connect underground. Provide. Provide. Be lovely and do no harm.*

From *Heart of the Land: Essays on Last Great Places*, edited by Joseph Barbato and Lisa Weinerman (New York: Pantheon, 1995).

LOUISE ERDRICH

Teresa Jordan is the author of several books about the American West, including her most recent, *The Year of Living Virtuously* (2014). She was raised in a multigenerational ranching family in the Iron Mountain area of southeastern Wyoming. There she experienced firsthand the joys and challenges of ranch life, and learned to love the western grasslands. While she was attending Yale University, both her grandfather and her mother died. A year after she graduated, the ranch fell victim to the economic upheaval of the 1970s and passed out of her family. Her memoir, *Riding the White Horse Home* (1994), tells the story of one ranch community in order to explore the fate of family agriculture in the twentieth century. Currently living in Utah, she continues to write and create visual art about Western nature and culture. In this essay, Jordan attends the release of bison on the Tallgrass Prairie Preserve in the Flint Hills of Oklahoma, just north of Osage author John Joseph Mathews's hometown of Pawhuska and not far from where Washington Irving toured in the 1830s. Managed by the Nature Conservancy, this is the largest preserved tract of native tallgrass on earth, containing thirty-nine thousand acres and a bison herd now numbering in the thousands.

✳ ✳ ✳

Playing God on the Lawns of the Lord

There are a thousand of us and three hundred of them. We are *Homo sapiens* and they are *Bison bison bison*. We have come to this spot in the Flint Hills of northern Oklahoma to watch them released onto five thousand acres of native tallgrass prairie. The prairie and the bison evolved together over thousands of years. They have been separated for well over a century. They are coming together now, some order is being restored, and we are in a celebratory mood.

Each of us feels honored to be here. This is an invitational event, open to members of the Osage Tribe, which owns the vast oil reserves under this ground but not the ground itself; to members and friends of The Nature Conservancy, which now owns the ground; to General Norman H. Schwarzkopf, Retired, of the U.S. Army, who has just a few hours before, at dawn, received his Osage name of Tzizho Kihekah, or Eagle Chief; and to a swarm of schoolchildren, who sit restlessly on the grass waiting for the bison to appear.

No one denies that we are part of a spectacle. CBS is here. NBC is here. CNN is here. The *New York Times* is here. They have the best seats in the house, on the elevated platform near the gate through which the bison will run. Excitement fills the air, something akin to what one feels at a homecoming game, and we all have our cameras in our hands. Now we see the bison in the distance, a bounding line of darkness above the tallgrass, heading the wrong way at a dead run.

The cowboys do not ride horseback, but in pickups and on four-wheel all-terrain vehicles. They turn the bison toward the gate; the animals turn back the other way. The cowboys turn the bison again; once more, they refuse. The cowboys try a third time, and this time the great shaggy beasts come through the gate on the run, three hundred head, a diversified herd, ranging from calves born only a few months earlier, in the spring, to massive bulls a dozen years old that weigh nearly a ton.

We have grown quiet, as we've been instructed to do in order not to startle the animals, and our silence turns quickly to awe. So it's true, what we have read in books and seen in movies. The sound of bison on the run really does travel over a long distance. This is a quintessential American sound, something we must carry in our genes, bred into Native Americans by thousands of years hunting and eating and living side by side with the beasts; bred into Euro-Americans as a haunting legacy of what our ancestors annihilated in a few short years only slightly more than a century ago. This is the sound of history and valor and triumph and squalor and sorrow, and it really does sound like thunder. But thunder would be borne by the air. We would feel it in our chests, our diaphragms. This sound comes up from the ground, through the soles of our feet. The sensation increases as the bison near, and they move like a dark roiling sea, swift and high bounding, closely packed. The damp overcast of the sky brings out the full spectrum of gold in their shaggy roughs and darkens the burnt sienna of their atavistic forms. They pass and we are no longer members of a crowd of spectators. Each one of us is alone, watching as if from great distance something primal and real. And most of us are weeping.

A FEW DAYS before the official release, the bison had been trucked to a holding pen on the Tallgrass Prairie Preserve so they could acclimatize before being set free. Minutes after being unloaded, several of them found old wallows, depressions in the ground hollowed out by their ancestors when they had rolled to rub off flies or shed their winter coats. No creatures had used the wallows for at least a hundred and fifty years, but the newcomers, fresh off the truck, instinctively made for the depressions and started rolling. It was as if both the animals and the land remembered.

It is easy to romanticize such a thing, and yet the bison and the prairie had evolved together since the last ice age, or for what would be, in Oklahoma, somewhere in the neighborhood of twelve thousand years. The prairie ecosystem is "disturbance dependent," driven by the interaction of climate, fire, and the grazing of vast nomadic herds. Before white settlement, it was one of the most far-reaching ecosystems in what is now America, with bison numbering between 30 and 60 million and prairie grasses—tall, short, and mixed—covering well over a million square miles.

It worked like this: as herds of bison, elk, and other grazers ate their way back and forth across the country, their hooves tilled the soil and their dung fertilized it. Parts of the prairie went ungrazed and those grasses aged and dried, becoming less palatable and more susceptible to fire as fuel loads increased. Fire occurred naturally from lightning strikes and was also set by the native tribes that used flame to help with hunting, defense, and ceremony. Millions of acres burned every year; fire could move so fast and devour so much territory that some Plains tribes called it the Red Buffalo. Fire removed old growth and returned minerals to the ground; once an area had burned, it greened up almost immediately. New, tender shoots attracted herds which were then more likely to ignore areas that hadn't been so recently burned—often those that had been grazed a year or two earlier. These areas would then accumulate plant material that would make them more susceptible to fire and the cycle would start all over again in the miraculous combination of leapfrog, synergy, and chance that we call nature.

Tall grasses covered the more humid, eastern parts of the prairie, some 200,000 square miles of Texas, Oklahoma, Arkansas, Missouri, Kansas, Nebraska, Iowa, Illinois, Minnesota, the Dakotas, and southern Manitoba. There, this constantly shifting balance of elements provided habitat for a staggering variety of plants, birds, and animals. But westward expansion eliminated the bison, suppressed fire, and plowed up the land. Today, the tallgrass prairie exists only in small and isolated patches.

The largest expanse survives in the Flint Hills of Oklahoma and Kansas where the soil is rocky and thin, and the land, for the most part, has never been plowed. Unfit for farming, the grasses have continued to be grazed, by cattle rather than bison. The cattle have been managed well, and responsible ranching has kept much tallgrass prairie in the Flint Hills in good shape. The Barnard Ranch that forms the core of The Nature Conservancy's 36,600-acre preserve is an exemplary case. According to Bob Hamilton, director of Science and Stewardship, all of the grasses and other plants in the original tallgrass prairie are present on the preserve, at least so far as range experts have been able to determine. The prairie *ecosystem*, however, that living give-and-take of nature, is essentially extinct. The suppression of fire and the somewhat different grazing habits of cattle from bison have changed the proportions of things; woody plants, once kept at bay by fire, have in some places invaded; seeds that need fire to germinate have laid dormant in the ground; broad-leafed plants that are more appealing to cows than to bison have been grazed back, and other, less tasty ones have come in. Whatever natural evolution and progression was in place before whites came has been interrupted; as naturalist John Madson has pointed out, this may look like a prairie, but without fire or bison, it can't act like one.

The hope on the preserve is to restore the functioning ecosystem. Fire has been reintroduced and about one-fifth of the bison unit goes up in flames each year in a number of small, controlled burns designed to mimic the original frequency of fires—several large burns in the dormant fall and spring, which historically would have been set most often by Native Americans; more frequent but smaller fires in the more humid midsummer lightning season when the prairie is actively growing. Burn-unit locations are designed not on the basis of a predetermined formula, but by the buildup of fuel. "The prairie," Hamilton says, "tells us what to do."

The newly introduced herd of bison will increase, over the next few years, to eighteen hundred head; the management team will cull them to simulate what predators once achieved. And, as their numbers increase, the bison will eventually range over the full preserve, in areas that are now being grazed by cattle managed to imitate the habits of bison.

There is irony that we need so much management to re-create what nature once did by herself. Once the ecosystem operated within boundaries defined by climate and geography; today, in an area populated by both Native Americans and Anglos, dotted by ranchsteads and towns, and crisscrossed by highways, it must operate within boundaries defined by man. These boundaries are too narrow to allow nature to follow its own course; untended fire, for instance, could wipe out neighboring homes or cause havoc with Osage oil reserves. When asked what the Conservancy will do about lightning strikes or accidental fires, Harvey Payne, the preserve's director, answers without hesitation: "Any fires we don't set, we will extinguish immediately." Payne comes from a family that has ranched in the Flint Hills for five generations, and he knows the importance of being a good neighbor. He also knows that the natural ecosystem had room for the law of averages to play out; here, an unplanned fire could upset the balance the preserve is trying to attain. Man has always played a part in this ecosystem; now the only hope to save it is to control it totally. Payne is pragmatic. "On this preserve," he says with sadness rather than hubris, "we play God."

SOME HOURS AFTER the bison were released, after the barbecue had ended, the band had left and the hundreds of guests had gone back to Pawhuska, Bartlesville, Tulsa, and points beyond, I took a long walk. The bison had settled down and were contentedly grazing; one old bull had wandered off by himself. I enjoyed my own solitude, walking in grass that towered over my head.

After the bustle of the day, the sounds of the prairie were soothing: the buzz of cicadas and crickets, the dee-dee-dee of a black-capped chickadee, the rustle of a rabbit I couldn't quite see, my own steps swishing through the grass.

John Madson once described the tallgrass prairie as the lawns of God, and it was easy to hear a prayer in the whispers around me. I topped a ridge and looked out over thousands of acres of gently rolling hills, a sea of grass turned auburn with autumn, interrupted here and there, along creeks and on rims of limestone, with dark lines of oak. The grasses seemed to stretch forever and yet there were variations—sagey green in wallows, brighter green on burns, darker greens and grays in gullies. The clouds broke for a moment and the grasses waved iridescent gold, eighteen karat; the sky closed back down and they turned smoky blue. I rubbed my eyes and everything was so rose-colored I couldn't imagine anything could ever be wrong.

The lawns of the Lord are different than the lawns we mortals have created, those greenswards of infinite sameness. When I sank down on my haunches, I could number a dozen plants within a single arm's reach, from the big bluestem that soared over eight feet tall to the low-growing violet wood sorrel. Biologists have counted over five hundred different plants on the acres around me and the names form a prayer of their own: jack-in-the-pulpit and fire-on-the-mountain, wild white indigo and soft golden aster, witchgrass and weeping love grass, switchgrass and twisted ladies' tresses, shepherd's purse and button blazing star. And that doesn't begin to name the more than three hundred birds and eighty mammals that thrive here as well. Surrounded by such riches and hidden from view, I could forget for a moment the power lines and roads and fences that delimit this prairie world. It was a pleasing fantasy, but passing. We live within such boundaries. The question is where we go from here.

"When we have something valuable in our hands, and deal with it without hindrance," wrote Pedro de Castañeda de Najera over 450 years ago, "we do not value or prize it as highly as if we understood how much we would miss it after we had lost it." A member of Coronado's expedition north from Mexico in search of the mythical Seven Cities of Cíbola, Castañeda was among the first Europeans to set foot on the western prairies, and he presaged a sadness that many of us feel today. "The longer we continue to have it the less we value it," he wrote, "but after we have lost it and miss the advantages of it, we have a great pain in the heart, and we are all the time imagining and trying to find ways and means by which to get it back again."

Sometimes we notice in time, and we can go some lengths toward repair. There would be no bison at all without the work of conservationists nearly a century ago. When a census turned up fewer than 550 of the animals, Theodore Roosevelt, ammunition king William Hornaday, and Molly Goodnight, wife of Texas cattleman Charlie Goodnight, joined others to establish the American Bison Society and preserve a small herd through the New York Zoological Society. The bison that now roam the tallgrass in Oklahoma are direct descendants of those that once lived in the Bronx. With luck, the prairie ecosystem itself will enjoy a similar salvation.

The Indian tribes that Castañeda encountered lived within the arms of God. They gave thanks for the life of each bison that fed them. Modern agriculture, on the other hand, has given us the sense that we *are* God. Developed over thousands of years, it has been the enthusiasm and genius of many of the world's peoples, and it has led us to believe that the earth should bend to our will. Science made us arrogant; now, as we begin to understand the ways in which we have damaged the systems that sustain us, it is making us humble. Perhaps, just perhaps, it can pave the path of our redemption. We have tried to tell the earth what to do; maybe now we can learn how to listen.

From *Heart of the Land: Essays on Last Great Places*, edited by Joseph Barbato and Lisa Weinerman (New York: Pantheon, 1995).

Raised on a small farm near Montevideo, Minnesota, Paul Gruchow would become one of the most respected literary advocates for the rural communities and natural environments of the Midwest and West. Through his personally engaging, insightful, well-researched, and often lyrical nonfiction books, such as *Journal of a Prairie Year* (1985), *The Necessity of Empty Places* (1988), and *Boundary Waters: The Grace of the Wild* (1997), he inspired a generation of writers and many others to explore, cherish, and protect what remains of the wild and fragile beauty of this region. He was also an experienced teacher, and in this essay from *Grass Roots* (1995), he demonstrates his gift for helping others to learn from nature, especially the tallgrass prairie ecology that meant so much to him.

What the Prairie Teaches Us

The prairie, although plain, inspires awe. It teaches us that grandeur can be wide as well as tall.

Young prairie plants put down deep roots first; only when these have been established do the plants invest much energy in growth above ground. They teach us that the work that matters doesn't always show.

Diversity makes the prairie resilient. One hundred acres of prairie may support three thousand species of insects alone, each of them poised to exploit—often beneficially—certain plants, microclimates, soils, weather conditions, and seasons. This exuberance equips the prairie to make the most of every opportunity, to meet every natural contingency. The prairie teaches us to see our own living arrangements as stingy and to understand that this miserliness is why they so frequently fall short of our expectations.

The prairie is a community. It is not just a landscape or the name of an area on a map, but a dynamic alliance of living plants, animals, birds, insects, reptiles, and microorganisms, all depending upon each other. When too few of them remain, their community loses its vitality and they perish together. The prairie teaches us that our strength is in our neighbors. The way to destroy a prairie is to cut it up into tiny pieces, spaced so that they have no communication.

The prairie is patient. When drought sets in, as it inevitably does, prairie grasses bide their time. They do not flower without the nourishment to make good seed. Instead, they save their resources for another year when the rains have fallen, the seeds promise to be fat, and the earth is moist and ready to receive them. The prairie teaches us to save our energies for the opportune moment.

The prairie grows richer as it ages. Our own horticultural practices eventually deplete the soils. The topsoil washes or blows away; without additives, fertility dwindles. But the soils beneath the protective cover of prairie sod deepen over time; their tilth improves as burrowing animals and insects plow organic matter into them; fires recycle nutrients; deep roots bring up trace elements from the substrate; abundant legumes and microorganisms help to keep it fertile. The prairie was so effective at this work that, more than a century after it was broken, it remains the richest agricultural region in the world. The prairie teaches us how to be competitive without also being destructive.

The prairie is tolerant. There are thousands of species of living things on the prairie, but few of them are natives. The prairie has welcomed strangers of every kind and has borrowed ideas from all of its neighboring communities. In doing so, it has discovered how to flourish in a harsh place. The prairie teaches us to see the virtue of ideas not our own and the possibilities that newcomers bring.

The prairie turns adversity to advantage. Fires were frequent on the unplowed prairies. The prairie so completely adapted to this fact that it now requires fire for its health. Regular burning discourages weedy competitors, releases nutrients captured in leaves and stems, reduces thatch that would otherwise become a stifling mulch, stimulates cloning in grasses, and encourages the growth of legumes, which capture nitrogen from the air and make it available to the whole prairie community. The prairie teaches us to consider the uses that may be made of our setbacks.

The prairie is cosmopolitan. On the wings of winds and of birds, in the migrations of animals and insects, down the waters of streams and rivers come the messages, mainly contained in genetic codes, that sustain the prairie. Its storms swoop out of the Arctic or sweep up from the Gulf; many of its songbirds are familiar with the tropical rainforests; its monarch butterflies winter in the highlands of Mexico; its ducks vacation on seacoasts and in desert oases; its parasites hitchhike upon all of them. We think that we have discovered the global village, but the prairie knew of it millennia ago.

The prairie is bountifully utilitarian. But it is lovely too, in a hundred thousand ways and in a million details, many of them so finely wrought that one must drop to one's knees to appreciate them. This is what, over all else, the prairie teaches us: there need be no contradiction between utility and beauty.

From *Grass Roots: The Universe of Home* by Paul Gruchow (Minneapolis, MN: Milkweed Editions, 1995).

✳ RICHARD MANNING, 1951– ✳

A longtime resident of Montana, award-winning journalist Richard Manning was raised on a Michigan farm. Since moving west, he has authored several books and numerous articles and essays on grasslands environmental issues, including *Rewilding the West: Restoration in a Prairie Landscape* (2009) and *Grassland: The History, Biology, Politics, and Promise of the American Prairie* (1995), from which this chapter is taken. Like much of his work, it represents an artful blend of journalism, history, environmental writing, and personal essay. Before seeking out the fate of a bull elk that roamed from Montana to Missouri, the former prairie home of this species, he visits a pioneer cemetery in Polk City, Iowa. There he draws a different conclusion than Aldo Leopold did about the future of the tallgrass. In providing seed for larger restorations like Walnut Creek (now Neal Smith National Wildlife Refuge)—which currently supports a herd of elk—that small, but carefully maintained prairie remnant raises "the possibility of a people no longer out of place" (286).

Enclosure

On two successive days in Iowa and Missouri in November of 1993, I was given alternate visions of the grassland's future. I present both here because I believe they are not exclusive; both will obtain.

On the first day, I was bouncing along the blacktop grid roads surrounding Des Moines in a four-wheel-drive pickup with Loren Lown. He was showing me his projects, a network of postage-stamp and roadside prairies he had resurrected in his role with the Polk County Conservation Board.

We drove to Polk City, an accomplishment not possible from some directions in the early days of settlement, even if four-wheel-drive pickups had existed. The area around Polk City had been a big wetland, impassable except when frozen in the winter; now it was drained by field tiles and farmed hard for corn and soybeans. The summer floods of 1993 had taken most of the corn, but it made no matter. The federal crop insurance payments would erase the memory of the floods, and the following spring, the planting would resume.

Outside Polk City there is a sand ridge, and the sand was the ridge's salvation. The ridge's flanks were too steep to plow. The ridge top, because it was sand, could be excavated all winter; sand holds no water and does not freeze. A place that can be dug all winter won't hold water for corn, but it will hold coffins, and so becomes the cemetery of a pioneer town. The ridge erects a hill, not much of one, but then it doesn't take much of a hill to stand out in Iowa. The tallest point between that graveyard and Minneapolis 250 miles to the north is the grain elevator at Bradford, Iowa. By the 1840s, the Indians had been chased from hereabouts and the cemetery began filling with pioneer graves.

The earliest generations of headstones mark single graves of children, with no family around. These were the kids of westbound settlers who had paused here only long enough to leave a child to the land's safekeeping. Also among the earlier stones, there is a family plot, one stone for a man whose life was used up by the land; next to him, three wives even harder used; and stretched out beside him in succession and around them eighteen children, like cordwood. Back farther still, in a rougher corner of the cemetery, there are rows of simple rectangular stones, uninscribed. The graveyard took the poor, too, but there was no money for carving the identity of the people whose lives on surrounding farms had not left them enough to ransom a name from time.

Today, the unburied of Polk City pay a man to keep the cemetery trimmed, and he does, by maintaining a clipped carpet of Kentucky bluegrass around the stones, a sort of monument in its own right.

Around the edges of the hilltop, however, on that forgotten ground too steep for corn or caskets, a prairie is rising. No one cared about the land on the flanks of that ridge for more than a century, and without fire, it went back to trees. The forest buried the tall-grass prairie at its feet. Then a few years ago, a woman of Polk City, a daughter of the stones, found the stunted, etiolated remnants of Indian grass, leadplant, hairy grama, side oats, and dropseed surviving in the trees. She organized and lobbied. Soon there was a fire, and the prairie returned.

A day's drive northwest of Polk City at Wounded Knee, South Dakota, there is another graveyard that the Iowa cemetery brought to mind. There, "rest in peace" seems a cynical joke. Among crosses and medals there are medicine bundles, tobacco offerings, and smudge, the markings of another prairie hilltop capped with the leavings of a brutal century. This hill holds the victims of the massacre at Wounded Knee in December of 1890 when the Hotchkiss guns of the U.S. cavalry massacred at least 153 and as many as 350 men, women, and children, Big Foot's band of Sioux.

Wounded Knee is full of angry graves bit by years of harsh Dakota wind, prickly pear, grit, and gravel. Like the rest of the West, the land around Wounded Knee is used hard. If these graves are to ever settle their anger, it will settle only when the prairie rises again.

The graveyard prairie of Polk City is sufficiently recovered now to raise seeds that are spread through the state's prairie network. Some go to the eight-thousand-acre restoration project at Walnut Creek, some twenty miles away. The graveyard prairie raises seeds of big bluestem, little bluestem, compass plant, leadplant, prairie clover, ground cherries, Indian grass, and in these are seeds of dick-cissel, Henslow's sparrow, horned lark, fox snake, elk, wolf, and Sioux.

The old farmers around here, the hard-heads, call the prairie "weeds." A botanist I know tells an old farmer that a weed is just a plant out of place.

"Are you telling me these plants have a place?"

Just so. The prairie raises a place.

The plains tribes, each in their own language, usually called themselves by a word that means "real people." We the educated have generally interpreted this as a sort of primitive arrogance, but hear this. As some of the tribes were forced from their home lands and onto reservations many times hundreds of miles and whole biomes away, many of the tribes ceased using their names. That is, once out of place they no longer were "real people."

The seeds of the Polk City graveyard resurrect the prairie and the prairie resurrected the Lakota, in the sense that we all ought to be Lakota, which is to say, we ought to be able to call ourselves, in our own language, real people. The seeds raise the possibility of a people no longer out of place, the possibility that we may again have a name, that we may ransom that name from the future.

ONLY LATE THAT afternoon after the graveyard did it occur to me that Kansas City was so close, maybe three hours by interstate southwest, and only late in that day did it occur to me that this was where I needed to go to finish my journey. I had to catch a flight the next afternoon, so I arose the next morning at 4:30 and beat a rental car down the deserted highway into Missouri. At Kansas City in the pouring rain and rush hour, I crossed the Missouri, the river whose other end had started me wandering through the grassland. Here, though, it looked nothing like the mountain streams I know, and I crossed it without much regard. It was not my destination. I was headed for Independence, a few miles east down another interstate. I already had passed through Liberty. Jefferson City lay a few miles east still, but I was headed for Independence and had a map.

I left the interstate there and was flung straight into a sprawl of suburban Quick Stops, fast food, and mini-malls that make the face of our nation that most people know. Then I drove a couple of miles toward the edge of the grid to where the newer rambling, wood-sided homes are chewing up the countryside with bluegrass lawns and cookie cutter architecture, and continued on around a winding road to Jackson County's Fleming Park. It was still early, and the park was empty of people, and besides it was raining a lot harder than a Montanan can imagine rain.

I found my way around Lake Jocomo, which is really a reservoir (contrary to the practice of the federal officials who build reservoirs and letter maps, we ought to call reservoirs by their names). The road topped a hill in the woods and some deer jumped across the road, then the woods petered out and surrendered the roadside to a long stretch of woven wire fence that contained a close-clipped exotic grass pasture. Inside stood standard issue, roadside-attraction bison, outside an elevated platform to provide the vantage to camcorders.

I had not driven all this way, though, to see these bison, so I drove on to the next fenced pasture of what a sign had just informed me was the "Native Hoofed Animal Enclosure." It really was a zoo, apparently another name too harsh for the park's sign. An adjacent sign said: "Please feed only apples, pears and corn. No other foods are allowed."

Inside were the elk I had come to see. Another sign said one of them, a bull, was from Montana and had been granted its life here by the governor of my home state, as if an elk's life was something a governor could grant. Briefly the sign told the story of the Sweet Grass Hills and an elk that wandered down the Missouri to be captured here, 1,800 miles from home.

Behind the fence in another close-cropped pasture fanned a herd of cows and calves, and, off alone serenely ruminating next to a pond, was bedded an enormous bull. He was the only bull in sight and had to be the elk I wanted to see, but I couldn't be sure and had to know. I drove back across the park to a restored colonial farmhouse that was park headquarters and asked.

"No, that's not him. That's the old bull, and he's just livestock by now. He's been raised all his life here and tame as [can] be. It's the rut, and we have to separate the Montana bull and a few cows from the old bull with his cows. Otherwise that Montana bull would try to kill the big guy. He'd kill him."

I'd always thought that the urge that brought that bull this far was the urge to spread his genetic information, and I believed he would indeed kill the big bull for the privilege of doing so.

I asked the guy if he didn't think it a shame to have a set of genes that can carry an animal 1,800 miles locked up behind a fence, and he said it was, but what else were they to do. Leave him wandering round the suburbs, and he'd be hit and killed on a road sooner or later.

Now I drive back to the Hoofed Animal Enclosure, determined to see the bull. I walk the perimeter of the front pasture back to where I can see the cross fence separating the two groups of elk. Every span of wire between every post in the cross-fence, once strung straight, has been beaten into an arc by the elk from Montana. The guy at park headquarters had told me the Montana bull spends the whole rut either breeding or attempting to batter down that separating fence so he can kill the other bull and take his cows.

The other bull had not responded to the upstart; no challenges issued from the opposite direction. Every U-shaped arc in every span of wire is pointed toward the old bull.

The back pasture is not cleared like the front, but choked with sapling trees, so I slowly crawl up and down the perimeter fence in the rain peering into the trees with my binoculars. Now and again I catch sight of a cow or maybe a set of legs that could be a bull, no telling.

Often in fall in Montana I will walk for whole weeks in the mountains, in late September when most everyone else is gone from the back country. It is the best time of year to be there for a panorama of reasons, but mostly I go to hear the bugle of bull elk. Like the cry of migrating geese, it is one of life's most sublime sounds, primal and haunting. One expects these barrel-chested, hormone-crazed beasts to issue something in the bass range, but the bugle of an elk, a song they sing almost only in the rut, comes from another place. It is a flute's whistle, shriller than anything one would expect, and it is without analog.

Only the most persistent, obstreperous, and biggest of the bull elk breed. After a month or so of fighting and copulation, a bull is worn out, almost literally. In a tough winter the big bulls die before any other animals in the herd, so weakened are they by their exertions. Maybe they know this, knowledge that tunes the shriek of their bugles.

The rutting cry of a bull elk is a seed song, a grass song, strung on the sad, tense line between mortality and immortality. I have always heard it as such, but never so sad as on that rainy morning when it broke through the sodden air of suburban Independence, Missouri, and filtered to my ears through a straight stretch of woven-wire fence.

From *Grassland: The History, Biology, Politics, and Promise of the American Prairie* by Richard Manning (New York: Penguin, 1995).

Current Poet Laureate of Iowa Mary Swander was born in Manning, raised in Davenport, and now divides her time between the Amish community of Kalona and Iowa State University (ISU), where she teaches creative writing and is Distinguished Professor of English. Much of her diverse literary work focuses on the interconnections between the environment and rural citizens. Recently, she collaborated with her ISU students to write *Farmscape*, for which they gathered first-person stories of farmers and food producers across the state and crafted them into a drama that is now being produced nationally. She is perhaps best known, however, as the author of the memoir *Out of This World*, which recounts her struggle with nearly fatal allergies (a condition called environmental illness), and her subsequent recovery among the Iowa Amish. In this excerpted chapter, she reflects on the relationship between her organic gardening—on which she depends for her health, but which is threatened by a grasshopper invasion—and the fate of the tallgrass prairies.

☀ ☀ ☀

The Grasshopper Year

[. . .]

During harvest, I learned the secrets and rhythms of preserving, the carrots root-cellaring in an outside pit, the canning lids pinging on the kitchen counter, the dehydrator whirling in the corner. I mastered the art of the pressure cooker, and everything I'd watched my grandmother do as a child suddenly took on a déjà vu–like quality. I remembered her selecting the best produce for canning, the blemishless tomatoes, the pristine green beans. I remembered her boiling water, sterilizing jars, inspecting the pressure cooker gasket, tightening the lid, watching the needle sweep toward the higher red numbers. The job became a ritual that connected me to generations of women. They stood in summer kitchens, aproned, their hands wet, their faces red, a stray strand of hair falling down into their eyes.

For my frozen goods, I sent away for cellophane bags, the plastic ones outgassing into my food, and rented freezer space at the local meat locker. After the plant burned one winter and destroyed months' worth of vegetables, I bought a small, used freezer of my own, only to have its motor die after several weeks' use. A day later, two strong men hauled its old carcass out of my basement and bounced a brand-new model down the steps. The new freezer lasted a year, then went out when I was away for a week one December. I managed to salvage a few soggy turnips, then purchased a dehydrator.

Whatever my method of preservation, for five winters in a row I had subsisted on my larder of root-cellared, canned, frozen, or dried vegetables. I worked hard, sacrificing other pursuits. On Saturday nights when my friends went off to a smoky bar to listen to banjo and fiddle music, I turned on the radio and spent the evening tapping my feet to the rhythms of washed beets and diced squash. Again, I connected to a previous era, when work and play were one. I had no leisure time, devoting most of my energy to my job and my garden, watching my health improve as my horticultural skills grew.

Then came the insect plague.

"You need some birds to eat those grasshoppers," Mahlon told me. His chickens kept his infestation in check. "But guineas are the best, really. If a person has a few of them around, you won't see a grasshopper again. Course, you have to keep the guineas around too."

Fannie and Max had guineas, beautiful black and white speckled birds with bony casques on their heads. They darted out into the road, shrieked when you pulled into the drive, and rattled across the roof with a *click* and *tack*.

"Yeah, we had some guineas once too," Mahlon said. "They'd set up in our tree at night, then in the morning, be gone. A week later, we'd find them at a farm a couple of miles over."

Maybe, if I just wait, a guinea will appear, I thought. But I'd never been lucky, so I called the feed store to place my order.

"We don't get guineas in after the first of June," the dealer said.

I chastised myself for my tardiness, for not anticipating this grasshopper problem, for waiting too long. I should have acted that day in May when I found the first hatched nymph on a cabbage leaf.

"Ducks'll do real good too," Mahlon said the next night after I'd hailed him on the road on his tractor.

I *am* lucky, I thought. I knew a man north of town who raised ducks, hundreds of ducks, his farmyard overflowing with their webbed feet. Don's place was such a duck attraction that in the fall unsuccessful hunters stopped there and bought mallards. After crouching all morning in a blind in the bitter cold, the hunters didn't want to go home empty-handed.

I phoned Don. "I'd like to buy some adult ducks."

As fast as the birds mature, I knew I couldn't wait for hatchlings to grow up and begin their instinctual bug scavenging.

"I'm sorry to say, this year I don't have a duck on the place. Coons have been so bad—must be the drought—not enough for them to eat. They just walk into the yard in broad daylight, bold as you please, and a few minutes later sashay out with a duck in their mouth."

"Do you know of anyone who has ducks?"

"No, just about every guy I know who had 'em doesn't anymore. Same deal. Coons. I have geese, though. Plenty of those. Want some geese?"

"Will they eat grasshoppers?"

"Well, I don't know."

Who would? The county extension service agent was very interested in recording the exact location of my infestation, but confessed she didn't know if geese ate grasshoppers. I'd have to call the state poultry expert for the answer. "But why don't you just spray?" she asked. Several days and long distance phone calls later, I finally reached the poultry expert. He, too, wanted to make me a pin on someone's state map, but finally informed me that no, geese weren't especially good predators of grasshoppers. "Why don't you just spray?" he asked.

I tried to explain my plight. "Think of me as a pioneer," I said. "How did the pioneers cope with grasshoppers?"

He told me that the pioneers did have their disastrous locust plagues. (A locust is just a grasshopper gone rogue.) But a little more than a hundred years ago, when my great-grandparents settled in Iowa, before farmers tiled and drained their fields, the state was dotted with bogs and sloughs that supported dense populations of birds. Canada geese, canvas backs, and ring-necked ducks, mergansers, pelicans, and sandhill and whooping cranes darkened the sky on their fall and spring migrations, and when they circled down to land, ate their share of grasshoppers. Year round, the native prairie chickens gobbled up thousands more of the insects, but the prairie chickens went the way of the buffalo, men hunting them for sport until near extinction. So here I was dealing with one human-made ecological imbalance on top of another.

I began reading the diaries and letters of prairie women, and found their tales of locusts ravaging their small settlements. 1874 came to be known as "The Grasshopper Year," when locusts swept across the plains like snowstorms, the insects dropping to the earth, covering every inch of ground, every bush or shrub like enormous-size flakes. Tree limbs snapped under their weight, and the grasshoppers ate right through sheets and quilts hastily thrown over precious tomato vines and celery plants. They devoured the leaves of fruit trees and ate through green peaches, leaving nothing but the pits hanging from the branches. They invaded homes and ate anything that wasn't encased in metal—even kitchen utensils, furniture, and wooden walls of the cabins. Window curtains were left in shreds, and young children screamed in horror as the grasshoppers wiggled through their hair and down their shirts. Men tied strings around their pant cuffs to keep the locusts from crawling up their legs. A Kansas woman even reported the grasshoppers eating a green stripe off the white dress she was wearing.

Pioneers learned that the grasshoppers would stop for nothing. The women's writings were filled with Biblical descriptions, accounts of locust plagues right out of Exodus that challenged the homesteaders' faith in themselves, their God, and the very idea of Manifest Destiny in the New World. The letters of Gro Svenson, a Norwegian immigrant, recorded her battle with locusts on the Iowa prairie near Estherville and explained that settlement's solution. Early in the spring, the pioneers doused their crude cabins and barns with water, then set fire to the surrounding fields and prairies in hopes of burning out the insects' eggs and young. Inevitably, the burning meant replanting and a late harvest, but the homesteaders were able to reap a crop from their one- and two-acre farms that year.

My garden isn't even a quarter acre, but the season was too dry, the plot too near the house to burn. Earlier that spring, brush fires had jumped across ditches, set fields smoldering, and threatened buildings all across the state. Day and night, the air smelled of smoke, and in April a county conservation commissioner told me, "Iowa's on fire." In early May, a friend in a small western Iowa town stayed up all night dragging buckets of water from her house to the edge of her yard to try to put out a wildfire that crept steadily closer, flames lapping toward her one-hundred-year-old family home, where her elderly and ailing mother slept inside.

Instead of the torch, I tried the phone and called an organic gardening magazine hotline. They told me to try some buckets of my own.

"Make grasshopper traps," the hotline woman instructed me, "by setting out buckets baited with molasses water. The grasshoppers will jump in and drown."

I combed the secondhand stores for small buckets, received donations from friends, and gathered up a stack on sale at the paint store, then covered their bottoms with goopy syrup. Sure enough, within minutes of placing traps out between my vegetable rows, grasshoppers leapt in, swam a few laps of breast stroke across the surface, then sank to their sugary graves. Dumbfounded, I stood out in my garden in the heat of the day among my tomato vines and watched hundreds of the insects plunge into the water. Hourly, the grasshopper carnage tallied up, yet more bugs kept coming.

I called the hotline again.

"Make a mixture of chili peppers and garlic and spray it all over your plants," the woman said.

The next day, spray tank slung over my shoulder, nozzle nosing the undersides of the leaves, I moved through the garden, coating all my vegetables with my homemade pesticide. At first, the concoction seemed to truly repel the grasshoppers, and I was jubilant. The insects still lurked near the garden, but had stopped eating it, their voracious hum quieted to a whisper. Occasionally, one would make a daring suicidal splash into a bucket, but most appeared slowed, almost listless, compared to their usual hyperactivity. I drove to the discount supermarket and bought a case of bottled chili peppers.

"Having some kind of party?" the clerk asked.

At home, the potion seemed to be wearing off, so I sprayed again, and this time the grasshoppers became excited, their wings vibrating, compound eyes sparkling with desire. The party had just begun.

"We got the salsa, you bring the chips." They buzzed, hopping and leaping from corn stalk to fence post, chomping and spitting more voraciously than ever.

I slumped down at the kitchen table. It was over. I spent the next few days foraging in the ditches and along the fences for lamb's-quarters and clover for my daily salads and tried not to even glance in the direction of my garden.

Then, a couple of mornings later, I woke early, remembering bug juice. Ah, yes! Bug juice, an old organic garden standby, a homemade pesticide that had never failed me. Bug juice, made from ground pest parts, was blended with water and sprayed back on the area of infestation. It worked, somehow, on the old holistic adage that like cures like. *Like* certainly cures you from ever wanting to get up in the morning and fix yourself a nice quart of orange juice in the blender again. I'd read one source that speculated that the very smell of dead pests was enough to drive brethren away from the fold, and another source that claimed that dead pests spread viruses throughout the flock that killed fellow followers. Amen, I said to both theories and prayed the juice would work.

That afternoon, I sieved off the grasshoppers from the molasses buckets and readied the juice. *Slish, slosh.* Masses of dead bugs swirled in the wire mesh, and just as I was about to whirl an Acrididae smoothie, Moses stopped by in his buggy.

He tied up his horse to the hitching post, bent over my buckets, and cocked his head. "Whatcha doing there?" he asked, and when I explained, he said, "Oh, you're cruel."

I may have been. When the blender gears began to grind mouths, claws, wings, ovipositors, and antennae together, I did have a moment of guilt, but as soon as I stood in the garden with the sprayer, I forgot my animal cruelty in hopes of saving the last shreds of my plants. The bug juice shot out over my cabbages, and once again I was optimistic, although guardedly so. I went to bed that night thinking I might have found a remedy to my problem, but at the same time was resigned to the possibility that I probably hadn't.

Days ticked by, and slowly I began to see some improvement. In Exodus, the Lord sent a series of plagues down on the resistant pharaoh who had enslaved Israel. The Egyptians suffered through infestations of ascending terror, from frogs, to maggots and flies, to grasshoppers. Then, finally, Moses parted the Red Sea, Pharaoh's army drowned, and all were delivered. My salvation was less dramatic—more like a rerouting of a small tributary—but little by little, the grasshoppers did begin to thin, and the cabbage, broccoli, and Swiss chard rallied to generate new leaves, fill in their flowerettes and heads, their fan-shaped leaves. After ten days, the garden had a different look and feel. No longer molested and limp, it stood up taller, stronger, more full-bodied.

When I reached the two-week point, I began to realize that I'd won, that the garden was definitely on the rebound, although it would never regain its original shape and hue. The grasshoppers still hovered thickly across the landscape, but I salvaged enough produce from my plot to make it through another winter. Yet, as I chopped okra (my only cultivated plant to make it through the plague unscathed), I began to wonder if my whole idea of gardening wasn't askew.

After years of organic gardening, I found it much easier to cultivate pest-resistant plants than to try to fight the pests themselves. So, to guard against future locust plagues, I thought I would delve into finding other species and varieties, like okra, which stood up well against the powerful jaws of the grasshoppers. Then I remembered that, in their journals, the pioneer women noted that the prairie grasses were the only thing that the grasshoppers left untouched. Plant a complete garden of okra and big and little bluestem grasses?

Was this *such* a crazy idea? Instead of prairie grasses, perhaps I could plant some of the prairie weeds that I had been reduced to foraging for my salads. Had I just said reduced to foraging? Foraging, like root-cellaring and dehydrating, which I ultimately found the cheapest and most satisfactory forms of food preservation, was our most ancient form of food production.

Agricultural cultivation, as we know it, is a fairly recent historical develop-ment. What if I didn't plant at all? What if I just freed myself from all the hard work of gardening altogether—the plans, the seed saving and buying, the seedlings and transplants, the cultivating and tools—and learned to forage for all my food?

I've come to rage at the way the prairie has been ravaged by the people who settled the Midwest, my ancestors and the pioneer women among them. In steamboats that crossed the Great Lakes, that made their way up the Mississippi, in covered wagons that forded the Missouri, then creaked and moaned behind oxen along the narrow paths that wound westward, the early settlers brought their European values with them, and most of America today retains their idea of the ideal landscape. Trees, mountains, and large bodies of water, the familiar in the pioneers' past, were sought out and treasured. For the most part, the prairie was revered not for what it was, an ecosystem of magnificent intricacy and beauty in its own right, but for what it could become.

Naturally, it could become farms. It could be cultivated, and so destroyed, to produce plants that were familiar. Plants that could be raised and sold to others who found them familiar. Never mind adapting to a new flora and fauna, of learning new ways of cooking and eating. Never mind the fact that this prairie had been sustaining populations of Native Americans for hundreds of years without much cultivation. The prairie was, and still is, devalued. With the introduction of the plow went the wildflowers and bluestems, the wetlands, the vast populations of birds, butterflies, and small mammals.

Sure, some understood. Gro Svenson marveled at the blooming prairie flowers and the changing colors and nature of the landscape. Others were awed by the woolly calyxed pasque with its delicate yellow center, the first flower to appear in the spring. As spring moved into summer, a succession of surprises sprang up, from the buttercup and Johnny jump-up, to the puc-coon, wild artichoke, wild indigo, spiderwort, wild garlic and onion, smart-weed, butterfly weed, wood betony, and purple cone flower. The sunflowers dominated the midsummer with the opening of the black-eyed Susans and compass plants. Then the goldenrods and asters finished off the fall.

Today, we've nearly finished off the prairie. Now, you have to really search to find small patches of it in parks and preserves. Never again will we, like those pioneers, have the opportunity to experience the majesty of its vastness. When I walk through Doolittle Prairie, my favorite preserve in central Iowa, I try to imagine what this plot would be like if its waist-high grasses and wildflowers extended to the horizon, if it were unscarred by power lines and the constant din of passing traffic. In Doolittle Prairie, I began to wonder if gardening, this activity that I thought had been keeping me in touch with nature, was only, in a weird paradoxical way, destroying it. Perhaps I should give up gardening altogether, restore my plot to prairie, and just forage; then I truly would be in balance with my native environment. For wasn't my small garden a part of the larger farming problem that had so changed the landscape?

Agriculture does have its advantages, though, I reasoned. It has allowed us to be freed from hunter-gatherer society and stay in one place, build a stationary community. Yes, I could wander up and down the road, looking for a meal, but I might need to wander farther and farther, trespassing on others' "property," if this were my only source of sustenance. At last, I decided I would get out my field guide and make more of a study of natural food sources, do more foraging, but keep my garden, reducing my dependence upon it.

Then the two ideas—foraging and gardening—began to come together in my mind. I thought of Wes Jackson's work at the Land Institute in Kansas to find a perennial variety of wheat that, like prairie grass, would be regionally adapted and eliminate the time, cost, and contamination of planting seeds and applying pesticides and herbicides. Why couldn't I move more toward Jackson's idea in my garden?

I developed a plan for a native garden. I began to allow some of the common weeds—dandelions, lamb's-quarters, plantain, and purslane—to remain in my spring garden and combine with different varieties of lettuce in my salad bowl. I reserved a section of my garden to become a patch of perennial edibles—sunflowers, milkweed, and mullein. These hardy flowering weeds brought not only beauty but an array of butterflies to my plot. True, as staples, weeds took some getting used to in my diet, but over time they provided a delicious, free, and organic source of food that never needed to be shipped across country, sprayed, or watered.

I don't know if I will ever convert entirely to a native garden, but I'm willing to experiment. In the end, my own northern European roots may run deeper than those of prairie grass, and I may not be able to give up the grow lights, the seedlings, the straight rows and strings. I may cling as tightly to the familiar tomato, potato, peas, and beans as did the pioneers. "Let food be your medicine, and medicine be your food," Hippocrates, the medical pioneer, said. As tense as it was, my locust plague got me to thinking beyond my own rake and hoe to a new perception of food. In a small way, this new vision may begin to heal me and my landscape.

And what of the larger population and landscape? Big changes start with small decisions. What if, for starters, more of us returned to gardening, then each midwesterner nurtured his or her own prairie patch? Not just a "meadow in a can," but a real selection of indigenous plants that could be used for breakfast, lunch, or dinner. What if, deep down in our own tissues, we began to feel the changes these native foods brought? The return of a portion of control over our food sources, our physical and mental health, and balance of our natural environment. As long as the prairie is populated mostly by Euro-Americans, it will probably never return to its natural state, but a network of native gardens might be enough to keep its identity before us, reminding us how much we've lost and, with careful living, how much we might gain.

The big picture of a world of clean food, clean water, the return of local flora and fauna began to emerge in my mind, yet there were further leaps of thought I hadn't even entertained.

"Now, some people like grasshoppers," Mahlon told me later that summer when my infestation was in control.

"Like them? The birds like them, you mean."

"No, people like them."

"Why?"

"You may miss those grasshoppers yet."

"Why?"

"All this time you complained about not having enough to eat, you could've had *them*."

Of course! Didn't I have a crunchy organic protein source right in my own backyard? Didn't I have the guts to try them? Didn't Pharaoh's army get drowned? Hadn't I been delivered to the promised land?

From *Out of This World: A Journey of Healing* by Mary Swander, 1995. Reprint (Iowa City: University of Iowa Press, 2008).

✳ DON GAYTON, 1946– ✳

Ecologist and author Don Gayton lives in British Columbia, but grew up in California and Washington. Whether writing about gardens, kokanee salmon, Puget Sound, or the vast grasslands of Canada and the United States, Gayton explores the complex, but always fundamental relationship between people and place, including his own. This excerpted essay, which comes from *Landscapes of the Interior* (1996), offers a brief history of the profession of ecological restoration, and portrays the life and work of one such scientist currently restoring tallgrass prairie remnants in suburban Winnipeg.

Tallgrass Dream

J ohn Morgan drops down to his knees in the grass, and gathers a bundle of stems in his hand. His is a natural, almost unconscious movement, a tool of his profession. We are at the Living Prairie Museum, an unexpected island of native grassland in the heart of suburban Winnipeg. It is a Sunday in early spring, and several families move along the trails. I kneel down too, feeling a little self-conscious, but no one pays any attention to us, grown men kneeling in the grass.

Living Prairie is a tiny forty-acre parcel, which the City of Winnipeg had the foresight to secure as a public ecological reserve. The vegetation we kneel in is the fabled tallgrass prairie, a grassland type whose fame and romance rivals that of the pampas of Argentina or the south African veldt. The vegetation around us is waist-high, and the growing season has not even started yet. This is a far cry from the short, droughty prairies of Saskatchewan, Alberta or Montana. At least thirty different plant species make up the complex thatch in front of us. By fall some of those plants, the statuesque grasses, will be taller than we are. It is not hard to see why Morgan, the botanist, nurseryman and prairie restorationist, takes a proprietary interest in this Living Prairie. As he methodically sorts through the plant stems in his hand, John tells me that scraps of unbroken prairie like this one are all that is left of a vast ecotype that once stretched from Chicago to Toronto, and from Lake Winnipeg to Texas. As we stand up again, I look around: a suburban neighborhood presses up on one side of us, and an industrial area looms on the other. I sense this visit will be an odyssey of loss.

[. . .]

From the small universe of his farm in nearby Argyle, Manitoba, John Morgan works to reclaim parts of our ecological patrimony, by saving and restoring tallgrass prairie. Here he catalogues the remaining scraps of Manitoba's tallgrass, produces seed and container-grown plants of tallgrass species, tests seeding methods, and undertakes actual restorations of prairie, right from the bare ground up.

The ambitious, almost quixotic profession of ecological restoration was born in the Midwestern U. S. in the late 1930s. The activities of restorationists fall into four general categories: identifying and protecting existing remnants; actually re-creating parcels of threatened ecotypes; devising management strategies for the enhancement of those existing and re-created parcels; and introducing the use of native plants into the world of home gardening and landscape architecture. The first restorationists, like Harold Green of Wisconsin, were isolated individuals, often considered either hopeless romantics, or outright heretics. A full-fledged restoration movement, with leaders and a philosophy, has emerged only recently, and the twin shadows of Thoreau and Aldo Leopold loom large over it. Morgan is part of a small group who pursue restoration full-time, but a host of others—farmers, grandmothers, schoolkids—pursue it as an avocation. In Calgary, neighborhood groups are rescuing native grassland sod about to be lost to housing developments, and transplanting it elsewhere; in Saskatoon, a university professor leads a successful citizen movement ("Rescue the Fescue") to save a small piece of relic prairie. Outside of Chicago, a conservation group uses a variety of agronomic methods to restore an abandoned farm field to the original tallgrass prairie. There is a grand irony in the fact that the original impetus for the ecological restoration movement began in the very cities that originally consumed prairie—Calgary, Chicago and Winnipeg.

The current impetus for preservation and ecological restoration seems to be the imminent threat of losing entire ecotypes. Manitoba's tallgrass certainly falls into that "threatened" category, if not beyond it: Morgan estimates the size of the province's tallgrass remnant to be 1/20th of 1 per cent of its original extent.

Ecologically and geographically, tallgrass stands as a buffer between the moist deciduous forests of eastern North America and the dry plains of the West. Like the other Canadian prairie zones of mixed grass, fescue and pacific northwest bunchgrass, the tallgrass occupies a young landscape, geologically speaking. The crushing presence of the glaciers created a matrix of fine-textured soils and level terrain, and their final departure, some 12,000 years ago, left a new niche to be colonized. The ancestors of prairie plants were originally forest understory species that crept westward onto the newly-formed plains as the ice age came to a close and the glaciers departed. Tallgrass prairie, being closest to the forests, was thus the earliest type of prairie to appear on the continent. By the beginning of European settlement, the Canadian range of the tallgrass prairie covered the Red River Basin of southern Manitoba, a broad, 100-kilometre-wide swath starting at the North Dakota border and trending northeastward almost to the southern shore of Lake Winnipeg.

Tallgrass prairie species produce massive quantities of tiny, hairlike roots, an effective drought survival mechanism. Ten thousand years of this root production in the fine-textured silts and clays left by the glaciers produced some of the richest, deepest soils to be found anywhere in the world. But the prodigious soil-building ability of tallgrass prairie led to its early undoing, as the vast, fertile lands within its range were ruthlessly plowed up and planted to corn, wheat and other crops. In contrast to the drier prairies of Saskatchewan and Alberta, where large-scale land conversion did not begin until the 1940s, Manitoba's Red River Valley was largely cultivated by the turn of the century.

Several small relict tallgrass prairies found in the Essex, Kent and Lambton counties of Ontario, separated from the Red River Basin by hundreds of kilometres of boreal forests, are a puzzle. Morgan tells me the consensus of expert opinion is that during the Altithermal, a period about 5,000 years ago when temperatures were warmer and drier than they are now, tallgrass prairie covered much of southern Ontario. As the climate cooled and forests prevailed again, a few small prairie "islands" were left. Fossil landscapes, if you will.

Like all prairies, tallgrass evolved in the presence of periodic wildfires, which, along with occasional drought, maintained a balance between grass and trees. Fire also improves nutrient cycling and prevents the buildup of excessive quantities of plant litter, which slows the normal solar warming of the soil every spring. As proof of the value of fire John cites the railroad rights-of-way, which railway workers traditionally burned as an inexpensive means of trackside vegetation control. These same rights-of-way, usually thought of as ecological disaster zones, are now some of his favorite tallgrass prairie seed collection sites, because of the vigor of the native vegetation. To be a restorationist, it seems, is to live with contradiction.

Fire was also critical to the maintenance of oak savannah, another vanished ecotype closely related to the tallgrass, which was found in the moister parts of the Red River Valley. Oak savannah consisted of widely-spaced, mature oak trees with tallgrass underneath, giving the ecotype a very pleasing, parklike appearance. Fire-resistant oak and the tallgrass species could survive the periodic fires that swept across the prairies, but fire-sensitive aspen could not. With the advent of modern fire suppression, aspen has crowded out and virtually obliterated any remaining oak savannah landscapes, and is now encroaching on many tallgrass remnants.

The early destruction of the tallgrass, and the distortion of the remnants by fire suppression, presents a real problem for Morgan; they have no blueprints upon which to model their restorations. In addition, the isolated remnants, like Living Prairie, are always suspect, because their flora is subject to attrition from within and invasion from without. Historical descriptions of tallgrass prairie are scarce and inadequate; the early explorers hurried across it on their way to search for beaver, the Northwest Passage, or other compulsions, and few stopped to describe what they saw. The earliest recorded mention of the Canadian prairies was made in 1690 by the rhyming Irishman Henry Kelsey, who dismissed it thus:

This plain affords nothing but beast and grass,
and over it in three days we past.

Even Samuel Butler, who was awed by the "ocean of grass," provided lots of Victorian grandeur, but little descriptive or botanical detail to go on. I was reminded again of the importance of the nonexistent profession of landscape archeology. Much great literature has come from the simple desire to understand how things *must have been*, for a person or an era; it is time for ecologists to understand the value of this desire.

Despite the absence of historical description, Morgan and others look to a single grass, big bluestem (*Andropogon gerardii*), as a major constituent, perhaps even the backbone, of tallgrass prairie. This majestic perennial bunchgrass is the tallest of all the dryland grasses, reaching a height of two metres and curing to a rich auburn color in the fall. Switchgrass (*Panicum virgatum*) and Indian grass (*Sorghastrum nutans*) are also thought to have been present in some abundance. All three of these grasses share a complex variation of normal plant metabolism that gives them excellent growth efficiency in the heat of summer, when other plants are forced to go dormant. It is also assumed that tallgrass prairie contained a great profusion of broad-leaved plants — goldenrods, asters, blazingstars and legumes, to name a few. So the modern restorationist might envision big bluestem as the matrix of this great, buzzing, midsummer prairie enterprise. Showy, weak-stemmed goldenrods would lean against bluestem's tangled growth and shafts of purple liatris would penetrate through it. There would be white yarrow and yellow sunflowers to please the butterflies, and leadplants to taunt the botanist with their greyish, dead-looking leaves. A hundred insects, some perhaps unknown to science, could be captured in a single sweep of the net. Forgotten birds that have long since left the prairies might work the air currents above the sward, and below it the soil would teem with complex microbial life. A worthy system to restore.

People like John Morgan re-create examples of tallgrass for many reasons: as a tribute to nature, as a simple gesture atoning for our historical misdeeds to the land, or as a legacy to hand to our descendants. Or perhaps the simple feeling that an ecotype has a certain right to existence that transcends human concerns. More pragmatic reasons, like the value of bio-diversity, and the maintenance of a threatened gene pool for possible future use in agriculture or medicine, may also be important. But the fundamental motive of the restorationist may be a very personal and spiritual one that is harder to define. "I can't put my finger on it," Morgan says, "but there is something very subtle and very powerful that happens when you have the privilege of kneeling quietly on a piece of tallgrass prairie."

The Morgan's farm, a quarter-section in the tiny hamlet of Argyle, north of Winnipeg in the heart of the Red River country, is where his dream and his business of tallgrass prairie restoration take shape. The cycle of restoration starts in early spring; snow has retreated to the aspen bluffs, and sheets of meltwater lie on the stubblefields. The air is full of migrating Canada geese. Doug Collicutt is a frequent visitor, as he and John discuss their plans for the coming field season. The heat has been turned on in the 1,800-square-foot greenhouse, and John and his wife Carol tend tray after tray of tiny young seedlings. Seed for these has been painfully collected from dozens of different locations: unplowed field corners, railroad rights-of-way, lightly grazed corners of cow pastures, even vacant industrial land inside of Winnipeg. The greenhouse is part production facility, part commercial enterprise (a native bedding plant business complements the Morgan's restoration consulting), and part laboratory. Here in the greenhouse's humid atmosphere the Morgans have confronted two fundamental aspects of native plant physiology: germination and root growth.

Tallgrass plant species are climax species. They are virtually all perennials, and their germinative systems are geared only to replace themselves within a very stable, long-term plant community. They are neither colonizers, like the weed species, nor highly uniform, like the crop plants. Although mature tallgrass plants can become quite large and aggressive, their seeds tend to be very small, with precise germination requirements. The seeds often germinate only after long periods of dormancy, and once they do germinate, early seedling growth can be quite slow. These long-term survival mechanisms are precisely those the plant breeder would select against when developing a crop, but the Morgans must work with these traits. They have searched the meagre literature for everything they could find on native plant germination, and have tried a full range of temperature and moisture combinations, seeding depths and seed pretreatments, in an effort to unlock the growth secrets of each tallgrass species.

Root growth of these same plants presents the opposite scenario: it is abundant, sometimes overwhelming. John laughs as he recalls their early seeding attempts using conventional two-inch-deep greenhouse cell trays. Root growth in some species was so abundant that it literally pushed plant and soil right up out of the container. John and Carol solved the problem by substituting the six-inch-deep cell trays used by silviculturists for tree seedlings.

June will find John tuning up his Ferguson tractor in preparation for seeding. In contrast to the large-scale implements his grain farming neighbors use, John seeds with a specialized, eight-foot grass drill that can deliver seed very precisely. Then he firms the soil around the seed by going over the field with a water-filled, 1,500-pound roller-packer. Morgan swears by extensive land preparation before seeding. When I suggest there is a paradox in using aggressive technology to re-create nature, he readily agrees. "No question. But right now, we use every technological trick in the book— repeated cultivation, chemical weed control, state-of-the-art seeding equipment; anything we can get our hands on that will give us good stand establishment. Otherwise, all we get is a horrendous crop of weeds."

If the planning, the land preparation, the seed, the timing, the weather and the weed control are all just right, this grueling, high-input establishment phase needs to be done only once. By the second or third growing season, the network of tallgrass perennials will have taken root and the system could, theoretically, go on forever.

Summer is a time for tending the seed nursery plots in Argyle. Classical four-by-eight research-style plots of switch-grass, stiff goldenrod, three-flowered avens, yellow coneflower and a host of other plants are laid out over two acres of sticky, gumbo soil. The Red River Basin was originally the bottom of glacial Lake Agassiz, which trapped centuries of fine, windblown silts and clays. Add to that the leaven of several millennia of tallgrass root growth, and you have a soil the color and texture of German chocolate cake. John's two young daughters, mistresses of the farmyard, wisely park their tricycles and don rubber boots before venturing into the nursery.

By crowding the nursery plants within closely-spaced rows, John has been able to induce a heavy seed set in many species. In another part of the farm, a seven-acre field has just been planted to big bluestem for seed. John and Carol have put most of their grain farming land on short-term leases, and plan to take it back as they are able to convert it to tallgrass seed production. One thirty-acre field is being prepared to be seeded to big bluestem next year. Seed harvesting of some of the early species in the nursery has begun already, and the hunt is on for good remnant harvesting sites around the countryside.

Seed harvesting begins in earnest in the fall. A pull-type harvester with a large rotary brush (similar to those seen on street sweepers) is the favored implement. This custom-made harvester, of Morgan's own design, can be set for various species by adjusting the height of the brush, the speed of its rotation and the groundspeed. John follows a strict protocol for harvesting natural areas, hitting them once only every few years, and maintaining a very broad genetic base in his seed stocks. He is also loath to send his seeds for use in distant locations, believing that the use of locally-adapted varieties is fundamental to ecological restoration. "I could easily sell my seeds to people in Calgary or Saskatoon," he says, "but that defeats the purpose of the whole exercise. The production of seed for restoration should be a local enterprise, for a local market."

After freeze-up, the farm's focus of activity shifts to the seed cleaning plant, on the second floor of a converted barn. At one end of the barn is a variety of sieves, shakers and fanning devices, all custom-made, since none of the conventional crop seed cleaning equipment works on the native seeds, which can be tiny, light, fluffy, fragile or all of the above. The rest of the barn is a motherlode of seed, stored in containers that range in size from gunnysacks to pill bottles, all neatly arranged in alphabetical order, from *Andropogon* to *Zizea*. John picks up a container at random, and shows me the tiny slivers of complexity inside. As he reseals the container, he shakes his head in bemusement. "It's funny," he says, "but I take real delight in knowing that we now have an entire ice-cream pail full of meadow blazingstar seed."

A high school playing field is an unlikely spot for a prairie restoration, but one of Morgan's most successful seedings took place at Winnipeg's Elmwood High School in 1991. Morgan and Collicutt were approached by an innovative geography teacher, Anne Monk, who was teaching a section on the Great Plains to one of her geography classes. "I am always interested in action-oriented classroom projects," she says, "and establishing a piece of tallgrass prairie seemed to combine action, learning and stewardship. It was a natural for us." The class obtained permission to rototill a little-used 2,200-square-foot chunk of the school yard. In early June, they broadcast a custom seed mix of seven grasses and twenty-five broad-leaved tallgrass species that had been collected from relic prairies. John, Doug and Anne then had the students "make like buffaloes" and press the seed into the ground. The catch was excellent, and now subsequent classes are learning to identify the true prairie species and weeding out any invaders.

Morgan and Collicutt have further honed their restoration skills in projects at Winnipeg's Kil-Cona Park and the downtown Forks redevelopment area, creating pocket prairies. They are also midway through a research project to determine the most effective means of cultivating, seeding and establishing tallgrass prairie. Some of the variables being tested are: fall versus spring seeding, drilling versus broadcasting seed, straw mulching and irrigating. They have also done some landscaping projects with prairie species, and welcome these departures from the classical tradition of mowed lawns and formal flowerbeds.

In trying to re-create prairie, Morgan is under no illusions. He is ever mindful of the ecological maxim that states "small relicts always degrade." What that means is that a five-acre prairie remnant may be forever doomed to invasion, and a five-acre restoration may never become a real, functioning prairie. Preserving or restoring tiny scraps in scattered locations may ultimately reinforce the odyssey of loss. The phenomenon of "edge effect" is one of the primary reasons; simple mathematics tells us that the smaller a parcel is, there is a greater proportion of "edge" area relative to the "inside" area, and a greater opportunity for the more aggressive introduced plants to invade from the surrounding area. This phenomenon can be seen at Living Prairie, where smooth bromegrass, a common introduced grass, can be seen invading from the adjacent land outside the reserve. Population dynamics are another problem; many species of plants and animals require a large, scattered and diverse population to maintain their gene pool. Add to this the complex and poorly understood web of interspecies dependencies, and the minimum size for a functioning, self-sufficient tallgrass prairie could easily be 500 acres, 5,000 or even more. As in the re-creation of historical fire regimes, ecological restoration work requires an awareness of our ignorance, and a good dose of humility.

Morgan himself takes a pragmatic view of what he does. "Sure, to call ourselves prairie restorationists is pretty arrogant. What we are really doing is replicating what we *think* was the original plant distribution of tallgrass prairie." He chuckles as he recalls the comment one observer made, who referred to what they were doing as "gonzo botany." "We're not even sure about the species mix, and we know there are a hundred other factors which make up a real, functioning ecosystem. But what we are doing is a start, and we are learning."

John and Carol are gracious hosts, and invite me to spend the night at their farm. Always the insomniac traveller, I was grateful that the guest bedroom was well stocked with reading material. As I read on into the still depths of this Manitoba night, I stumbled on to an essay by William Jordan of the University of Wisconsin, one of the restoration movement's theorists. As I read my interest mounted, to the point that I realized my mouth had gone dry and my palms were slightly damp. I was in the rare grip of a compelling idea that, to paraphrase Jordan, is as follows:

* Human beings have always felt a certain tension between themselves and nature.
* One of the prime functions of culture is to mediate this tension.
* Culture cannot mediate the tension between humans and nature if it operates only in the literal dimension. Successful mediation occurs when the realm of imagination, through the technologies of ritual and the performing arts, is involved.
* To the extent this is true, the real root of modern alienation from nature is not science or technology, but a loss of belief in the efficacy of ritual.
* The re-establishment of a satisfactory human bond with nature will depend on the development of the technologies of imagination and ritual.
* The process of ecological restoration provides an ideal framework for the development of a system of rituals for working out the terms of our relationship with nature.

This was grand logic: seditious, unsupported, inflammatory, but compellingly correct. Jordan doesn't ask us to back up, to some pretechnological state, nor does he ask us, as so many do, to adopt the rituals and worldviews of another culture. He asks us to go forward into our own culture, embracing both science and ritual, to devise a new paradigm for our relationship with nature. Midway through this landscape re-exploration, I knew I had sighted a major landmark.

The cycle of ecological restoration is complex and multifaceted, and Jordan sees it as a kind of reprise of human history: seed collection recapitulates hunting and gathering, as seeding does agriculture. Research activities are a capsule of science, archival work a microcosm of history. By doing restoration, we participate in and review these human advancements, their effects on us, and their effects on our landscapes. But the culmination of the restoration cycle is a humble new-old assemblage of plants, in some schoolyard, old right-of-way, or park. Restoration is a small piece of our ecological patrimony regained, a grain of sand recaptured. The attempt to actually *reproduce* an ecosystem may be an act of incredible arrogance, and yet it is also an act of profound faith in the future of nature and human beings.

I left the Morgan's farm the next day, reluctantly, and John drove me back to Winnipeg. Before going to the airport, I asked if we could stop to tour the Living Prairie once more, so I could fix that tiny shred of tallgrass landscape in my mind. As we walked the narrow footpaths, John offered an insight about his occupation. "You know, I'd like to think that the work I do is not so special or unique, that there will come a day when ecological restoration is recognized as a common profession, like medicine or law. It's good, honest work, and there's a lot of it to do." Good work. As we head back to the parking lot, I take a last look back across the prairie; lilac-tinted clouds stack up in the massive sky overhead, and the wind works its way through the tallgrass. To re-create this, or even to fail in the attempt, is certainly good work.

From *Landscapes of the Interior: Re-Explorations of Nature and the Human Spirit* by Don Gayton (Gabriola Island, BC: New Society, 1996).

✳ LANCE M. FOSTER, 1960– ✳

Lance M. Foster is an enrolled member of the Iowa Tribe of Kansas and Nebraska, a member of the Bear Clan, and a descendant of White Cloud, an Ioway chief. Raised in Helena, Montana, he has earned graduate degrees in anthropology and landscape architecture from Iowa State University, and attended the Institute of American Indian Arts in Santa Fe, New Mexico. He has served as a consultant on cultural sites for state archaeology and historical agencies in Iowa and Minnesota, as well as working on heritage management for the National Park Service, collaborating with many tribes and Native American groups. He has been active in language revitalization and cultural preservation efforts in his tribe, and has written numerous articles, essays, and the recent book, *The Indians of Iowa* (2009). In this excerpted essay, Foster explores the history of his tribe's relationship to *Tanji*, the prairie, and his own quest "to discover the sleeping form of the Iowa past under the monotonous blanket of the present" (178).

✳ ✳ ✳

Tanji na Che
Recovering the Landscape of the Ioway

Baxoje min ke. I am an Ioway, a member of the Iowa Tribe of Kansas and Nebraska, located near White Cloud, in northeast Kansas. The state of Iowa owes its name to our ancestors, who lived in the lands that became Iowa before their removal to Kansas through the Treaty of 1836. We are legally known as the Iowa, although we pronounce it "Ioway."

When I came to Iowa to attend graduate school, I hoped to see a land that breathed the stories of old times. Maybe I expected too much, since I was raised in Montana, where so much of the land is as the Creator made it. When I came to Iowa, I could taste the chemicals that ensured the monoculture of corn and soybeans that swept from horizon to horizon. Iowa, I learned, is the state that has been most transformed from its primeval condition. Almost all its wetlands have been drained, its forests decimated, and its animals eradicated, and less than 0.1 percent of its prairies is left. Thus the pastoral countryside of Iowa is not native, and my ancestors would hardly recognize it.

However, I was determined to understand this place, this Iowa I found, and to discover the sleeping form of the Iowa past under the monotonous blanket of the present. Our landscape and culture was like a great and graceful earthen vessel, molded over ages by loving hands, that was shattered by an angry visitor into a thousand pieces. My time has been one of searching for the scattered shards of memory and place. Some shards have been recovered and assembled by specialists in new ways—as literature, archaeology, history, or mythology. But I believe that the essence is still there. Sitting Bull, the Hunkpapa Lakota, once said that whatever you have lost, if you go back and look hard enough, you will find it. Some shards may be disguised as an old story; some, buried at an archaeological site; some, hidden in an archaic phrase. Some pieces may even be in the song of a meadowlark or in the swirl of snow that comes in as you shut the door. As my uncle Herman Bear Comes Out, of the Northern Cheyenne, once told me, all the answers are still out there, in the Land, where they came from in the first place. So I have come to discover *Tanji*, the prairie.

[. . .]

Reimagining the Prairie: *Tanji*

The tallgrass prairie was not a featureless expanse of grass to the Ioway. Here and there, features natural and cultural marked a land that was familiar and comforting. The lands were under the great dome of the Sky, which was conceived of as the roof of the Earth-Lodge, with the Island-Earth as its floor and the Four Winds as the Protectors and Earth-anchors.[1] The Directions were named by the Winds: *Byuwahu*, the East, "The Sun Comes from There"; *Mansje*, the South, "Warmth"; *Byuwari*, the West, "The Sun Goes There"; *Umeri*, the North, "The Cold Side." The South Wind was considered beneficient, while the rare Northeast Wind brought killing blizzards and was thought to be maleficient.

The other great Powers, besides the Winds, were the Day, the Night, and the Winter. Although the Big Male Winter, the time of glaciers, had been killed by Mischinye, the Great Hare, long ago, the land still bore the scars of that glacial age, and the Small Female Winter returned every year with her bitter winds.

The cycle of the year was marked by events occurring during the months. There were variations, but the cycle collected by anthropologist Alanson Skinner is the most widely known today.[2] The new year began in March, with the greening of the land, encouraged by the singing of the frogs; thus March was called the Frog's Moon (*Pesge etawe*). April (*Mak'anye*, Cultivating Moon) and May (*Bi wa'un nyinge*, Nothing-to-Do Moon) marked the beginning and end of preparing the fields for corn. The changes in the appearance of the landscape as summer approached were noted in the names of June (*Wixra shuwe*, Little Flowers) and July (*Wixra xanje*, Big Flowers). Late summer, fall, and winter were the time of hunting. August was the Buffalo Rutting Moon (*Che Kiruxe*); September, the Frost in Animal Beds Moon (*Doxina Gremina*). The bugling of the bull elk as he gathered his harem gave October its name, the Elk Whistling Moon (*Huma Yochinya*). November was the Deer Rutting Moon (*Ta Kiruxe*); December, the Raccoon Rutting Moon (*Mingke Kiruxe*). January and February were, respectively, Little Bear Jumping (*Munje'tawe Shuweinye*) and Big Bear Jumping (*Munje'tawe Shuweinye Xanje*). The names reflect the growing size of bear cubs as they emerge from their dens on the warmer days of winter.

The summer was also a time of the Thunderers and tornadoes. Tornadoes were called *t'a-t'anwe*, which described the incessant and erratic jumping movement they made while skipping along their path of destruction. I have heard some nations call them the Finger of God. The Thunderers traveled the river valleys during fierce storms, skimming along the bluffs and stopping to rest on mounds that marked their routes or on cedar trees, which they had great affinity for. Sometimes the Thunderers fought the Underwater Spirits, the Ischexi. The allies of the Thunder in their status as Guardians of the Sacred were the Cedar and the Snakes.

[. . .]

Tanji na Che   271

The wetlands and marshes that covered roughly 20 percent of the present state of Iowa were places left from earliest creation, remnants of the time of the Flooded-Earth from which Trickster had created the present Earth-Island, with the assistance of Muskrat. Muskrat was instructed to dive to the bottom of the Flood and to bring up mud, which Trickster spread around to make this Earth-Island. The Marsh (*Jegixe*), with its swarm of invisible and visible life, was the still-powerful remnant of Creation, and the prairie pothole region continued to be a part of the process of ongoing Creation. In fact, some of the burial mounds built high on river bluffs and prairie knolls have rich marsh earth added to them. This may be symbolic of the connection between the rebirth of the dead through the mound and the energy of creation, issuing from the primal marsh.

During the fasting or vision quest, other sacred places included canyons, bluffs, caves, and isolated high hills and rock formations and glacial erratics, some with inscribed petroglyphs. Over the years, I have heard elders tell stories of haunted and mysterious places on the prairies of northwestern Iowa and southwestern Minnesota. Oceyedan Mound was avoided by all tribes as the home of a race of small and angry spirits. Pilot Knob was said to be haunted by the spirits of Ioway and Sioux warriors who died there. Spirit Lake was the home of one of the Ischexi, the Underwater Deities that ruled the Underworld. The Loess Hills were said to occasionally have a "Sun Bridge," which was a jumping-off place for the dead in their travel to the west and the Land of the Dead.[3]

Bordering the Big Sioux River in northwestern Iowa and southeastern South Dakota, Blood Run was a large ceremonial site at which oral tradition says many tribes—including Ioway, Omaha, Ponca, Otoe, Arikara, and Dakota—camped together. Tradition also holds that this was where the Arikara introduced the intertribal adoption ceremony known as the Pipe Dance to the other tribes.[4]

The great rivers were connected by overland trails through the upland prairie, with villages and sacred places at their meeting points. The Ioway chief No Heart of Fear prepared a map in 1837 to defend the Ioway's claim to aboriginal possession of the lands that would become Iowa.[5] The map reveals the rivers, villages, trails, and migration routes of the Ioway across their homelands.[6] In marshy country, the glacial eskers and ridges provided relatively dry travel for people and buffalo. The Ioways were known as great walkers, and many of their names contain the word *mányi*, which means "to walk or go along."[7] When the French explorer Nicolas Perrot first met them in 1685, he noted that they often killed deer or buffalo while running after them.[8]

Moreover, the Ioway language reveals a great familiarity with overland travel through the prairies and, at the same time, the difficulties encountered from the varying vegetation. For example, *nathdáge* means "to leave a path while walking through tall grass." *Wahánré* means "to walk under and through brush or tallgrass while pushing it aside." *Wathánje mányi* means "to walk through brush or a thicket, shaking it," and *waxrán mányi* means "to walk through brush where there is no path."[9]

Most of the great trails in Iowa that were later used by the Sauk and the Europeans were made by the Ioway and their ancestors. One such trail that is recorded in historic French maps led from the Ioway villages on the Upper Iowa and Mississippi Rivers, across the marshy glacial plain of northern Iowa, to their villages on Okoboji and Spirit Lakes, and thence through the Loess Hills to their village on the Missouri River.[10] Although a detailed study has not been made of this route, I strongly suspect that it followed glacial landforms such as eskers, which provided relatively dry passage across marshy terrain.

The prairie in a very real sense defined the Ioway as a people, and they, in turn, helped define and maintain the prairie. By contrast, when a group of Huron and Ottawa fled from Iroquois aggression and sought refuge in the prairie lands of the welcoming Ioway, they felt vulnerable and dismayed by the endless open expanses. They were a woodland people and feared the overwhelming spaces as much as the European settlers did two centuries later. The Huron and Ottawa returned to the eastern forests and meadows. They preferred to face the wrath of the Iroquois rather than the loneliness of the treeless lands.[11]

The magnet that drew the Ioway away from their rivers and bluffs to the open prairie was *Che*, the Great One, the Buffalo, although at the same time increasing pressures from other native populations to the north and east began pressing the Ioway to the west and south.

Che: The Great One of the Prairies

Today, most Americans associate the buffalo with the Great Plains, rather than with the tallgrass prairies of Iowa and Illinois. But buffalo were definitely important to the Ioway and the tallgrass prairie, too, as proved by archaeological evidence, historical records, and the traditions and language of the Ioway. The Ioway dependence on the buffalo seems to be as old as the tribe itself.

At the archaeological sites of the Ioway and our Oneota ancestors are buffalo bones as well as artifacts with buffalo motifs.[12] Archaeologists have also found evidence of long-range communal bison hunts from Iowa into Kansas.[13] During dry periods, when the prairie expanded at the expense of forested areas, the buffalo range also expanded. At the same time, Oneota settlements spread and prospered.[14] With their surplus of buffalo, the Ioway were suppliers of buffalo robes to their relatives the Winnebago (along with that other prairie resource, pipestone from Minnesota).[15] Sites of trading centers have been located along the Mississippi, with hides, artifacts, and dried meat.[16]

In the historic period, trade in bison hides and pipestone between the Ioway and the Winnebago was observed by the French missionary Father André in 1676.[17] Even as late as the early nineteenth century, Lieutenant Stephen Kearny recorded a herd of 5,000 buffalo on the upper Raccoon River.[18]

Culturally, the interdependence of buffalo and Ioway shows up in many more ways. The Ioway had two principal clans, the Bear and the Buffalo, and our stories tell that when the clans of the eastern woodlands, under the leadership of the Bear Clan, met with the Buffalo Clan on the western prairies, the Clans formed one People. While the Bear Clan led the tribe in the fall and winter, the Buffalo Clan led in the spring and summer. The Buffalo Clan name, *Nawo Dayi*, means "Road Maker," referring to the wide buffalo trails worn across the prairies, which were used as roads when moving camp.[19] (These buffalo roads were converted by later Euro-American settlers into wagon roads.)[20]

Buffalo Clan traditions also tell how buffalo wallows collected water, which was used by other animals for drinking. Medicinal plants, like the

LANCE M. FOSTER

Male Buffalo Bellow Plant, also grew by the wallows, and the buffalo some-times were seen using these wallows for healing their sick. In 1851, Topo-muk, an Ioway Buffalo Doctor, said:

Sometimes when a Buffalo is shot in the breast, he becomes very sick, staggers from one side to the other . . . then two other Buffalo get one on each side, and support it and urge it along, perhaps, and helping it until they get it away. When they doctor it . . . they have a basin of earth [wallow] in which they place the water. This they use in blowing on the wound. If there is no water nigh, one runs off to a stream and drinks water and runs back again and pours it out of his mouth into the basin. They bring the weeds also on their horns and using the water in the basin they blow as we do on the wound.[21]

In hunting buffalo, the Ioway occasionally used the fire-surround method, in which hunters encircled a group of buffalo and fired the prairies around the herd, leaving some areas unburned. At those points, the hunters lay in am-bush.[22] Although I have not yet found any traditions of systematically man-aging the prairie for better game range through fire, Duke Paul Wilhelm of Württemberg did write in 1823 about the intentional firing of prairies by the Ioway: "In the autumn the Indians set the dry prairie on fire and so lay whole forests in ashes, the wind driving the fire on until it reaches the river."[23]

As the botanists tell us today, there was a constant war on the prairie margins between the grasses and the advancing woodlands of bur oak and hawthorn. The firing of prairie grasses and margins, favorable to the deep-rooted grass communities and less favorable to the woody plant communi-ties, ensured that plentiful pastures near favored campsites would encourage the buffalo to remain in the area. Buffalo are notoriously fickle, and their movements are hard to predict. For such a large animal, they are amazing in their ability to be inconspicuous on the landscape, and the scouts often had difficulty finding the herds. Buffalo also stampede, and are notoriously destructive when they do, so precautions had to be taken to keep hunters in line and prevent them from frightening the animals. In pre-horse days, the Ioway, like other Indians, often stalked buffalo under decoy wolfskins because wolves commonly prowled the margins of the herds and were paid little attention as long as they did not get too close.

In 1859, Robert White Cloud, an Ioway, was interviewed by the ethnologist Lewis Henry Morgan: "He says they also regard the buffalo as a god. . . . He said the Indian believed the buffalo after being killed and eaten by him had power to cover his bones again with flesh and come to life again, and that he did thus come to life again after being killed."[24]

While the buffalo was the most vital life-giver to the Ioway, many other beings' lives were entwined in the sacred nest of grass, *Tanji*. The elk (*Huma*) supplied meat and hides, and his shoulder blade was used for hoeing corn and beans. The antlers were worked into hide scrapers and quirts. Elk Medicine was powerful for attracting the women, much as when the bull elk bugled, and the cow elk came running. October, as I have said, was known as the Elk Whistling Moon (*Huma Yochinya*).

The wolf (*Shunta*) was loved and respected, not feared. The wolf was the scout, the one that kept watch. My ancestor Mahaska wore a white wolfskin as a cape, which was said to give him the power of invisibility.

White-tailed deer (*Ta*) were hunted along the riverbanks and were a staple. They were so important that special hunting medicines were sometimes used. The skins of unborn fawns were used in the most sacred ceremonies, such as during the tattooing of young girls of high status to bless and protect them.[25] Archaeological evidence indicates greater dependence on buffalo in western prairie sites, and more on deer in eastern wooded sites.[26]

Although the black bear (*Munje*) was not a regular visitor to the prairies, his bad-tempered relative the grizzly (*Mahto*) was occasionally seen. He had a fierce reputation, but *Mahto* also pitied the people and gave them the Grizzly Bear Society, which was complementary to the Buffalo Doctor Society in the healing arts.

There are so many animals of the prairie that were vital in the lives and minds of the people. The meadowlark (*Postinla*) sang about the true nature of approaching strangers. In its nest, the mouse (*Hindunye*) hid ground beans, which might prove invaluable survival food; it was this mouse that raised one of the Hero Twins, who rid the world of monsters and made it safe for people. The gopher (*Mayinye*) shot dangerous medicine at the unwary from its burrows.

There are many more stories that reveal the richness of the prairie landscape and the Ioway intimacy with it. The regathering of them, and their attachment and embodiment in living form and place, is a sacred journey.

Last Thoughts: New Flesh on Dry Bones

In my years in Iowa, looking beneath the blanket of modern agriculture, I have found many connections between the prairie and Ioway customs. For example, the next time you go to a powwow, watch for the Grass Dancers. The yarn and fringe you see hanging from their regalia is a reminder of the old times, when warriors stuffed bunches of tallgrass into their garters and belts or sashes and danced vigorously with low, sweeping motions of the feet, to flatten the prairie grasses and make the dance grounds ready. Among the Ioway, this was a part of the *Herushka*, or *Iroshka*, the primary warriors' society. The warrior dances of the *Iroshka* survive in the deep-voiced Southern Drums of groups like Yellowhammer and in the searching and dignified steps of the Straight Dance.

Whenever I see the Grass Dancers and the Straight Dancers and hear those Southern Drums, I think about the old tallgrass prairies stretching out like wave upon wave of an endless ocean. And when I think about those prairies, I think about the Old Folks and about the buffalo.

So in April 1997, I decided to visit Walnut Creek (now Neal Smith) National Wildlife Refuge near Prairie City, southeast of Des Moines, because I had heard that a small herd of buffalo had been brought there to graze the restored prairie. I went on a Monday, but to my disappointment, I found that the Prairie Learning Center was not open on Mondays (or Sundays, a word to the wise). Because it was not open, however, I had the refuge almost to myself.

Ironically, the rolling slopes were scarred by the machines being used to replant prairie. Another area was so completely burned that it seemed carved from coal, even though I knew that fire is a necessary tool in establishing and maintaining the health of the prairie. But it was amazing to see the amount and variety of life that was returning. I saw delicately colored birds that I did not recognize, and even a small lizard scooting across the road.

I was there to see the buffalo, though, which are kept in 800 enclosed acres, with a so-called buffalo-proof fence, so I took the road that warned of "Buffalo on the Road." Well, the road went right through the enclosure, but I saw no buffalo. I looked and I looked. I wondered if somehow I was not worthy. While the terrain was rolling and broken, and some portions were wooded, it still amazed me that such large animals could remain so well hidden. I drove through once, turned around, and drove through again. Still nothing.

I was disappointed, and I drove down by the bur oaks, down below the charred slopes. I sat there for a while, listening to the tree frogs peeping near Walnut Creek, and enjoying the smell of growing sweetgrass somewhere nearby.

Then I thought about how the Old Ones believed that the buffalo could come alive again, by clothing its desiccated bones with a mantle of living flesh. And the thought came to me that this prairie was coming alive in the same way. It had once been killed and the flesh torn away, the soil with its coneflowers and Indian grass, so that the prairie was invisible, nothing left but a sheath of hybrid corn and scattered patches of weeds and scrub. But now the skeleton was again becoming clothed in its old flesh of prairie life — plants, birds, and animals. Like the old stories of the buffalo returning to life, this prairie was returning to life. And somewhere, just over that ridge, unseen as in a great cavern beneath the earth, the buffalo were waiting.

From *Recovering the Prairie*, edited by Robert F. Sayre (Madison: University of Wisconsin Press, 1999).

NOTES

1. Lance Foster, "The Ioway and the Lost Landscape of Southeast Iowa," *Journal of the Iowa Archeological Society* 43 (1996): 1–5.

2. Alanson Skinner, "Ethnology of the Ioway Indians," *Bulletin of the Public Museum of the City of Milwaukee* 5 (1926): 181–354.

3. Lance Foster, *Mayan Jegi: The Ioway Indians and the Lost Landscape of Iowa* (Ames: Native Nations Press, 1997), 53.

4. Foster, *Mayan Jegi*, 78.

5. For further interpretation of the map, see Mark Warhus, *Another America: Native American Maps and the History of Our Land* (New York: St. Martin's Press, 1997), 34–43.

6. William Green, "Ioway Cartography, Ethnohistory, and Archaeology," MS in preparation.

7. Foster, "Ioway and the Lost Landscape," 2.

8. Martha Royce Blaine, *The Ioway Indians* (Norman: University of Oklahoma Press, 1979), 20.

9. Foster, "Ioway and the Lost Landscape," 2.

10. Mildred Mott Wedel, "Peering at the Ioway Indians through the Mist of Time: 1650–circa 1700," *Journal of the Iowa Archaeological Society* 33 (1986): 50–77.

11. Mildred Mott, "The Relation of Historic Indian Tribes to Acheological Manifestations in Iowa," *Iowa Journal of History and Politics* 36 (1938): 227–314.

12. Joseph A. Tiffany, "An Overview of Oneota Sites in Southeastern Iowa: A Perspective from the Ceramic Analysis of the Schmeiser Site, 13DM101, Des Moines County, Iowa," *Proceedings of the Iowa Academy of Sciences* 86 (1979): 89–101.

13. Richard Fishel, *Excavations at the Dixon Site (13WD8): Correctionville Phase Oneota in Northwest Iowa* (Iowa City: University of Iowa, Office of the State Archaeologist, 1995); Brad Logan, "Oneota Far West: The White Rock Phase" (Paper presented at the fifty-fourth annual meeting of the Plains Anthropological Society, Iowa City, 1996).

14. Guy Gibbon, "Cultural Dynamics and the Development of the Oneota Lifeway in Wisconsin," *American Antiquity* 37 (1972): 166–85.

15. Blaine, *Ioway Indians*, 17.

16. Dale R. Henning, "Oneota Evolution and Interactions: A View from the Wever Terrace, Southeast Iowa," in *Oneota Archaeology: Past, Present, and Future*, ed. William Green (Iowa City: University of Iowa, Office of the State Archaeologist, 1995), 65–88; Tiffany, "Overview of Oneota Sites."

17. Mott, "Relation of Historic Indian Tribes," 241.

18. Mildred Mott Wedel, "Indian Villages on the Upper Iowa River," *Palimpsest* 42 (1961): 586.

19. Foster, *Mayan Jegi*, 55.

20. Blaine, *Ioway Indians*, 1979, 22.

21. Quoted in Blaine, *Ioway Indians*, 225.

22. Tom McHugh, *The Time of the Buffalo* (Lincoln: University of Nebraska Press, 1972), 69–70.

23. Paul Wilhelm, Duke of Württemberg, *Travels in North America, 1822–1824*, trans. Robert Nitske, ed. Savoie Lottinville (Norman: University of Oklahoma Press, 1973), 307.

24. Leslie A. White, ed., *Lewis Henry Morgan: The Indian Journals, 1859–62* (Ann Arbor: University of Michigan Press, 1959), 70.

25. Skinner, "Ethnology of the Ioway Indians."

26. Lynn Marie Alex, *Exploring Iowa's Past: A Guide to Prehistoric Archaeology*, 2nd ed. (Iowa City: University of Iowa Press, forthcoming); Joseph A. Tiffany, "The Milford Site (13DK1): A Post-Contact Oneota Village in Northwest Iowa," *Plains Anthropologist* 42 (1993): 205–36.

THE
TWENTY-FIRST
CENTURY

✳ ✳ ✳

Born and raised in Burlington, Iowa—Aldo Leopold's hometown—Lisa Knopp received her education at the University of Iowa, Iowa Wesleyan College, Western Illinois University, and the University of Nebraska–Lincoln, where she earned a PhD in creative writing. Her widely published essays often explore the intersections of place, home, nature, and spirituality. She is the author of five collections of essays, most recently *What the River Carries: Encounters with the Mississippi, Missouri, and Platte* (2012) and a new book-in-progress titled *Like Salt or Love: Essays on Leaving Home*. Currently living in Lincoln, Nebraska, she is a professor in the English Department's Creative Nonfiction Writing Program at the University of Nebraska at Omaha. In this essay from *The Nature of Home* (2002), she examines the history of Arbor Day, and the legacy of those who "failed to see that," on the native prairies, "treelessness and barrenness are not equivalent states" (19).

Far Brought

On May 14, 2000, Ian and I strolled through the arboretum near Arbor Lodge, the Nebraska City mansion where Carrie and J. Sterling Morton and their four sons once lived. When I had visited this state historical park over a decade earlier, I had sought nineteenth-century Nebraska history. But this time I sought natural history: the lay of the land, evidence of what the land, flora, and fauna had been, what it is now.

On this May morning, the hilltop arboretum was cool, damp, dark, and rather foreign-looking. And for good reason. Some of the more than 260 species of trees and shrubs included in the arboretum are native to

Nebraska. But most species are not. *Ginkgo biloba*, Yulan magnolia, Norway spruce, sassafras, Chinese chestnut, Scotch pine, swamp white oak, golden rain tree, Japanese pagoda, tulip poplar, London plane, white pine and bald cypress are native to other parts of the continent or world and thus are exotics or aliens.

My son and I paused before a grove of American chestnuts. In 1904, imported Asian chestnuts hosted a fungus to which they were immune but the native chestnuts were not. Within forty years of its introduction, the blight had completely eradicated American chestnuts, once the dominant tree in northeastern deciduous forests. The Mortons' grove escaped the blight because of its location in a part of the continent where American chestnuts don't naturally grow and where people had not yet begun planting the blight-bearing imported chestnuts. But the Mortons' grove of white pines did not fare as well. Sterling had planted them in 1891 to prove that the eastern white pine, native to parts of eastern Canada, the northeastern United States, the Appalachian Mountains, and the Great Lakes region, could grow in Nebraska. This pine, accustomed to more moisture and less extreme temperatures, perished in the drought of 1937. Shortly thereafter, the grove was replanted. The stories of the groves of these two imported species contrasts with that of the native Osage orange hedgerow that the Mortons planted almost a century and a half ago. Four of these spiny trees persist to this day.

Ian and I completed our stroll at the Prairie Plants Garden, which was added to the arboretum in 1979 by the Nebraska Game and Parks Commission. I suspect that Sterling, who preferred what he called "far brought" over native species, would have approved such an inclusion in his arboretum since it reminded visitors what the land was like before he had improved it.

FROM THE MOMENT they arrived in Nebraska Territory, the Mortons had designs upon the land. In 1854 Carrie and Sterling claimed and purchased a quarter section of land just west of Nebraska City, on the highest ground in the area, now the site of Arbor Lodge. When facing the Missouri River just a few miles east of their claim, they saw bluffs covered with the western edge of the oak-hickory forest and flood plains wooded with native ashes, willows, box elders, and cottonwoods. Because the then untamed Missouri flooded frequently and fiercely, scouring the lowlands of saplings and drowning older trees, the Mortons saw a riparian forest that was forever young. When they turned their backs on the river, they saw grassland, broken by woodlands clumped near rivers and streams and in the lowlands. Once this grassland extended, more or less, from the Rockies into central Illinois, from southern Saskatchewan into north-central Texas. Once prairie was our continent's largest and most characteristic biome. When the Mortons settled in Nebraska, they saw prairie that was relatively healthy and intact. Periodic wildfires, droughts, the integrity of plant communities, and grazing by bison, pronghorns, mule deer, and elk kept the prairie safe from the encroachment of trees and other woody plants.

Apparently the Mortons did not consider the grasslands comforting or homelike. Sterling and Carrie spent their childhood and early adulthood in and around Detroit, Michigan, an area forested with aspens, beeches, birches, elms, maples, oaks, cedars, firs, hemlocks, white pines, and spruces. But their earliest memories were set in even more eastern forests: Sterling was born in Adams, New York, on April 22, 1832; Carolyn Joy French was born in Hallowell, Maine, on August 9, 1833. The bluffy, timbered Missouri River, that easternmost edge of Nebraska Territory, might have resembled the Mortons' geography of home. But the landscape that the Mortons saw when they faced west was entirely Other.

Because they broke the prairie on their claim in 1855, the Mortons knew something about prairie plants. In his 1871 address at the opening of the University of Nebraska, Sterling said, "One of the grandest of material labors is the reduction of untried lands to tillage" (Olson 157). By "grand," I do not know if he was referring to what he perceived to be the lofty task of converting any landscape to productive cropland or orchard or if he meant that the job was "grandly" arduous. Certainly, the latter was true. Those who used wooden or iron plows had to stop frequently to clean the sticky prairie soil off the moldboard. In *Where the Sky Began: Land of the Tallgrass Prairie*, John Madson writes that the roots of the bluestems and prairie clovers were so tough and wiry that sometimes they damaged plows and injured draft animals. And breaking the tallgrass prairie was noisy work. The sodbusters called the leadplant "prairie shoestring" because when its strong roots were cut by a plow, they popped like breaking shoestrings. Nor was one plowing enough to subdue the prairie. The aerial and subterranean parts of the dominant tall grasses—Indian grass, big bluestem, and switchgrass—are roughly equal, with the seed heads waving six to ten feet above the ground and roots burrowing as far into the earth. While the aerial parts of little bluestem rise only a couple feet, the roots plunge twice that far. The roots of prairie forbs are even more remarkable. The roots of members of the sunflower-aster family can extend eight to twelve feet. The fleshy taproot of the purple coneflower extends ten or more feet. Roots of the leadplant extend sixteen or more feet. In healthy, undisturbed prairie, the sod is so crowded with roots, that alien species—blue grass, leafy spurge, maple trees—can't gain a toehold. Because most of the prairie's biomass is beneath the earth's surface, the plants are safe from droughts, fires, harsh winters, hordes of insects, and big native grazers. Thus the Mortons and other sodbusters found that the tough, deep roots of prairie plants had to be broken again and again before the land was safe for corn and wheat, planted fence row to fence row.

Because the Mortons farmed and kept orchards, they knew something about the extraordinary fertility of prairie soil. But they probably didn't know what accounted for it. When the last ice sheets of the Pleistocene melted about twenty-five thousand years ago, vast stretches of dried mud remained. High winds carried the fine dust particles (*loess*, an Old German word meaning "light" or "loose") south into what is now Nebraska, Iowa, and western Missouri. Because glacial drift contains a larger variety and greater quantity of the soluble minerals that plants use for food, the soil formed from loess is superior to soil formed from native bedrock.

The fertility of "prairyerth" is also due, in part, to an abundance of humus, the dark, organic residues formed when bacteria and fungi break down dead plants, animals, and insects: plant food. Madson says that in forest soils, the twenty to fifty tons of humus per acre are concentrated on or near the earth's surface. In unbroken prairie soils as much as 250 tons of humus per acre is distributed from the surface to the subsoil. Prairie plants are well nourished from seed head to root tip.

Yet no matter how much the rich, fertile soil yielded, Morton failed to see that treelessness and barrenness are not equivalent states. No matter how much money Morton made from the hogs, corn, and fruits he raised on prairie soil, he spoke of and acted with hostility toward the native landscape. In the March 12, 1870, issue of the *Nebraska City News*, of which he was the editor, he wrote of the need to "battle against the timberless prairies." In his 1872 "Fruit Address" to the Nebraska State Horticultural Society, Morton proclaimed that his goal was to make Nebraska our nation's "best timbered state." Sterling's biographer, James C. Olson, writes that on Christmas 1876, Paul Morton gave his parents letterhead stationary that bore an engraving of their house, Arbor Lodge, in the upper left-hand corner. Stuck in Sterling's farm journal is a sheet of this stationery on which he had scribbled: "From a photograph of house taken by Dr. Smith of Nebraska City in summer of 1876. Had the Dr. set his camera in the same place twenty-two years before it would have been a picture of barren prairie so far as the eye could see without a tree in sight" (245).

Of course Morton wasn't the first to be distressed by the "timberlessness" of the plains. When the first white immigrants arrived in what is now Nebraska, 3 percent of the state was forested, or rather, 97 percent of the area was grassland—tallgrass prairie in the eastern third of the state; mixed grass in the Sandhills and south-central part of the state; short and mixed grasses in the Panhandle. This landscape didn't satisfy the overwhelming majority of the immigrants who were what Kansas historian James C. Malin called "Anglo-American forest men." Part of the baggage these newcomers brought with them from Europe or the eastern United States was, in Malin's words, the belief that "the presence of forests was natural and the absence of trees was an unmistakable sign of deficiency or abnormality of nature" (2). So, when immigrants, who rarely if ever had seen the sun rise or set on an unobstructed horizon, arrived on the Great Plains, they felt overwhelmed, frightened, or diminished by the open land and sky. Many of them were so fixated on the lack of trees that they could not see the grasses, the rich soil, the integrity of the prairie communities. Shortly after his arrival in Nebraska, Judge Edward R. Harden of the territorial supreme court wrote to his wife in Georgia that he was catching the first boat home and would never return to Nebraska Territory: "It is poor country no Timber, sickly, and out of the world and settled up with Savages" (27). One of those savages, Omaha chief Big Elk, told members of Major Stephen H. Long's 1819 expedition, "If I even thought your hearts bad enough to take this land, I would not fear it, as I know there is not wood enough on it for the use of the whites" (Rice B2).

Nor was Morton the first immigrant to dream of remaking the prairie. In "Women and the 'Mental Geography' of the Plains," Sandra L. Myres writes that many newcomers to the plains "believed that the real physical world which they saw about them could be transformed by an increased population and the application of modern science and technology" (42). The science and technology offered by Samuel Aughey, a land speculator, chairman of the Natural Sciences department at the University of Nebraska and first director of the University of Nebraska State Museum, was simple though arduous. Aughey believed that one had only to plow the grasslands to produce a more agreeable, more eastern landscape. In his 1880 *Sketches of the Physical Geography and Geology of Nebraska* Aughey asserted that by breaking "primitive" prairie soil, one could increase absorption of rain-fall, which in turn increased evaporation, which in turn increased rainfall. Aughey's views, popularized by Charles D. Wilbur, were repeated in the diaries and journals of diverse grassland settlers. For instance Elizabeth (Mrs. George) Custer wrote, "The cultivation of the ground, the planting of trees, and such causes, have materially modified some of the extraordinary exhibitions that we witnessed when Kansas was supposed to be the great American desert" (Myres 42).

Morton did not come to Nebraska Territory because he loved the land; rather, he came, according to his biographer, "for the express purpose of achieving fame or wealth, preferably both." Immediately upon his arrival Morton went to work on his dual goals of becoming rich and famous and of creating a home place that resembled the home he had left. He suc-ceeded on both counts. Among Morton's worldly accomplishments were: the founding of Nebraska's first newspaper; election to the territorial legis-lature within a year of his arrival and several times thereafter; appointment as secretary of the territory by President Buchanan in 1858; appointment as secretary and acting governor of the territory for a term running from 1858 to 1861; selection as the Democratic nominee for governor in 1866, 1882, 1884 and 1892, as well as to the U. S. Senate and House of Representatives (he lost each race); an attorney for several Chicago corporations; coauthor with Albert Watkins of the three-volume, posthumously published *Illustrated*

History of Nebraska; and appointment to the cabinet as U.S. secretary of agriculture during President Cleveland's second administration (1893–97). According to his biographer, Morton was "a conservative from a section of the country that seemed for a time to produce only radicals; a man, who, though virtually always in the minority, was ever a force to be considered." Morton supported free traders, the gold standard, the rights of labor, and the Confederacy. As a well-paid lobbyist-publicist for the Burlington and Missouri River Railroad, Morton opposed the Granger Movement, the farm organization that secured the passage of laws limiting railroad rates.

As Sterling's wealth and fame grew, so did Arbor Lodge, from a four-room cabin in 1855, to a thirty-room, neocolonial mansion at the time of his death on April 27, 1902. Sterling and Carrie's son, Joy, the founder of the Morton Salt Company, continued his parents' expansions and renovations. When Joy and his family donated the property, which had served as their summer get-away, to the state of Nebraska in 1923, Arbor Lodge included fifty-two rooms.

Sterling was also successful in his goal of re-creating his estate in the image of his old home in the East. Immediately upon arriving on his quarter section, he began planting trees. Dr. George L. Miller, founder and editor of the Omaha *Herald* (now, the *Omaha-World Herald*), wrote in the May 13, 1868, issue of that paper of his friend's landscaping activities: "The farm itself bears the most gratifying evidence of Mr. Morton's early appreciation of what was needed to make it yield the solid as well as the luxurious comforts of Home. His orchards, numbering hundreds of apple trees, remind one of those a century old in the East. . . . All around that splendid farm may be seen proof of the constancy with which Mr. Morton has given direction to fruit and tree culture. He is constantly sticking the 'cuttings' or roots or fruit of forest trees into the ground" (Olson 156).

Morton was not content to transplant native cottonwoods from the riverbanks and bottoms. When County Commissioner Oliver Stevenson brought seed potatoes back from a trip to Pennsylvania, Morton wrote in the February 2, 1867, issue of the *Nebraska City News*: "If every Nebraskan who visits the East would look after matters of this kind, and emulate Mr. Stevenson by introducing new and improved kinds of grains, vegetables, and fruits the whole State would be much benefited" (Olson 158). Morton hoped that tobacco and hemp would become "staple products of Nebraska." Likewise he introduced Suffolk pigs into Otoe County, adding imported fauna to the imported flora. When Carrie returned from a trip to Pike's Peak, she brought an Engelmann spruce seedling in a tomato can to add to the arbor. Sterling's only opposition to alien species was that "far brought" trees didn't grow as easily as native trees, which added ammunition to the arguments of those who believed that apples and pears could not be grown on Nebraska soil.

Nor was Morton content to forest just his own property. On January 4, 1872, he offered a resolution to the State Board of Agriculture, of which he was a member, to establish a tree-planting holiday. The board accepted the proposal and offered one hundred dollars to the Nebraska county agricultural society and twenty-five dollars' worth of books on farming to the individual Nebraskan who properly planted the largest number of trees on the holiday. Two board members wanted to call the holiday Sylvan Day, which Morton rejected since it only referred to forest trees. Arbor, on the other hand, referred to all trees. The name Arbor Day was unanimously accepted.

The next step for Morton in reconstituting the landscape was to compel individual Nebraskans to plant trees. In his famous "Fruit Address" before the State Horticultural Society, also on January 4, 1872, he linked the planting of trees with home, culture, and morality:

There is comfort in a good orchard, in that it makes the new home more like the "old home in the East," and with its thrifty growth and large luscious fruits, sows contentment in the mind of a family as the clouds scatter the rain. Orchards are missionaries of culture and refinement. They make the people among whom they grow a better and more thoughtful people. If every farmer in Nebraska will plant out and cultivate an orchard and a flower garden, together with a few forest trees, this will become mentally and morally the best agricultural State, the grandest community of producers in the American Union. . . . If I had the power I would compel every man in the State who had a home of his own, to plant out and cultivate fruit trees. (Olson 163)

Such sentimentality and boosterism were readily accepted and repeated. On Nebraska's first Arbor Day, Lincoln's *Daily State Journal* charged every

LISA KNOPP

property-owning Lincolnite to "put out a tree or two, if not more, with his own hands, if necessary." Moreover, the editor wished that "business of all kinds could be suspended" so that every person "capable of making a hole in the ground" could plant something — tree, shrub, or even a rose bush — in recognition "that this is a treeless country, and that what nature has left unfinished, the enterprise of Nebraskans will complete. Nebraska only lacks trees to be the Elysium of the continent" (Rice B2).

On April 10, 1872, the first Arbor Day was celebrated. Much to Morton's disappointment, the eight hundred trees that he planned to plant on his farm didn't arrive in time. James S. Bishop, a farmer who lived four miles southwest of Lincoln, planted ten thousand trees — cottonwoods, soft maples, Lombardy poplars, box elders, and yellow willows. He won the state fair premium for the finest grove of cultivated timber in the state. On the first Arbor Day, Nebraskans planted more than one million trees, though the editor of the *Nebraska Farmer* claimed that twelve million was a more accurate count. On that first Arbor Day, Morton was positively effusive. In a letter to the *Omaha Daily Herald*, he wrote: "Then what infinite beauty and loveliness we can add to the pleasant plains of Nebraska by planting forest and fruit trees upon every swell of their voluptuous undulations and, in another short decade, make her the Orchard of the Union, the Sylvan queen of the Republic" (Olson 165).

Arbor Day sponsors wanted a celebration every year and a forest for every farm. In 1874, the legislature set aside the second Wednesday of every April as Arbor Day, a legal holiday. In 1874, Morton's sometimes friend, sometimes foe, and fellow orchardist, Governor Robert W. Furnas, issued a proclamation encouraging Nebraskans to celebrate the new holiday. That Arbor Day, Morton noted in his diary that he set out two hundred elms, ashes, and lindens on his farm.

In 1885, the Nebraska State Legislature moved the date of Arbor Day to April 22, Morton's birthday. (Ironically, Earth Day, first celebrated in 1970 coincided with the celebration of Arbor Day.) By 1892, forty-one of the forty-two United States (Delaware being the one exception) as well as several other countries celebrated the tree-planting holiday. In 1895 the state legislature passed a resolution that Nebraska be known as "the Tree Planter's State," a nickname that persisted until 1945 when Nebraska was redubbed "the Cornhusker State." By 1900, seven hundred million trees had been planted in the Midwest alone. In 1972, the centennial of the first Arbor Day, Nebraskans planted seven million trees. Now the holiday is celebrated in every state on a date or dates established by each state's legislature. Now the green road signs welcoming travelers to Nebraska identify our state as the "Home of Arbor Day," asking us to identify ourselves by a movement that sought to make Nebraska look like something other than Nebraska.

The movement to forest the grasslands has had almost a century and a half of political backing in the form of cash incentives from both the state and federal governments. In 1869 the state legislature exempted one hundred dollars' worth of property for every acre of trees planted and maintained. Since most farmers paid their entire tax bill by planting trees, the incentive became too costly for the government and the law was repealed in 1877. Under the federal Timber Culture Act of 1873, anyone who qualified for a homestead could acquire an additional quarter section by planting forty acres "of the same trees" and tending them for ten years. Later the law was revised so that one had only to maintain ten acres' worth of trees for eight years.

The Kinkaid Homestead Act of 1904, sponsored by Congressman Moses P. Kincaid of O'Neill, Nebraska, provided 640 acres for homesteaders in northwestern Nebraska and free tree seedlings to any homesteaders or farmers living west of the hundredth meridian. In 1924 this legislation was replaced by the federal Clarke-McNary Act, by which two million trees per year were sold to Nebraskans at cost for farmstead and feedlot windbreaks. Through this program more than sixty million trees were planted in Nebraska. New Deal projects planted many thousands of miles of shelterbelts in the Midwest and the Great Plains to reduce erosion. In 1902, University of Nebraska botanist Charles Edwin Bessey persuaded President Theodore Roosevelt to provide 206,028 acres of land in the Sandhills, an area dominated by mixed-grass prairie, for what would become the largest, human-made forest in the United States: the Nebraska National Forest.

Planting the prairie with pine forests, orchards, windbreaks, and arbors may seem benign compared to turning it under and either asphalting it or replanting it with nonnative grasses and crops. Yet each of these activities were and are inspired by the same mind set: seeing the land and those who dwell upon it as a commodity or resource. According to this philosophy, land that lacks immediate and practical use is without value and can be remade according to the owner's desires—desires that are usually inspired by economic self-interest, such as Morton's. And so forests continue to be leveled, wetlands drained and filled, deserts forced to bloom like a rose, and the grasslands broken.

To the Anglo-American forest people, the planting of almost any tree was and is desirable, since it converts the ugly and barren into the beautiful and productive. Yet tree planting can have disastrous consequences when the species planted is inappropriate for a place. Alien plant species steal moisture, nutrients, and sunlight from native species. In the absence of natural predators, "far brought" species have no checks on their ability to reproduce and can quickly overtake a habitat. Since ecosystems are comprised of complex, intricate interdependencies, the insertion of an alien species into an exquisitely balanced ecosystem or the displacement of a native species from its exquisitely balanced belonging-place affects countless other organisms, in great or small ways.

The Russian olive, a native of southeastern Europe and western Asia, was introduced in the United States in the late 1800s and widely planted as a windbreak. Since then, this aggressive species has invaded riparian areas that were historically open and has lowered the water table. Those riparian areas dominated by dense stands of Russian olives do not host the same rich diversity of bird species (piping plovers, least terns, and sandhill and whooping cranes, to name but a few) found on open sandbars, fields, and prairies dominated by native plant species. Likewise, the Siberian elm, native to northern China, eastern Siberia, Manchuria, and Korea, was introduced in the United States because of its fast growth and ability to withstand summer droughts and cold winters. Consequently, this species was the most widely planted shelterbelt tree in the 1930s and the dominant species in Nebraska's urban forests. This elm produces many hundreds of samara (one-seeded, winged fruit) per tree. The wind-disseminated seeds sprout quickly and easily, forming dense thickets of seedlings on disturbed prairie, making it even more unlikely that the land will return to a vigorous, healthy state. In short, tree planting is desirable only if one selects species that preserve the integrity and stability of the native ecosystem. In some cases, the best way to contribute to the well-being of an ecosystem is not by planting trees but by removing "weed" trees.

I do not hold Morton, Furnas, Kinkaid, Bessey, and the other forest people accountable for what they could not have known. Since the prairie was one of the last biomes to be studied, there was little information to challenge the cultural stereotype of prairie as a deficient, monotonous wasteland. Nor could they have known the value, nay, the necessity of genetic, species, and ecosystem diversity. Wallace Stegner writes that most people in new environments were and are driven by "the compulsion to impose themselves and their needs, their old habits and new crops on new earth. They don't look to see what the new earth is doing naturally; they don't listen to its voice."

But I do hold accountable Morton and anyone else who refuses to become acquainted with the place they call home and who fights so hard against what is natural and right for a place. Nothing in Morton's writings indicates to me that he grew to appreciate the economy, productivity, diversity, and complexity of the tallgrass prairie that once covered eastern Nebraska, or the mixed-grass prairie that he encountered on his political and business trips into central Nebraska. Nor have I found evidence that he grew to see the beauty of the grasslands. No accounts of bronze-purple stalks of big bluestem or the golden plumes of Indian grass nodding in the wind. No accounts of finding pleasure and solace in the seam where sky and land meet. No mention of the leadplant with its silver-green leaflets and its cones of purple blossoms and gold stamens or the abundance of orchids (lady's tresses, prairie fringed, showy orchis, bracted) or the scent of prairie roses or the brilliant red-orange of the Illinois bundleflower. It appears to me that the high spot near the Missouri River where Morton lived, worked, and dreamed remained for him an estranging-place. In his efforts to fill the emptiness that he projected onto the land with what did not belong there, he squandered his time, energy, intelligence, and prosperity. If Morton had not had such contrary designs on the land, he might have experienced the freedom and discipline that comes from living in and with nature.

In one respect, Morton and the other forest people were a magnificent failure. Despite their tree-planting zeal, despite Morton's legacy of 128 years of tree-planting incentives, at the turn of the millennium, only about 3 percent of Nebraska is forested—the same percentage as was forested a couple of hundred years ago. What has changed, however, is where one is likely to find trees. The native deciduous forest on the west bank of the Missouri has been cleared for agriculture and the expansion of such Missouri River cities as Omaha, Bellevue, Papillion, Plattsmouth, and Nebraska City. Native red cedars have moved out of the river valleys and onto range and woodlands. Dams and reservoirs have reduced floodplain forests; yet flood control has permitted trees to clog rivers and streams, robbing sandhill cranes and migrating waterfowl of their preferred habitat. At the same time, tree planting in cities— the only tree-planting efforts on the Great Plains that I support, since trees reduce the heat island effect, clean the air, and create the rather sheltered, rather private, outdoor enclosures that we want near our living spaces— is far below what it should be. According to Assistant Nebraska State Forester Dave Mooter, Nebraska's urban forest is less than 50 percent stocked.

But in another respect, Morton and the other forest people were a magnificent success. A recent *National Wildlife* report states that only .2 percent of the prairie remains, making it the rarest and most fragmented ecosystem in the United States, the one in gravest danger of disappearing altogether. John Madson observes that "[t]oday, it is easier to find virgin groves of redwoods than virgin stands of tallgrass prairie." Even when one does find a stand of never-broken prairie, it is usually but a fenced-off plot, a museum piece, too tiny and isolated to accommodate those species with large territorial needs, too tiny and isolated to even suggest its former range and glory. Now we can only imagine what once lay outside the Mortons' back door—dark bison moving through bright, seemingly endless acres of big bluestem, golden sunflowers, white asters, and purple gayfeather.

From *The Nature of Home: A Lexicon and Essays* by Lisa Knopp (Lincoln: University of Nebraska Press, 2002).

WORKS CITED

Harden, Randolph. "Letters from Pioneer Nebraska by Edward Randolph Harden." *Nebraska History* 27 (1946): 18–46.
Madson, John. *Where the Sky Began: Land of the Tallgrass Prairie* (Boston: Houghton Mifflin, 1982).

Malin, James C. *The Grassland of North America: Prolegomena to its History, with Addenda and Postscript* (Gloucester, MA: P. Smith, 1967).

Mooter, Dave. "Urban Forest." *A Walk in the Woods* (Lincoln: Nebraska Game and Parks Commission, 1993), 66–69.

Myres, Sandra L. "Women and the 'Mental Geography' of the Plains." *The NEBRASKA Humanist* 9 (1986): 39–45.

Olson, James C. *J. Sterling Morton* (Lincoln: University of Nebraska Press, 1942).

Rice, Ginger. "Tree Planters Gave Land a New Shape." *Journal and Star* (16 April 1972): B2, B3.

Stegner, Wallace, and Page Stegner. *American Places* (New York: Dutton, 1981).

✳ CINDY CROSBY, 1961– ✳

Raised in Plainfield, Indiana, Cindy Crosby graduated with a degree in journalism from Ball State University, where she worked on their award-winning university newspaper, and later operated an independent bookstore in Bloomington, Indiana, with her husband, Jeff. She currently lives in the Chicago area where she teaches natural history and gives public talks on many subjects, including writing, prairie ecology, and faith. Winner of the 2013 Paul Gruchow Essay Contest and in 2005 selected as Isle Royale (MI) National Park's artist-in-residence, Crosby is the author or compiler of seven books and has written numerous essays about the environment. She has also volunteered hundreds of hours to ecosystem restoration and wildlife monitoring. This chapter from *By Willoway Brook* (2003) is set at the hundred-acre Schulenberg Prairie in the western suburbs of Chicago, where she is currently a steward supervisor. Originally farmland, in 1962 this tract of land became one of the first areas in the United States that was intentionally re-planted as native prairie. The Schulenberg Prairie is part of the Morton Arboretum—established by Joy Morton, the son of J. Sterling Morton, founder of Arbor Day. Here she describes the transition, in her relationship to prayer and prairie restoration, from solitude to a new appreciation of the necessity of community.

Pulling Weeds
Community

That land is a community is the basic concept of ecology.—Aldo Leopold, *Sand County Almanac*

Christian prayer is above all the prayer of the whole human community, which Christ joins to himself.— *The Liturgy of the Hours*

I know that the prayers of those who have gone before me are part of the structure that is informing my new life of prayer. Yet I have experienced little community in a way that stimulates me to understand prayer. Not knowing quite how to begin, I search for community on the prairie.

The arboretum is closed. The hot, humid day is handing its keys to cool evening, and out on the Schulenberg Prairie a group of volunteers are planting swamp saxifrage and prairie plantain. My hands are muddy, and the moist area I'm working in is full of water draining off the upper fields from the long, gray month of rains that quit only yesterday. Overhead, goldfinches are chattering in high-pitched voices to each other about the events of the day, and across the gravel road, a few volunteers laugh as they plug rare plants into the ground. A red-tailed hawk swirls through the air currents overhead, looking for anything edible our efforts might have flushed through the foliage.

It's May, and the first meeting of the year for the arboretum's prairie volunteers. Almost twenty-five people have turned out to help weed, plant, and keep the prairie in good shape. Long after the arboretum's iron gates have clanged shut, we'll keep on working until sunset. The previous summer I often saw the group getting ready to work, and I vowed I'd show up next year and see what it was all about. This summer I came, I saw, I planted; I'm hooked.

"Do you want the shovel?" calls Julie from across the road. I assure her my trowel is working fine, and keep digging. The prairie plantain has a slightly bigger root system than the swamp saxifrage, and the bank in the water run-off area is blanketed with erosion mat and stones. Bracing my boots against the bank, I keep chiseling away at the hole.

For one who loves to come to the prairie to reflect, to rest, and to find silence, these summer evenings are a cause for readjustment. I've thought of the prairie as a place for solitude. Now I'm seeing it as a place for community. Together, we're doing in two hours what it would take an individual weeks to accomplish. Although some of these folks have been coming out for years to volunteer on the prairie, for about half of us this is out of the ordinary. Through working side by side in the mud, we're already forging a common bond. United by tallgrass.

The power of community is new for me, both on the prairie and in prayer.

For a long time, I was unwilling to volunteer and associate my quiet place with a place of work and people. After I've spent a summer investing some sweat into it, I've found that working with a community of those who also care about this particular prairie has intensified my relationship to the landscape rather than spoiling its niche for me as contemplative ground.

This is no small breakthrough. I used to be put off by community, especially in prayer. I don't like to pray out loud in a public setting, and I'm usually silent in group prayer as others spontaneously speak their petitions and their praises. Putting thoughts together orally eludes me. More likely than not, I am mentally rehearsing what to say.

When I do pray spontaneously with certain people—who pray as the spirit moves—I find myself, eyes closed, inadvertently speaking at the same time as someone else. An awkward silence follows, then a hissed, "Go ahead." "Oh, no, please, you go." "No, it's okay, go ahead." Not exactly an encounter with the divine.

Recently my family joined a church community whose regular morning service includes reciting the Lord's Prayer together. As I chant the well-worn words "Our Father, Who art in heaven, hallowed be thy name . . ." something unlocks inside me. When I close my eyes and repeat the words of centuries, I experience a connection with community that has eluded me for many years.

It's not only me. A friend of mine tells me she is feeling estranged from God these days, and unable to pray on her own. When she tries to run through the liturgical prayers she's learned by heart, the words scramble up in her brain. Yet when she sits with her church family on Sunday morning and the prayers are recited together, the words effortlessly trip off her lips. The dynamic of the whole, the synergy of the group prayer, is what allows her to connect with the words that confound her in private. It's a regular affirmation each week of a need for each other.

As prairie volunteers, we meet at the appointed time, we sweat, we bleed, and we get muddy together. Weeds are pulled, paths are cleared, and plants are tucked into the dirt, snug in their permanent locations. Sometimes it feels like progress; other days we look at each other at the end of the evening and wonder if we've accomplished a thing. But our actions build on the work accomplished by countless laborers before us, all with a vision of what could be done, a faith that this landscape could be reconstructed to resemble its original splendor. Working together.

Praying the Lord's Prayer, the psalms, and other traditional prayers alone at set hours during the day allows me to feel the invisible community of others all over the world, praying around the clock words that are two thousand years old. I feel the pull of this community, a supernatural strength of numbers, an energy force greater than my words and myself. I can lean on the prayers that have gone before me, the prayers said alongside me, the prayers that will continue long after my lips have ceased to utter another sentence. I'm a link in an unbroken chain that is greater than myself; building on the work of centuries and unfinished in my lifetime.

EACH TUESDAY EVENING before darkness closes in, the veteran prairie volunteers in the group take us on a little educational walking tour. Tonight it's a swing through the late-May blooming plants — pink shooting stars, spiky cream wild indigo, sky blue Jacob's ladder, and gangs of violets in purple and white. A breeze ruffles the knee-high plants, while in the west the sky flies its silken sunset banners, reflecting the colors of the flowers below.

I thought I knew the prairie plants well, but already I learn I've misidentified one plant as golden Alexander, *Zizia aurea*, and it's prairie parsley, *Polytaenia nuttallii*. "Look at the leaves," urges Marj, whose acquaintance with the prairie spans several decades. "See the difference?" Now that she points it out, I guess I can. I tuck this information away and then ask her about a one-inch high mat of white flowers with alternate leaves that has eluded my identification, despite all my field guides. "Oh that," she says, waving her hands like it's a no-brainer. "Bastard toadflax."

Well. That's one mystery cleared up.

As the summer progresses, we move from planting diverse prairie specimens to ridding the prairie of the plants that will eventually crowd out the desirable ones. July evenings are spent pulling sweet clover, *Melilotus alba*, a long-term offender. At one time it was a positive, intentional presence in the landscape. When the Schulenberg Prairie was a farm, sweet clover was a good nitrogen fixer for the soil and was planted as a rotation crop.

When the prairie restoration began in the early sixties, the sweet clover swiftly transformed into a raging liability. Left to its own devices, it quickly blankets an area, choking out native plants and insolently waving its white lacy wands. It's particularly interested in exploiting the edges of the mown paths and any "wounds" that appear on the prairie, gouged out by natural elements or human interference.

Bent over in the waist-high bluestem and switchgrass at the end of July, I concentrate on one small area around me, ridding it of sweet clover. The smaller shoots and knee-high plants come out easily. Larger plants, especially those up to my shoulder, are more stubborn. Their taproots grasp the dry prairie soil with a tenacity that defies my steady pull. I dig in my heels, grip the base of the plant, and lean back, all my weight aligned against this menace. My hands slide along the thick stems, and I feel something begin to give. Then, *snap!* I almost fall backward as the clover comes off in my hands. But not the root.

Removing only the part of the plant that grows on the surface is an aesthetic, but temporary, solution. Where one stem was, three will spring up. The tap-root has a survivalist mentality. Now I have to dig out the taproot or clip the stem at least at the soil line to forestall its return. I opt for cutting it off. A little dirt scraped away, a couple of snips, and it's finished.

I pick up the clover plant and toss it in the pile, then arch my aching back and mop my forehead with my bandanna. The sun is setting, and the white flowers on the prairie are lit up with a last-light brilliance. Including the clover. As I look across the prairie toward the savanna, it appears in waves and waves . . . *ad infinitum*. It's enough to make me despair.

But I don't, because I'm not in this alone. Working with me, forming a broken line, are twenty-plus other volunteers, all wrestling with their own patches of sweet clover, all bent on obliterating the enemy. By myself, I could pull clover all summer, twenty-four hours a day, and likely not make a dent in it. Our combined labor brings about substantial progress.

The prairie itself is an interconnected network of grasses and flowers, soil and water, fire and weather, birds and mammals, insects and people. It is at its best when everything is working together. When one piece of the whole is out of kilter, it triggers a domino effect. Let the wild lupine, *Lupinus perennis*, disappear from some prairie remnants, and the Karner blue butterfly, whose larvae prefer this particular plant, loses one of its primary sources of life. Sweet clover creeps in on the Schulenberg Prairie, and soon the richness of native plant life has vanished. The annual burn changes the vole and mouse populations. Willoway Brook runs low, and the dragonflies and damselflies swarm to lay their eggs in the shallows. It's a community of change. Each action influences another; each member of the community builds upon the efforts of the whole.

Charm in the singular, beauty in the aggregate. Significance in the single cardinal song that pours out into the landscape, and glory in the flock of cedar waxwings' soft music as they congregate in the evenings by Willoway Brook. Without the individual, the opportunity for landscape is lost. Without the landscape, context for the individual vanishes.

A dependency that I once resisted forms in my life. Now it becomes strangely alluring. Community prayer is powerful. The prayers of many form a woven net to catch us, to hold us, and to encourage us when we falter.

"We know well enough that every finger of the hand has a use that is separate and that no finger could do so well," William Bryant Logan writes in *Dirt*. "Yet even to do that, the community, the ensemble, is primary. And through the neighborly relation of parts, the hands perform those functions of which prayer is the plainest manifestation: to dig down, to grasp, to lift, and to let go."

Through the efforts of many: planting, replanting, and pulling weeds, the prairie will continue to sustain community.

The single red-tailed hawk I see regularly on my prairie walks showed up with two others in tow this week. Although each hawk spiraled on different levels of a thermal, they soared together over the prairie, shrieking to one another, flying over the tallgrass mingled with white sweet clover that defied our group efforts. Sharing the thermals that whorl them ever higher, they have learned how to live in community. I'm learning to do the same.

From *By Willoway Brook: Exploring the Landscape of Prayer* by Cindy Crosby (Brewster, MA: Paraclete Press, 2003).

CINDY CROSBY

✳ THOMAS DEAN, 1959– ✳

Thomas Dean was born and raised in Rockford, Illinois, before attending Northern Illinois University, the University of Wisconsin–Madison , and the University of Iowa (UI), where he earned a PhD in English. Currently, Dean is senior presidential writer/editor at the UI, and teaches writing, literature, and humanities courses. Dean is also certified as a Land Ethic Leader through the Aldo Leopold Foundation. In addition to his publications in national journals and anthologies, Dean's books include the edited collection, *The Grace of Grass and Water: Writing in Honor of Paul Gruchow* (2007), and *Under a Midland Sky* (2008), the latter of which explores how "the atmospheric drama" of weather is "intimately intertwined with our characters and the nature of our days as Midwesterners" (1). In this excerpted essay, originally published in Ice Cube Press's anthology *Living with Topsoil* (2004), Dean traces the roots of prairie into the topsoil and deeper, through geologic time, to the "Iowan Devonian sea" (69).

Finding Devonian

In the hot summer sun, I stand on the fossilized stratum of an ancient seabed. I am in Iowa, my chosen home. I stand near Coralville Lake, only a few miles from my house. Yet my feet, and as I bend down my hands and my fingers, touch the calcified remains of life millions of years old. My mind reels back to that time as best it can. Out of my imagination flow ancient salinate waters, temperate and relatively shallow, an equatorial sea. At my feet are beautifully waving crinoids, "sea lilies," plant-like animals attached to the sea floor, their delicate tendrils bending languorously in the waters, capturing food particles as they float by. The round pitted sponges and the ridged brachiopod shells complete the delicate geometry of this ancient ocean. My skin feels the disturbance of the current not too distant. A long shadow cuts the waters. Perhaps it's an armored arthrodire, looking for its next meal. I am on the Devonian sea floor.

The water drains from my reverie. I am back in the present, standing once again on top of fossilized limestone. My son and daughter are peering through small karst holes on the surface of a flat rock shelf. The holes belie millions of years of subterranean water flow, and Nathaniel and Sylvia stare through them at the cool, dark pool of water in the tiny cave underneath. We are on the Devonian sea floor—its fossilized remnants that a disaster of the century gifted to eastern Iowa over a decade past.

Deep winter snows and endless spring rains spelled disaster for the Midwest in 1993. My wife Susan and I were living in southeastern Wisconsin at the time, Nathaniel not a year old. Near the shores of Lake Michigan, the impending tragedy of the Mississippi River Valley was little more than news reports to us. With both relief and guilt, having moved from Iowa two years prior but leaving our hearts there, we watched the inexorable swelling of the Minnesota River on TV. We knew the rising waters to the north were a wave to hit Iowa within a few days. The Des Moines and Cedar Rivers in Minnesota were pushing down into Iowa, the bloodstreams of the Mississippi Valley engorged, spilling too much too fast toward the Big Muddy. The Des Moines, the Cedar, and then the Skunk, the Raccoon, the Iowa, the Maquoketa—the rivers of the Iowa we loved were soon all overflowing their own banks, often taking whole towns with them. As the floodwaters receded, the U. S. Geological Survey crunched the numbers to announce the most devastating, costliest flood in U. S. history.

[. . .]

In July and August of 1993, the spillway [at the Coralville Lake dam] fulfilled its safety valve mission, but in a way hardly expected and certainly unprecedented. As the enormous waters pounded down and flowed into Iowa that summer, nature called in its debt. The waters rose to as high as five feet above the spillway and smashed downward toward the campground. For 28 days. The torrential deluge continued unabated, sometimes nearing the 20,000 cubic feet of water per second discharge rate of the dam itself. The horrendous flow was emblematic of nature's fury writ large across the landscape of the middle land that summer.

Of course, the waters eventually receded. And the devastation of the flood below the spillway was awesome, in all senses of that word. The road and campground beneath were obliterated. Layer upon layer of river silt and topsoil were washed downstream. Three-ton blocks of bedrock limestone were carried like matchsticks hundreds of feet.

But the flood's marvel lay not only in the evidence of the waters' power. The river's fury had washed away, as literally as it could, hundreds of millions of years of history. The incredible force of moving water had managed to dig a trench 20 feet at its deepest. In so doing it laid bare a sea bed never seen by humans, and not seen by any living creature for 375 million years. The flood had discovered, in the literal sense of that word, the Devonian.

I, like most children before and after me, and certainly like my own son and daughter, was enchanted by fossils. The romance of ages long past and the seductiveness of fantastic creatures, especially those of enormous size and ferocity, spark visions of paleontological careers in most youngsters at some time or another. I was the same, carefully examining stones in the alley behind my home and on the school playground for signs of tyrannosaurus rex. I dug a lot as a kid. Even in my own back yard, I had visions of finding arrowheads of the natives who had lived on that northern Illinois ground. Whenever we picnicked or camped at Atwood Park—the former Camp Grant military training compound of World War I—I would always end up at "Bullet Hill," where there had been a shooting range, and where some dedicated digging would yield plenty of spent ammunition.

Yet I think our impulse toward the petrified and buried past lies in something deeper than mere fancy. My backyard archaeological and paleontological expeditions originated in a desire for connection to place as much as a grasping toward mystery, even if I didn't realize it myself at the time. The ineffable resonances of times past, of history—human, floral, faunal, geological—were what I sought. The people, events, extinctions, and tectonic shifts that, for decades, and centuries, and millennia, have accreted to create a place—those, I realize now, are what I sought. I now understand better why, as a child, I stared so silently and intently at the couch where Abraham Lincoln sat in the Tinker Swiss Cottage museum, and why, though realizing now the disrespect of my gaze and the exhibit itself, I couldn't tear my eyes from the "Indian skeleton" on display at the Burpee Museum of Natural History, both in my hometown of Rockford, Illinois. These were not only pieces of history in general, or even remnants of my place. They *were* my place, the available matériel, however piecemeal, of the layering of experience that is the essence of place.

When you fall in love with another person, you desire the totality of shared experience. Living a shared life means sharing history, memory, experience, and body. The connection to place should be just as intimate, if it is to be real and meaningful. Finding place is a quest for physicality as much as it is the pursuit of knowledge and memory. Yet this physicality of living deeply also requires the deepest possible historical connection, right down to geologic bedrock. I have found such a connection amazingly available in the North Woods of Minnesota, my second beloved land. At the southern edge of the Canadian Shield, in a climate that supports only the shallowest of forest duff, some of the oldest rock in Earth's history—greenstone—juts out to our waiting eyes and fingers. I can literally touch the deepest depths of that place.

And so, here in my home in Iowa, I want to find the Devonian. Iowa's natural marvel is its topsoil, the rich, dark, fecund organic compost that birthed the prairie and feeds the world. Yet our topsoil is Cenozoic, a geological youngster. Its Quaternary roots are less than two million years long. Living with topsoil, caring for and respecting it, is one of our most important obligations as stewards of our place. But I am impelled to go deeper, to dig further into older, harder first things.

The Devonian is obviously not Iowa's oldest age. In some form or another, our land has always been part of Earth's existence. Through the foggy mists of geological time, we would have to plumb Pre-Cambrian Archean eons to arrive at the first land. But the Devonian is back there, between 300 and 400 million years. It is Paleozoic rather than Mesozoic, predating the compelling dinosaur-laden Jurassic and Triassic by at least 200 million. Yet the Devonian is the era that has tapped most persistently into Iowa's imagination, among geologists, paleontologists, and ordinary Iowa-ologists.

A lot has to do with availability. Our wonderful Iowa limestone is Devonian, and it is rife with fossilized life. The Devonian in our beautiful land is easy to see. But "because it's there" doesn't fully explain our fascination. Although Iowans pride themselves on suffering Midwestern extremes —scorching, humid, tornadic summers, and frigid, snowy, biting winters— I think most of us secretly long for equatorial paradise. And, in fact, Iowa *was* equatorial—in the Devonian. As part of the supercontinent of Pangea millions of years ago, before tectonic violence ripped us asunder from the proto-European and African continents that we snuggled, we lay south of the equator. And we were underwater—a shallow sea teeming with exotic life. When our geographical defensive hackles rise, and we say things like, "Even though we don't have oceans . . . ," we should step back and retract such apologies. Regardless of whether or not oceans are somehow inherently superior to meadows, we *were* an ocean, and we should keep that as part of who we are now. We are quick to accept that, as individual human beings, we are not only the sum of our personal experiences, but also our familial and tribal histories. We should say the same of our geological selves, so essential to our connection to place. We should subsume the Devonian, and everything that came before and after, into our prairie roots.

The essence of the prairie root is deep life. Our rich soil, in its original state as host to millions of acres of grasses and wildflowers, held, fed, and nurtured the complex and prolific root systems that barely suggested their vitality above ground. Our region's distinguishing characteristic is its gift of growth, turned toward the world's most robust agriculture in the modern day.

We can hold tight to that line of affiliation with the Devonian of millenia ago. The Northern Hemisphere's oldest known vascular plants are Devonian, and by the end of the age, the first trees and forests birthed from the first ferns and seed plants sprouted toward the sky. But here in Iowa, of course, we were submerged. Still, this sea was an especially rich repository of prehistoric life, and serves as an apt underwater ancestor to the prairie. The remnants we see in today's limestone have a name that suggests life's richness in this period, as well as suggesting a wonderful whimsical character: "fossiliferous." Trilobites scooted abundantly across the sea floor. Brachiopods dotted the underwater ground with their clam-like shells offering life-filled pointillism to the landscape. Delicate crinoids held fast to the bottom, their waving strands of multiple arms . . . seeking food, resembling plants waving in the wind. Corals—horn, hexagonaria, and others—and sponges abounded, accented by a variety of bryozoans, variegated mosses with branching, lacy, and encrusting fingers holding fast to rock and coral.

The romance of ages past always contains an element of danger, however, and our Iowan Devonian sea was no exception. We must have our large and menacing creature to threaten paradise, and we have it in the arthrodire placoderm, a frightening armored fish, the largest creature in this ancient ocean, growing up to 10 feet long. The model arthrodire is the centerpiece of the Devonian diorama of the University of Iowa's Museum of Natural History. With yawning open mouth, frightening long sharp teeth, and bizarre armored plates, the king of the Devonian placoderms defers only to the model giant sloth in popularity to the thousands of children who visit the displays each year. The prize find of the gorge is an intact arthrodire dorsal plate. Our reverence for this fossil brought it to a special display inside the dam visitors' center.

The persistence of the Devonian in our geological imagination is long-standing. Famed naturalist Louis Agassiz sparked the flame in 1866. Traveling to the frontier, where the railroad terminated at Iowa City, he delivered one of his famous lectures on multiple glaciations and drew everyone's attention to our great prehistoric treasure, the ancient coral reefs that gave our companion community Coralville its unique name. Agassiz was able to articulate for us our inexorable attraction to the Devonian.

The coral deposits of Coralville are not truly "reefs." More accurately, the gorge displays "biostromes," or layers of fossilized life, rather than the slowly accreting mounds of skeletal remains that form coral reefs. As a result, exploration of the Devonian Fossil Gorge is as literal a walk through time as you can get, a staircase to prehistory.

[. . .]

Especially since the prehistoric marvel nature provided in the wake of tragedy in 1993, I, and many other Iowans, have been able to find Devonian more easily than ever. Not only can we catch glimpses of it in rock cutaways that our roads pierce through, and not only can we touch its surfaces and

THOMAS DEAN

subsume its spirit in the limestone edifices that surround us, but our imaginations and bodies can touch reality in the gorge that lays bare the life of Iowa's ancient sea. The fossil gorge facilitates other important elements of place-making as well: we can share our pride of place with visiting friends through an enjoyable field trip, and we can sink taproots of connection into place with those closest to us through regular family visits. The gorge also provides great stories to tell; how the gorge came about through the Great Flood of 1993 has entered into local legend, and the tale itself is almost as important as the material result we enjoy today.

The ultimate lesson that comes from finding Devonian is not what I originally sought. We reach through the long past for a sense of permanence, but what we grasp is change. The permanent is elusive, a phantom, and what seems to represent forever is really only temporary, delicate and fragile. Although my quest for finding Devonian intended to sink permanent footings into historical, imagined, and literal bedrock, connection to place is always shifting, if not ephemeral. No place is static. Yet place's constant alterations make for beauty and new fascinations. At the Devonian fossil gorge, blazing stars of the geologically young prairie shoot up from the soils ringing the ancient fossil floor. At the foot of the gorge, at the lowest depths of the stairway to prehistory, waters pooling in the depressions are establishing a new swamp, replete with a forest of cattails and croaking frogs. During one visit with friends, we were treated to the beautiful sight of two deer bounding across the gorge, our modern forest denizens loping gracefully through a petrified ancient seabed.

Yet the constancy of change also reminds us of how delicate and fine are the gossamer threads holding us in place. As soon as the floodwaters revealed the fossils that opened up new old worlds to thousands upon thousands of visitors, those stone remnants of life, preserved for millions of years in the dark recesses of bedrock, began to succumb to a swifter process of erosion than they had ever known. The gorge is a transient, not permanent, phenomenon. Despite attempts to prevent pilfering, some people will extract fossils from the gorge. Curious hands touching the ancient remnants, tracing their beautiful yet delicate lines and patterns in understandable tactile awe, will wear those same patterns down. But probably an even greater threat is nature itself. The exposed rock surfaces will weather. The extremes of Midwestern atmospherics will cause the stone to flake. Modern plant life, as it has already started to do, will take hold in the crevasses of rock and on its surfaces. Although the Devonian Fossil Gorge will help us connect to our place for many years to come, in the longer term, it is just as mortal as our own short lives. That itself is not a tragedy, nor an invitation to mourn the fate of place. The gorge's transience is in many ways a cause for celebration, at least a lesson in joyful acceptance. Place, which is really nothing more than our own relationship to our environments, natural and constructed, is a creative dynamic that is most interesting in its interplay between permanence and change, between the museum of tradition and the frontier of the new, between revering what has come before and participating imaginatively in creating what comes after. In Iowa, I have found Devonian. And one of its main lessons for me is the marvel of hurtling through time, space, imagination, and land *beyond* that faraway age I have finally just come to know.

In the hot summer sun, I stand on the fossilized stratum of an ancient seabed. I watch my son and daughter gleefully exclaim over a newfound sea lily. "Here's a big one!" they shout. They are in the Devonian, and they are now of Iowa. A decade or two in the future, they will have careers, their own homes, maybe their own families. I'm not sure what the Devonian Fossil Gorge will look like then. Maybe its prehistoric three-dimensional stone pictures will be faded and worn. Maybe the eastern Iowa prairie, forest, and swamp will have re-established themselves in this space. Maybe the gorge will remain intact, and we can relive its wonders through the eyes of grandchildren, renewing our bonds of place and family. But regardless of what the future years will bring, and how much situation and circumstance will have changed, I will always have an anchor in this place, for I have found Devonian.

From *Living with Topsoil: Tending Spirits, Cherishing Land*, edited by Steve Semken (North Liberty, IA: Ice Cube Press, 2004).

✳ JOHN T. PRICE, 1966– ✳

John T. Price is a native of Fort Dodge, Iowa, and earned an MFA in non-fiction writing and a PhD in English at the University of Iowa. A recipient of a fellowship from the National Endowment for the Arts, his nonfiction about nature, family, and home has appeared in many national publications, including *Orion* and *Best Spiritual Writing 2000*. He is the author of three memoirs with a focus on the prairies, including *Not Just Any Land: A Personal and Literary Journey into the American Grasslands* (2004) and most recently, *Daddy Long Legs: The Natural Education of a Father* (2013). Currently the director of the Creative Nonfiction Writing Program in the English Department at the University of Nebraska at Omaha, he lives with his family in Council Bluffs, Iowa. In this title essay from *Man Killed by Pheasant and Other Kinships* (2008), Price recounts his unlikely awakening into a new intimacy with the landscapes of home.

✳ ✳ ✳

Man Killed by Pheasant

So I'm driving east on Highway 30, from our new home in Belle Plaine to Cedar Rapids, Iowa. It's a four-lane, and because I'm an eldest child, I'm driving the speed limit, around fifty-five, sixty miles per hour. I'm listening to Jimi Hendrix cry "Mary"—imagining, as usual, that I am Jimi Hendrix—when in the far distance I see some brown blobs hovering across the highway: one, then two. By the way they move, low and slow, I suspect they're young pheasants. As I near the place of their crossing I look over the empty passenger seat and into the grassy ditch to see if I can spot the whole clan. Suddenly, there is a peripheral darkness, the fast shadow of an eclipse, and something explodes against the side of my head in a fury of flapping and scratching and squawking. In an act of extraordinary timing, one of the straggling pheasants has flown in my driver's side window. And being the steel-jawed action hero I am, I scream, scream like a rabbit, and strike at it frantically with my left arm, the car swerving, wings snapping, Hendrix wailing, feathers beating at my face until, at last, I knock the thing back out the window and onto the road. I regain control of the car, if not myself, and pull over, undone.

That's the time I should have been killed by a pheasant. For reasons peculiar to that summer, I recall it often. It occurred, for one, while I was on my way to teach a technical writing course at a nearby community college, a summer job to help me through grad school. This "distance learning experience" took place exclusively by radio wave, with me in an empty room on campus and my fifteen students scattered at sites within a hundred-mile radius. The technology was such that my students could see me, but I couldn't see them. To converse we had to push buttons at the base of our microphones, so that each class felt like an episode of *Larry King Live: Judy, from Monticello, hello, you're on the air.* "The future of higher education" my supervisor called it. And I never did get the hang of the camera. I'd turn it on at the beginning of class and there, on the big-screen monitor, would be a super-close-up of my lips. I'd spend the next few minutes jostling the joystick, zooming in and out like one of those early music videos until I found the suitable frame. Sometimes my students would laugh at this, and I'd hear them laughing, but only if they pushed their buttons. If there was an electrical storm nearby, I wouldn't hear them at all.

On the way to such a displaced, bodiless job, a near-death experience had some additional currency. As did the larger natural disaster unfolding around me. It was the summer of the great Iowa floods, 1993, and the reason I was on Highway 30 to begin with was that my usual route to campus had been washed out by the swollen Iowa River. This was a serious situation: People had been killed and Des Moines had been without water for over a week. "Nature Gone Mad!" was how the national media described it.

Although aware of the widespread suffering, I was privileged to watch the whole thing unfold more gently from the roadways of my rural commutes. And what I saw was a wilderness of birds. Bean fields suddenly became sheer, inaccessible places where herons stood piercing frogs in the shallows, where pelicans flew in great cyclonic towers, where bald eagles swung low to pick off stranded fish. Perched on soggy, neglected fence posts were birds I hadn't seen since early childhood, bobolinks and bluebirds and tanagers. Their color and song drew my eye closer to the earth, to the ragged ditches full of forgotten wildflowers and grasses—primrose and horsemint, big blue and switch—safe, at least for a while, from the mower's blade. The domesticated landscape of my home had gone wild and I was mesmerized by it.

Toward the end of the summer flooding, when the dramatic presence of wild birds dwindled, I thought a lot about Noah, about those end days on the Ark between the release of the raven and the return of the dove, between knowledge of a decimated landscape and faith in one that, through decimation, had become reborn. When it was all over, I thought I understood Noah's first impulse, once on dry land, to get drunk and forget. I'd lived my entire life in Iowa, the most ecologically altered state in the union, with less than one-tenth of 1 percent of its native habitats remaining. "Tragic" is what the ecologist Wes Jackson has called the plowing up of this prairie region, "one of the two or three worst atrocities committed by Americans." Not that I'd ever cared—it's hard to care about a wild place you've never seen or known. Yet in those short, flooded months of 1993, I witnessed a blurry reflection of what the land had once been: a rich ecology of wetlands and savannahs and prairies, alive with movement and migration. Alive with power. Under its influence, I felt closer to my home landscape than ever before. So when that power slipped from view, I was surprised to find myself longing to chase after it. Having spent most of my life wanting to leave the Midwest, where might I find the reasons to stay, to commit?

Death by pheasant didn't immediately come to mind. Although, in the wake of the floods, death was part of what I longed for. Or rather the possibility of a certain kind of death, the kind in which you become lost in a vast landscape and die, as Edward Abbey has described it, "alone, on rock under sun at the brink of the unknown, like a wolf, like a great bird." This had nearly been my fate on that cliff in Idaho, during our honeymoon, and in my mind it helped define that wilderness as a place worthy of respect, a place of consequence and a kind of fearful freedom. My German friend, Elmar, calls this freedom *vogelfrei*, which loosely translates into "free as a bird." Far from the positive spin we've put on this phrase, *vogelfrei* refers to the state of being cast out from the tribe, so free you'll die in the open, unburied, to be picked apart by birds. It's a state of fear and vulnerability and movement, one that might, especially here in the agricultural Midwest—a place seemingly without fang or claw or talon—make us more attentive to the natural world, more humbled by its power to transform us.

At first flush, my collision with the pheasant didn't seem to hold that kind of possibility. But it could have. If, for example, this had happened to me as a child or adolescent or as a member of a New Age men's group, I might have made something more of it. When I was a boy, some of my favorite comic book characters were mutations of man and animal—Mole Man, of course, and Spiderman and Captain America's ally, the Falcon. Imagine the comic book story that could have developed this time: A mild-mannered English professor is struck in the head by a wayward pheasant, his blood mingling with the bird's while, coincidentally, a cosmic tsunami from a distant stellar explosion soaks the whole scene in gamma radiation. Emerging from the smoldering rubble: Pheasant Man. No, *Super* Pheasant Man! As Super Pheasant Man, our mild-mannered professor finds he has acquired the bird's more powerful features—its pride and daring, its resilience, its colorful head feathers—learning to use them for the good of humanity while at the same time fighting the darker side of his condition, namely, a propensity for polygamy and loose stools.

But I was not a boy when I met that unlucky pheasant on Highway 30, which is too bad, because for a long time afterward I found nothing particularly uplifting about the experience. Instead, I saw my life, and death, made a joke. Imagine the regional headlines: *Iowa Man Killed by Pheasant; Mother Files for Hunting License.* Imagine the funeral, where, in the middle of "I'll Fly Away," one of my more successful cousins whispers to his wife, *You know, it wasn't even a cock pheasant that killed him. It was just a little baby pheasant.* Imagine the members of that hypothetical men's group, who, in their wailful mourning of my death, botch up the spirit animal ritual and condemn my soul to be borne not on the wings of an eagle or a falcon, but on those of a pheasant, stubby and insufficient, struggling to get us both off the ground, never getting more than maybe fifteen feet toward heaven before dropping back down to earth with a thud and a cluck.

JOHN T. PRICE

No, thank you. I do not wish to become one with the pheasant, in this life or in the next. Yet seen through the history of the land, this bird and I have been colliding for centuries. Having evolved together on the grasslands of distant continents, we were both brought to this country by the accidents of nature and technology and desire. As Americans, the pheasant and I have come to share certain important historical figures, like Benjamin Franklin, whose son-in-law was one of the first to attempt to introduce the ring-necked pheasant, a native of China, to this country—an unsuccessful release in New Jersey. Its introduction to Iowa over a century later was by accident, taking place during a 1901 windstorm near Cedar Falls that blew down confinement fences and released two thousand of the birds into the prairie night. They've remained here ever since, sharing with my people an affinity for the Northern Plains to which we've both become anchored by the peculiarities of the soil. This soil, loess and glacial till, is migrant and invasive, like us, having been carried here from ancient Canada by wind and by ice. Its rich, organic loam, black as oil, brought my farmer ancestors to the region and has, at the same time, held close the range of the ring-necked pheasant, lacing the bird's grit with calcium carbonate. Because the ringneck requires an abundance of this mineral, it doesn't stray far, not even a few hundred miles south into the gray prairies of, say, lower Illinois.

So the pheasant and I have remained settlers in this region, watching as others of our kind move on. As such, we have come to share some of the same enemies, like the fencerow-to-fencerow, get-big-or-get-out agricultural policies of the 1970s and '80s. These policies, enacting yet another vision of migration, dramatically expanded agricultural exports and, at the same time, led the region to the farm crisis of the 1980s, to the flight and impoverishment and death of thousands of industrial and farm families. For the pheasant as well, despite set-aside programs, this fencerow-to-fencerow world has held its own kind of impoverishment, a destruction of habitat so thorough that two hundred pheasants have been known to crowd a shelterbelt only a hundred yards long. In such a bare-naked world a good blizzard, like the one in 1975—while I was honking truck horns with my grandfather—has the power to wipe out 80 percent of a local pheasant population in a single evening.

Yet, in sharing enemies, we have also been, together, the common enemy.

To the prairie chicken, for instance—one of the many native citizens that had the unfortunate luck to precede us here in the "heartland." For almost a century European settlers hunted and plowed down prairie chicken populations in Iowa. But the ring-necked pheasant has also played a role, destroying the prairie chicken's eggs, occupying its nests, and interrupting, seemingly out of spite, its dancing-ground rituals. Some argue that restoring prairie chicken populations will have to coincide with significant reductions in pheasant populations. Reducing pheasant numbers around here, however, is about as easy and popular as reducing our own. The difficulties are partly economic: In Iowa, we hunt and eat this bird to the tune of about a million and a half a year. It's one of our biggest tourist attractions. But I wonder, too, if we don't see in this bird, at some unconscious level, a dark reflection of our own troubling history in the American grasslands, our role as ecological party crashers, as culture wreckers. Our role, ultimately, as killers and thieves. To question the pheasant's claim on the land is, in some way, to question our own.

It's unfair, of course, and dangerous to project our sins onto another species. When tossing around ethical responsibility, the difference between us, between instinct and intent, is significant. But the pheasants needn't worry about taking the blame. Hardly anyone around here gives them a second thought. That indifference was part of my problem when I finally began searching, while living in Belle Plaine, for reasons to care about my home landscape. In relation to that bird, as to most of the familiar, transplanted wildlife around me, I felt nothing. The pheasant was common, and the last thing I wanted to feel as a Midwesterner was common. Since early adolescence I'd been fleeing a sense of inadequacy shared by many in this region, a sense of self marked—as Minnesotan Patricia Hampl has said—by "an indelible brand of innocence, which is to be marked by an absence, a vacancy. By nothing at all." For Midwesterners like me, the complex, the worthy seemed always to be found elsewhere. Not here in this ordinary place, this ordinary life.

Not surprisingly, in the years immediately following the flood, I didn't seek that new relationship to the Midwest in the familiar land immediately around me, but by traveling to distant and, in my imagination, more exotic landscapes such as the Black Hills and Badlands in western South Dakota.

What I discovered in those places did indeed transform me. Near Wind Cave, I saw for the first time elk bugling and mating at home on their native prairies. While sitting in a fly-plagued prairie dog town, I saw for the first time a bison bull wallowing in the brown dust. On the grasslands near Bear Butte, where Crazy Horse once sought vision, I saw for the first time a falcon stoop to kill a duck, the native cycles of predator and prey, of wild death, still lingering. Even *vogelfrei*, that fearful freedom—I felt it for the first time in this region while walking lost in the deep earth of the Badlands. My journey through these places toward commitment has been awkward, fragmented, and at times pathetic, even comic. Yet the significance of those experiences cannot be underestimated, how they have worked to cure a lifetime of ignorance and indifference. How, to use spiritual terms, they have filled what once was empty.

But if the spiritual journey to a place begins, as some claim, with mortal fear, then it was not the bison or the falcon or the Badlands that first drew me closer to the region in which I'd been raised. It was the pheasant, that particular baby pheasant—there on a highway in eastern Iowa—which almost, as my sister Allyson would say, rocked my world. In a sense, that's exactly what it did. It made me wake up, become more observant of what's lurking in the margins. What's lurking there, despite the rumors, is the possibility of surprise, of accident, of death. And if it's possible in this over-determined landscape for a pheasant to kill a man, then why not also the possibility of restoration, renewal, and, at last, hope?

That's a romantic stretch, I know, and at the time of the incident itself I didn't feel particularly worshipful of its surprise. As I sat in the car, wiping my face, I just felt lucky—*Thank God it wasn't a two-lane!*—and then ridiculous. The whole thing was so absurd it might've been a dream. I carefully leaned my head out the window to see if the pheasant was still on the road. It wasn't. I thought about going back to see if it was injured, but decided against it—after all, it was only a pheasant. Besides, I was late for work, where in a few minutes I would be taking my own precarious flight through the airwaves, across the flooded land, to students I would never see, never truly know.

I started the car and eased back onto the highway. As I approached cruising speed I saw something move out of the corner of my eye. I jerked, swerving the car a little. A feather. An ordinary brown feather. Then another and another—there must have been a dozen—floating in the breeze of the open window. They tickled and annoyed me. Yet for reasons I still can't explain, I kept the window open, just a crack, enough to keep the feathers dancing about the cabin. And that's how I, the man almost killed by a pheasant, drove the rest of those miles, touched by its feathers in flight, touched by an intimacy as rare and welcome, in my tragic country, as laughter in a storm.

From *Man Killed by Pheasant and Other Kinships* by John T. Price, 2008. Reprint (Iowa City: University of Iowa Press, 2012).

In 1981, at age twenty-six, Steven I. Apfelbaum—a northern Illinois native and trained ecologist—first laid eyes on the rural land in southern Wisconsin that would become his life's passion. He eventually purchased the eighty-acre property, and like Aldo Leopold before him, spent the next decades restoring its prairies and other ecosystems with the help of his partner, Susan (who grew up in Walnut, Iowa). What Apfelbaum learned has informed his work as senior ecologist for Applied Ecological Services (AES), a firm he founded, which is known for its international science-based ecological design and restoration work. The experience also became the subject of his book, *Nature's Second Chance* (2009). In this chapter, he tells the story of his evolving relationships with rural neighbors and community members—relationships that, as he discovered, speak to the complex, but vital interconnections between natural and "social" ecosystems (146).

Getting to Know Your Neighbors

The knock at the door caught me off guard since I hadn't heard a vehicle drive up. As I opened the front door, I recognized the young man as the son of a local farmer I'd once met at a chili cook-off. This farmer and his sons worked a pretty tired farm a few miles away that had been in one family longer than most of the other local farms. His son was of medium build with a strong brow and burly forearms. I tried to invite him in, but he declined politely and stood on the porch. He said he wanted to speak to me about my fields.

"That sure is a weedy mess," he said, looking out over the field.

It didn't bother me that he was so direct. I enjoyed these kinds of encounters. They gave me the opportunity to explain my work. So I tried to engage him, as I had done with many other visitors and neighbors those first couple years after planting the prairie, and would continue to do with many more.

"Yes, it's a mess," I agreed, "but it's been planted this way on purpose. Those aren't weeds, actually. They're young wildflowers and other seedling prairie plants."

He looked at me with an uncomprehending smile.

"For the next year or two it will look like that. Random. That's what it's supposed to look like. It's just getting started."

He kept grinning. He looked back over the field nervously and then smiled from ear to ear, as if he'd just remembered a punch line.

"Say, I'm not too clear on what you're saying, but I was just driving around looking for some land to rent and noticed that your fields appear to be fallow and weren't planted to corn yet this year. If that's so, would you rent that land to me for planting corn?" he asked, punctuating "corn" with a broad friendly grin.

"The entire farm has been planted. Those are young native prairie plants and wildflowers," I repeated, pinching my thumb and index finger in his line of sight like crosshairs to draw his attention to the nascent seedlings. "See how they're coming up under those agricultural weeds?"

"So, you'll just let the weeds grow instead of planting corn?" He asked this slowly as if perhaps I was the one who didn't understand the situation.

"Actually, the entire farm is fully planted, and none of it's for rent. Sorry." I had a full day's work ahead of me, and this tutorial didn't seem to be producing much fruit.

"Oh, that's too bad. Well thanks. I just figured that if it was sitting idle, I'd be a fool not to ask."

"Wait a year or two and the wildflowers will be lush and beautiful."

He stopped a second to consider that, as he patted his vest for his keys.

"What good are they? Can't sell 'em, can you?"

"Actually we do. They're a crop. We produce wildflower seeds."

"So your crops aren't in rows?" he asked. "Don't you cultivate and spray to keep the weeds down?"

"Actually, we burn the fields to keep the weeds down. Fire encourages growth of native prairie grasses."

"We used to burn our fencerows to keep the weeds down," he said, visibly pleased to finally share a point of reference. "Well, thanks, and good luck." He shook my hand and pivoted around for one last look at the weedy mess before he walked around the house to his truck.

This was a scene that would be repeated with many variations for years. Neighbors simply don't know what to make of the way we view and use our land.

A weedy mess is what most of them saw in the beginning. But after a few years the mess became so spectacular it couldn't be written off so easily. The green growths of giant ragweed and foxtail grass slowly gave way to big bluestem, coneflowers, and fifteen-foot compass plants. And each subsequent year the prairie showcased its variegated display of wildflowers from late May through early November on par with the most colorful children's kaleidoscope.

As the land changed, so did perceptions. Even the farmers took notice when the wildflowers bloomed. Neighbors, AES clients, and the occasional school group started coming around to take in the full spectrum of color and get up close to the wildflowers. We welcomed them. When we could we conducted informal nature walks. Otherwise we just let people explore on their own. In time we were conducting conservation demonstrations, proposing that interested visitors plant prairies on their marginally productive and erosion-prone farmlands, and showing how this helps restore ecological health and foster abundant wildlife.

WORD GOT AROUND, not only about the diversity of plant species on the farm but also about wildlife flourishing in fresh new habitats full of foraging opportunities. And where the dubious farmers had vilified the weedy mess, the hunters got excited about the prospect of shooting game. The farm quickly became the largest block of wildlife habitat in the community, so the fields were coveted for their bounteous hunting potential. I recognized the draw. I was no stranger to hunting and had occasionally invited colleagues over for pheasant hunts. But there weren't enough game animals to make the place a hunting ground for the masses. And frankly, I hadn't been encouraging all of this wildlife simply to see it killed for sport.

After a heavy frost, the prairie's autumn hues brighten the rolling hills. At about the time the hay crops and corn foliage fade to tawny yellows and browns, the prairie grasses explode in glowing waves of reddish blue and gold. Like kids in a candy store, neighboring hunting enthusiasts inevitably come knocking on our door, hoping for some of the spoils. Inevitably we'd open the door to reveal a group of men decked out in traditional plaid hunting jackets and modern fluorescent orange vests. Out the window we could see friends and dogs in the driveway, waiting for the affirmative nod that the hunt was on. But I rarely granted permission, because giving someone the green light guaranteed that dozens of friends, and friends of friends, would show up the following day asking permission.

"It's the first hunt for my eight-year-old son" was a typical refrain from a door knocker in a sporty orange vest. "Sure looks like a great place for a young guy's first hunt, what with all the rabbits and pheasants out there."

Sometimes it was coyote hunters coming round, hell bent on lecturing me about predator–prey relationships.

"Neither the pheasant populations nor the Republicans are gonna survive if the predators aren't trimmed," they'd argue.

"Sorry, but hunting just isn't allowed. The ecosystem is still quite fragile," I'd explain. "Besides, the pheasant population doesn't seem to be in decline, even though coyote and fox populations are on the rise."

They'd still try to hunt coyotes in winter anyway. Trucks would follow a squadron of radio-collared foxhounds. When the radio signals indicated that the dogs had cornered a coyote, the trucks would swoop in, driving as close to the farm fields and yards as possible. The hunters would park on the access roads overlooking Stone Prairie Farm, while their dogs trespassed in search of coyotes seeking shelter in the prairie habitat. Again the hunters would appear at my door asking for permission to shoot the coyotes. I'd always say no, explaining that I'd only reconsider if they needed subsistence food or were addressing a wildlife disease issue that threatened ecological health.

"After all, the coyote and fox pups play with our dog, Max," I'd offer. "I can't deprive him of his friends."

I'd let them retrieve their dogs, but only on the condition that they leave their guns in their trucks.

I didn't like these run-ins. The hunters were usually disgruntled, but fortunately they always honored my requests. I did have one particularly bad experience, though, with the farmer who'd rented my land during the conversion from corn to prairie. He came over one late fall day in 1992 and asked to hunt pheasants. I said no, telling him I'd already invited some colleagues and a neighbor to hunt that same afternoon.

I thought no more about it until he reappeared later that year during deer-hunting season, determined to take advantage of the abundance of game. In an audacious move, he cut a fence and drove onto my land in an all-terrain vehicle, crisscrossing the fragile prairies in an attempt to flush out the animals. Other curious hunters had parked around the perimeter and were poised to shoot should a spooked deer flee into the open. At my wit's end, I called the county sheriff's office. Officers showed up with the game warden, who took photos of the hunters and their license plates to discourage future poaching. To be neighborly I didn't press charges for the blatant trespass. But I did get him to repair the fence, thanks to the sheriff, who facilitated our detente. After this incident the requests to hunt on the farm tapered off.

These days only a few hunters ask. It's gotten around that hunting is not allowed at Stone Prairie Farm. Susan and I field the calls that do come in and take the opportunity to explain our philosophy. The hunters get detailed explanations of the restoration and are told that our focus is on producing high-quality habitat—and hunting experiences for our family and close friends. We even suggest they could do the same on their farms.

IN AN EARLIER CHAPTER I mentioned the patriarch of a local farm family who disapproved of our land use. "A noxious weed patch!" he'd call the farm, only half joking. I met Carl around 1981, not long after moving in, when he was tending to his livestock on a section of the farm that he was still renting. He'd drive through my yard to get to the back pasture where he kept heifers, dry cows, and one large bull. He came twice daily from his farm a few miles south in Illinois to feed and water the livestock.

Occasionally we'd meet on the driveway and talk. He'd sing the praises of his straight, weedless crop rows and the well-maintained habitat-free fence lines. And I'd argue against his policy of leaving no acreage uncultivated, as well as his insistence that the value of every inch of ground needed to be quantified in purely utilitarian terms. I didn't try to insult his philosophy or suggest that one should get nothing out of the land except enjoyment. But I would remind him of specific examples in which farm use damaged the land, as with the silt-clogged streambed and the crumbling, tattered banks where his cows had done their damage.

I think we respected each other, even then. But our debates frustrated both of us, especially Carl. I tried to appreciate the critical reality that every corn plant, cow, or gallon of milk was an asset in a portfolio that he had inherited from his father. He made a clear distinction between subsisting off the land and producing products for markets, as his family did. I'd ask him if there could be no stronger land ethic in the face of market forces, and he'd look at me with a blend of pity and disdain. I failed to convince him, and time and time again we returned to this theme in our discussions.

Despite Carl's distaste for our land use his daughter Patricia developed a passion for wildflowers. When I first met Patricia she was eight or nine years old, shadowing her mother in their garden, carrying a child-sized basket and tending to the vegetables. While she helped with milking chores and everything else expected around an operating dairy farm, she always gravitated to the garden. Then one year when she was in high school I was surprised to find her working as a summer employee at the AES nursery. From that point on she spent a lot of time on the farm, visiting with us and walking the trails, guidebook in hand, to learn about the plants.

Eventually Carl's lease ended and our regular encounters ceased. We lost touch with Patricia as well. She grew up, as they do, and went off to college.

One evening, six or seven years later, she reappeared. She told us she was getting married and asked if they could hold the ceremony overlooking the blooming wildflowers.

"You don't think your father will find that sacrilegious?" I asked.

"Well, first of all it's my wedding!" she said, smiling. "But you know what? He's really had a change of heart over the years. Now that he's retired from farming, he's a different person. I think over time the flowers won the argument for you. Sometimes he'll even stop the car when he's driving by and get out to admire the land."

STEVEN I. APFELBAUM

Susan and I agreed wholeheartedly. We even gave them permission to mow a small plot near the wind generator so that they'd have a clear area after the ceremony for a dining tent. My only complaint was that we'd planned a family vacation then and wouldn't be able to be there!

The wedding took place on a sunny Saturday afternoon and the prairie was ablaze in color. Some neighbors who'd attended got word back to us about the handpicked black-eyed Susans, yellow coneflowers, and bergamot that the entire family, including Carl, had picked the preceding day for the lush bouquets and the wedding party's corsages.

Thank-you letters from the bride greeted us upon our return. In them Patricia and her husband, Mark, waxed poetic about the acres of beautiful flowers and related how moved their relatives and friends had been. It was clear that the landscape had been considered a part of the ceremony.

Years afterward some of their family members were still making visits to walk the trails and learn about the flowers. Carl and his wife, after finally retiring from farming and moving into town, asked if they could collect seeds to grow wildflowers in their new garden. I suppose that besides the meaningful experience of his daughter's wedding, Carl's change in attitude was solidified by his retirement from farming. Now he no longer needed to correlate the maintenance of acreage with the generation of money, or calculate the toil that went into each bushel of corn.

Of course we helped him find the best combination of seeds for his limited garden and offered to be of assistance should he have any questions in the future, or should he wish to replenish his seed stock. We were pleased that he could stop and smell the flowers and enjoy his retirement.

Getting to know Carl and his family helped Susan and me to better empathize with the struggles of the surrounding farmers, some of whom we'd come to see at times as the opposition. Through this relationship, we saw how our experiment related to the larger landscape and how our influence was being felt in the greater community. While young startup farmers were almost entirely production oriented, the older operations began to dedicate more resources toward the care of the land. Older farmers in general seemed to be more accepting of conservation initiatives and took an interest in issues of land use and corresponding productivity.

Our prairie is actually quite cost-effective. With minimal financial investment, it provides something for every season, from roots and berries to rabbits and pheasants, not to mention an abundance of fiber. Granted, the cornfield beats the prairie in a conventional marketplace that measures short-term financial return and sees the land merely as a means of production. However, against wildflowers, or organic vegetable and fruit production, corn actually lags in value on a per-acre basis. In our current marketplace this fact is skewed by the massive government subsidies of corn, but I suppose I can't knock the farmers for playing by the rules that allow them to make a living.

Add in current climate change concerns and the imperative to reduce greenhouse gases, and the prairie wins hands down in its contribution and value per acre. On a per-acre basis, with no fertilization, irrigation, or cultivation, prairies sequester almost five times more carbon dioxide than cornfields. In fact, most cornfields are net emitters of carbon dioxide and contribute directly to soil erosion and impaired water quality. And they provide little or no wildlife habitat value. Which would you rather have next door, a thriving wildlife habitat, or a corn-flake factory?

AS OUR REPUTATION spread, we got a fair number of unannounced drop-ins. Many folks had "windshield" experiences. They were too shy to ask if they could walk the land and just drove slowly by. But some were so inspired that they'd come back yearly to experience the changes.

We received numerous requests from schools for field trips, particularly involving biology classes wanting to collect plant specimens. Around 1996 one unusually articulate sixteen-year-old named Justin demonstrated stunning observation skills for someone his age. We spent the whole afternoon with him on the prairie, helping him identify plants so that he could gather some for class. A week later he came over for supper and told us how impressed his teacher had been with the diversity of species he'd collected. This began a curious relationship.

Justin had grown up nearby on his father's farm. After his sophomore year his folks sent him to a rigorous college preparatory school in Kansas City, so he was only around during the summer months. As a youth I had thought about little other than ecology. In contrast Justin was more of a tech geek who loved music, computer technology, and games. We encouraged him and were always glad to have him over, in part because Noah was already off to college in Iowa by then, and we found comfort in being able to help another young man as he was beginning his journey.

Justin liked to impress us with his growing knowledge of plants. For instance, he could explain the subtle differences between various sunflower species, identifying stem characteristics and flower structures that were not easy to distinguish. Ecology was perfect for his detail-oriented mind. If he wasn't in the prairie or sitting at the table, he'd come over to dog-sit or do some gardening. Naturally, when we bought new stereo equipment, we hired him to wire up the new system.

He ended up becoming a radio station DJ in Kansas City. I'll admit I was surprised that he'd chosen such a physically circumscribed profession, given how much he enjoyed the outdoors. Still, whenever he came home to visit his folks he always found time to drop by and walk the trail. And if we'd missed him he'd leave a note wedged in the door listing the newest plants he'd identified. I was a bit disappointed that he didn't pursue ecology or a related field, but I knew that his experiences on our land had stayed with him, providing a link to the land that endured even as his daily life moved away from the prairie. I believe he'll always carry that love for nature, and share it someday with a family of his own.

NEWSPAPER JOURNALISTS also found their way to Stone Prairie Farm. After the local Janesville and Brodhead papers published articles in the mid-1990s, we received wider attention, eventually attracting major outlets like the *New York Times* and the *Los Angeles Times*. These were odd interactions for us. Even though the reporters posed thoughtful questions about the value of restoring prairies, they did so like trial lawyers who think they already know the answers. And like Carl, they tried to narrow the discussion to the question of return on investment, as if that were the only indication of success. They showed little interest in our thoughts on the nonfinancial value of our efforts and the joy we'd derived from the restored farm. But when I explained the economics of native seed production, such as the high value per pound of big bluestem grass seeds or even more valuable species, they took copious notes. Most journalists seemed resigned to satisfying their readership with simplistic results-based reporting.

We did meet some writers and photographers who seemed to understand that it was difficult to convey the scope of our achievements in a short article. Fortunately, the land often spoke for itself, punctuating an anecdote with the sudden arrival of bobolinks, bluebirds, an upland plover, or a meadowlark. When the reporters experienced nature directly in this way, they listened more intently to the pulse of life on the prairie, and their writing was influenced by the wealth of information they absorbed.

After reading a *Los Angeles Times* article, an elderly woman living in a retirement home in California sent a handwritten letter telling us how heartening it was to know that we were restoring prairies. She congratulated us and said that she wished she were young enough to visit the farm, because our work had renewed her faith in the future. It had shown her that restoration benefited both nature and humanity by correcting overdependence on monoculture and creating abundance through diversity. One doesn't have to be a revolutionary to see that the current market paradigm equating financial wealth with happiness is fundamentally antithetical to planetary health. It's common sense. Our elderly friend merely hoped that future generations would be able to enjoy the natural order that had been an inalienable right of humans for millennia. Whether we grow corn or wheat or native prairie plants, we have to recognize that we have the choice.

WE HAVE MET many of our neighbors at various summertime community events. We visit the occasional county fair and thresheree, as well as other rural cultural events, such as historical "mountain man" and Civil War reenactments. Among our favorites are the local "Cheese Days," which are much more than a farmer's market. Local businesses display their wares at this festival, but the main focus is the many dairies and cheese makers that provide samples of their products. People come from far and wide to taste regional delicacies, drink locally brewed beer, and generally to celebrate the Swiss cheese-making heritage that defined the culture of this part of America and its relationship to the land.

Neighbors meet to gossip, talk about the pending corn crop harvest, and discuss politics while eating cream puffs and dancing the polka. A percentage of the money raised at these events goes to scholarship funds. For weeks afterward, the local newspapers and club newsletters run stories on the lucky scholarship winners, the overall financial success of the event, and the prize-winning parade float. AES has even joined the fun by putting together a mobile prairie display. One year, AES employees and their children decorated a truck-drawn trailer filled with thousands of pots of blooming prairie plants for the Cheese Days parade in Monroe, Wisconsin. No one had ever seen such an eco-friendly float, and it was a big favorite with just about everyone.

Overall our relationship with our neighbors and the broader culture is complex and ever evolving. The social ecosystem and our position within it change with time, finding one point of equilibrium, shifting as new factors are introduced, then establishing another, much like any natural ecosystem.

From *Nature's Second Chance: Restoring the Ecology of Stone Prairie Farm* by Steven I. Apfelbaum (Boston: Beacon Press, 2009).

✳ BENJAMIN VOGT, 1976- ✳

Benjamin Vogt was born in Oklahoma City, grew up in western Oklahoma, and at the age of ten moved to the Twin Cities in Minnesota. He received a BFA from the University of Evansville in Indiana, an MFA from Ohio State University, and a PhD from the University of Nebraska–Lincoln. His poetry, essays, and photography have appeared in *Crab Orchard Review, Creative Nonfiction, ISLE, Orion, Subtropics,* and the *Sun.* Benjamin is the author of the poetry collection, *Afterimage* (Stephen F. Austin University Press). He is at work on a memoir about his 1894 homesteading family in Oklahoma, the prairie ecosystem, Cheyenne and Mennonite integration, and his "love/hate relationship" with the Great Plains. Benjamin also owns Monarch Gardens, a prairie garden consulting firm, and speaks regionally about gardening with native plants. In this essay, he recounts the "electric-shock moment" of discovering a single western prairie fringed orchid in a tallgrass restoration near where he currently lives in Nebraska.

✳ ✳ ✳

Platanthera praeclara

I'm in a recently restored tallgrass prairie in east central Nebraska, just steps away from a tractor moving soil along the banks of the Platte River, recreating marshes and open water for shore birds and migrants—piping plovers and least terns. It's sunny but cool, late October, the breeze cancelling out the sun one minute, then the reverse the next.

Trailing a group of prairie nerds I'm taking my time. Our guide, Chris Helzer, works for The Nature Conservancy and is showing us their operations. His thin frame has drifted one hundred feet away into the stubble of a prairie they grazed hard that year, so not much is more than a foot high.

I'm enjoying the sound of the tan grass and blackened flowers against my jeans—each step like a washboard in some bluegrass band. The crunch beneath my soles is as satisfying as popping bubble wrap as I try to not think about butterfly cocoons our small group might be trampling.

Then I hear an excited call, a quickening pace of two or three people. I glance over to see Chris on his belly, angled between stubble and an open patch of soil. Soon several are laying flat, their bodies petals splaying out from a dim center. When I arrive I hear Chris saying how he hasn't seen one in months, that he was hoping there'd be some out here since the last he'd seen were in this field a year or so ago.

It's not that noticeable, just six inches high with a few burnt-edged, white blooms, petals like ridiculously fake eyelashes. I don't want to ask what it is because I don't know—I want to keep the illusion that I'm a part of this knowledge, of land like this, of prairie. Finally, I hear the word "orchid" and have images of Nicholas Cage in Florida swamps, stalking ghost flowers for an illegal trade.

This single western prairie fringed orchid is a miracle, I learn. They prefer well-watered upland and untouched tallgrass. They are pollinated by a certain sphinx moth. They can survive only by having a symbiotic relationship with specific fungi in the soil that help their limited roots take up water and their seeds to sprout. A prairie orchid is like serendipity, I think, or just dumb luck. Here in the middle of corn fields a patch of prairie is being managed like a vast garden, hundreds and thousands of acres. And here at the heart of that infinitesimal prairie a half dozen people sun themselves around a few blooms, careful to keep their distance. When they get up they back away on all fours then slowly rise, careful to walk around and to the west, making a wide arc of reverence.

I know it sounds strange to be like this about a flower. We don't do this with nature, we don't do this outside. Books and smart phones, yes, of course—we even sleep with these. And only on birthdays or greeting card holidays or in the beginning of amorous relationships do we gift flowers, often greenhouse-grown and bred to the point of lacking any scent. Sometimes, when we stop to know the world, it gifts us electric-shock moments calling us to re-awaken, and those moments are always sprinkled with symbols and objects, metaphors we carry with purpose into our memory. It would be easy enough to walk by an orchid and not know, even easier to push through a tall stand of bluestem or indian grass whose roots go ten feet down and stretched out would reach miles. I still don't appreciate that little orchid all alone in the prairie, pure luck to exist, as much miracle as human design. I wonder if the geese flying overhead look down and remember. Maybe next year there will be more orchids. Maybe there will be none.

Unpublished essay, 2012.

✳ ELIZABETH DODD, 1962– ✳

Award-winning poet and essayist Elizabeth Dodd was born in Colorado in 1962 and raised in southeastern Ohio, though she descends from Oklahomans on both sides of the family, and so has an ancestral connection to the central plains. At Ohio University she majored in English and French, and from Indiana University she earned an MFA in poetry and a PhD in literature. She is University Distinguished Professor of English at Kansas State University, in the Flint Hills, where she teaches creative writing and literature and is an Affiliated researcher with the Konza Prairie Biological Station, where this essay takes place. She is the author of six books, including *Prospect: Journeys and Landscapes* (2003), and *In the Mind's Eye: Essays across the Animate World* (2008)—which won ASLE's Environmental Creative Writing Book Award. In this chapter from her memoir, *Horizon's Lens* (2012), Dodd looks to the stars to reclaim familial connections to landscape and language—a fitting end to this anthology, which began with an Osage story about the celestial origins of the prairie and its people.

Constellation

One fall near dusk I watched a flock of goldfinches feeding in a stand of sunflowers. When I stepped toward them from the trail, my body moving among the rustling stalks, the flock turned to smoke and rose into the air. The cloud then settled in the trees along the creek while I stood silent among borders and conjunctions—the adjacent hill leveling out into recumbent flood plain, crepuscular air resting in bunchgrass and deciduous thicket. Moments later the birds reversed their flight, returning from the

333

nearby treetops to continue feeding on seeds, stoking the inner fire of their metabolism for the coming night of cold.

As I tell the story of the finch flock, the nouns and verbs stand out, lovely—and maybe lonely—in their own particularity, framed in my mind and on the page by the pale matting of space-between-words. It's the way I experience the world, I think, sensitive always to the distances among things, the distinctive skin of separation so many objects present to all the elseness that surrounds them. Is that a deer or a shrub, I wonder, that darker patch I see halfway down the tawny autumn hillside? Either way, my attention frames interest in the general rhythm of subjects and complements, things I might keep handy in the pockets of thought, fingering them as I move through the landscape—one I know well or one where I've never before set eyes or foot.

This week I've been reading a nineteenth-century publication on onomatopes, word-roots derived from sounds made by the things they signify, so my head is abuzz with unfamiliar consonant clusters. The author had collected examples from Siouan speakers, Native American words that could have once gestured toward this very topography. So when I leave the quiet house for a walk along that same stretch of trail, I remember the finches and think of their flight, and my own movement, like echoes of the place itself. An hour into today's hike, snowflakes lunge from the northwest, countless tiny wind-sped points of cold. It's the first snow of the season, though winter hasn't even arrived. When I turn to face the oncoming weather I get a little disoriented, as if I'm speeding along to meet the wind.

Zu'de, that wind might be saying. *Ga-zu'-zu-de*, whistling or roaring as it touches whatever lifts skyward from the ground. The first means simply a whistling sound; the second, the frequency of whistling (practically incessant!) that is wind in the Great Plains. *Ga-zu'-zu-de*. Sometimes it drove newcomers nearly to distraction—one wealthy Kansas rancher's wife in the 1880s was said to hide in her fine home's underground root cellar just to get away from the sound. But I love the insistence of the prairie wind. When I listen I hear vowels as the wind hits the gallery woods by the creek, consonants when it shivers whole hillsides of grass. *Su'-'e*, the sound of walking through grass.

In Siouan speech the things of the world and the things that they do

ELIZABETH DODD

might all adhere to one another; in many languages indigenous to the Americas, verbs tend to absorb both their subjects and their objects in a little ecology of signification. Attention cinches them all together in agglutinating syntax-turned-word-formation. It's a way of speaking so different from my own that I have to think hard just to imagine it, and just when I think I've caught the concept, it slips away.

I've brought a topographic map of these hills and draws, but nothing is labeled on it. With a finger I can trace that darker line of trees by the creek, approximating where I think I might be standing. If anyone has given names to the intermittent streams that wrinkle these hillsides, I've never heard them, so I remember the landmarks with narrative pegs: here's where the turkeys will cluster all winter; here's where I once saw a badger push through the brush. These are recent inflections of the watershed, a memory-base that can't be more than a few microns thick. If, in some sort of remote-sensing linguistics measurement, conducted with several geographers and plotted into a software database, I could survey the word-maps laid on the prairie landscape over centuries, I'd need to populate at least two fields with the findings. Siouan, the speech of the Kansa people; Caddoan, the speech of the Pawnees.

For me, both tongues exist on the page rather than within the ear, though I sound them out slowly, experimentally. Right now I'm trying to shape words that, a century and a half ago, might have been breathed by someone —shall I call her a Kansan?—on this same hillside. *Kh'-dha*, she'd say to describe her own movement. I try it, too. The last vowel should be slightly nasalized, softening the percussive quality as though sound itself drops into the grass and is lost. *Kh'-dha*: the sound made by pushing one's way through grasses or sunflowers. That's what I did, that late afternoon more than a decade ago, when the birds turned to smoke. That's what I'm doing right now as well.

"They used to tell stories": a translated sentence I find in a book of Pawnee grammar. The linguist picks the phrase apart, showing how these little morphemes—stick-tight seedlets of thought—have coalesced into what looks like a single word in the text. *Ar ri ir ur i uks ra:i:wati hus u:ku.* Oh, it's a constellation of meaning. In the typescript, colons indicate where I should extend the vowels, voice them a little longer than the others. I try, reading the explanations as I go. *Ar*, an evidential, so we know the speaker heard the old stories himself. *U:ku*, showing the habitual aspect, how the old people used to tell the stories over and over. And between these, a dense cluster of syllabic signification. *Ur . . . ra:i:wati* means to tell the story, but *wati* is actually a metaphoric, transitive verb, "to dig." (I think of how archaeology and oral history both sift the past, lifting the evidence free from the dust.) *Uks* is a marker of aspect, indicating the completed quality of the verb. They used to tell, it hints, but not anymore. Those days are gone. My own grandparents are gone, and with them some of the hints and whispers of my bloodline's native connection with the continent. Choctaw—late arrivals to the southern plains, some of my father's people; they spoke Muskogean, a language of the Southeast. And my mother's people—well, their story's been garbled through generations. Algonquin? Athabaskan? The answers had been lost before I knew the questions I'd like to have asked.

Pawnee "is primarily a language that is remembered," the scholar laments. "In no family is it any longer used on a day to day basis." And when I read this I think how he, too, seems unconsciously to have slipped into a kind of rhetorical formality, frontloading his sentence with the negative, arranging the verb and its adverbial modification so that the pacing is slow, almost stately, delaying the moment of silence at the sentence's end. The grammar resembles a funeral procession, carrying the departed away across the page. My brother, pallbearer, carried my grandmother's casket when the family gathered for her death. At my grandfather's funeral I followed my father in speaking of his father's life—nearly a century lived in the Oklahoma shortgrass landscape. When my turn came I read a poem aloud and then returned to my seat among the mourners, dressed again in silence.

If you could cinch the written page closed like a pouch, wrap up a bundle of language, a memory bag, what then? If you reopened it slowly, could you imagine the language still there waiting for you, reemerging like stars once the sun has gone down?

Rahurahki. That's the Deer, a constellation I can make out even in my front yard through the scrim of hackberry branches and the competing haze of small-city lights. They're the stars that make up Orion's belt, and they don't look like the profile of an individual nor the outlines of the animal's antlers lifting in figurative form. The Skidi Pawnees said the stars would "follow one another up" in the east. To see what they saw, you have to imagine the hardly glimpsed creatures picking their quiet way forward—a small group of deer, three or four, moving slowly as if through the oaks by the creek. Those stars might be their heads, raised and listening; they might be white tails poised before flight.

For the Pawnees, the Big and Little Dippers were *raruka'i:tu'*, stretchers used to carry the sick or the dead. The four corner stars are the litter bearers, and a short line of stars seems to trail along, mourners or worried family members carrying the weight of mortality through timeless skies. The Milky Way is the road of the dead, *ru-ha-rú-tu-ru-hut*, the white path above, where spirits are driven by wind. But it has less lofty identities as well. It could be dust raised by the feet of innumerable bison; it could be a stream with foam pushed by the current. *Raki:rarutu:ru:ta*. I like the word as I read it aloud. It sounds liquid, I think; it sounds like movement.

On a mild night you can lie awake, half hiding yourself in the windbreak of your own sleeping bag, watching the lid of the sky until your eyes close, only to wake later and see how the stars have moved. East to west, the constellations travel the sun's vacated pathway. Treeless horizon, cloudless sky, you can try to glimpse the hints of form those stars imply.

Among my favorites from Pawnee astronomy are the Snake and the Swimming Ducks. The Snake is part of Scorpius; the serpentine body points its head toward the Milky Way, as if it has paused in the moment before it will twitch and slither away. The Swimming Ducks appear to float companionably together near the Snake's tail. The celestial birds are migratory: at my latitude, the constellation appears heliacally in the sky in the spring and disappears in the autumn. That's a good way to arrange your concept of the year, of course; it all begins anew in spring, crescendos through the months of summer, progressing toward senescence as autumn slides toward winter's grip. That's how the Pawnees perceived the order of time's passing, and they answered its movement with ceremonies that opened in spring and closed in fall. As James Murie explained just over a century ago, this was so because "all the animals are hibernating and the birds have gone south. Even the stars have changed their places." In phenology's rhythmic pulse, whole rafts of ducks move through the equinoctial skies of the central flyway, but I like the intimacy of those two small points of light, a pair who have made the trip together and are settled in to their own small niche of sky, like some small pothole lake left by the long-since melted age of ice, until once again it's time to move on, and together they depart.

Much as I like the Swimming Ducks, I have no certainty about their Pawnee name. By the early twentieth century, when Murie, whose mother was Skidi, began collecting information about traditional Pawnee culture, he came up against loss after loss. He had no Caddoan word for the constellation, only its English translation. And other waterfowl figures complicate the question. Another scholar explains that two months of the year are named for avian constellations, but I can't find clarification of when such months might fall in our own calendar year. And here I find Little Duck, Kiwaksi, and Big Duck, Kiwaks-kutsu. It's easy to hear the floppy chatter of ducks in both those names, close as they are to our own *quaaack-quack-quack*. Were the Swimming Ducks pictured individually, not collectively? Were they the same as another constellation called Ku:hat, the Loons? Loons do pass through this part of the continent in spring and fall migration—I've seen them once or twice—but they don't stay to nest through the summer months. I'd imagine the birds' travel would be tagged—call them the Flying Loons—if they were the chosen harbinger.

By now I want to know what part of the prairie's rhythms were indexed in the Pawnees' astral lexicon; I'd like to find any correspondence they saw between the worlds above and below. But the gaps in my own knowledge and the historical record move in like high cloud cover, obscuring the patterns I might have glimpsed. And after all, I live in town: the urban glow smears the sky above.

From the Lakotas, the Kansas' cousins, I learn Wakinyan. To me the constellation looks like a giant dragonfly with flat-tipped wings and narrow body; Polaris is near enough I can imagine that it's an insect the dragonfly pursues. But this is the Thunderbird, and I have to admit, it utterly commands the place where I first see it paused in flight just above the prairie hills. Across that astral landscape another animal moves through the autumn nights, named simply by the Lakotas Tayamnipa, Animal. The Pleiades comprise its head; the bright tip of its tail is Sirius. In the middle of its body the stars of Orion's belt form its backbone, from which poke a few starry ribs. Is the creature a bison? Is it a bear? I consider both suggestions but prefer not to choose. This time I like the purely abstract quality of animal. With neither hoof nor paw, the creature presents just a hint of a body. If you consider perspective, it's as if the earth has traded places with the sky, and we gaze down on the formation from above.

IN PHOTOGRAPHS I've seen stars arrayed on an oval of elk hide, four-pointed blazes inscribed on the pale leather in scatters and clusters you could hold in your lap. Called the Skidi Pawnee star chart, it's a handsome thing, laid out lengthwise with a sense of symmetry. Like open pages from a vellum album, the chart's two halves meet in a vertical band of indistinct stars said to represent the Milky Way. East and West redden the oval's far edges with traces of paint. Constellations bunch together, crowding one another in the chart's open field. There are the Swimming Ducks and the Snake that's painted near them. The Big Stretcher pauses in its presumed motion, the trail of mourners following in the four stars' wake. But this chart offers no celestial verisimilitude, no navigational atlas for the world above. Astronomer Von Del Chamberlain suggests its role is to "capture star powers that could then be used by the Skidi in maintaining a secure life." He notes that bits of leather lacings remain along the hide's perimeter— it's easy to picture the drawstring pulled closed, effectively bagging the facsimile sky inside the elk hide.

I've read accounts of how a Pawnee elder named Running Scout pointed out the various star images to his younger tribesman, James Murie. Murie was by then working for the Field Museum collecting ethnographic materials—stories and artifacts—and Scout was willing to share his knowledge of the old ways with an interested listener. Murie explained to the museum ethnographers how he'd likely be able to procure the elk hide star chart and the other items from the tribe's traditional bundles. One of the elders "was telling me the other day that he sees that the boys who he is trying to teach the old religion seem to have no interest in the ceremonies. . . . It hurts the old man," he reported, and the elder had decided to leave his artifacts "in your care."

Scout's words now rest in old wax cylinder recordings, notes written in both English and Pawnee, and quotations in scholarly texts. Fragments, handwritten by Murie, suggest the way Scout wanted to indicate the incompleteness of his knowledge; I imagine, if I could read the original Pawnee speech, there'd be linguistic evidentials hinting at what Murie casts as whole phrases of uncertainty. "Well brother the stories I lost—forgot—of what the old man used to tell about . . ." Scout, a survivor of the great dislocation when the tribe was removed from Nebraska and confined in Oklahoma, knew all about loss.

The transcriptions indicate that he must have been gesturing to the painted figures on the hide as he identified a star's name and story. "Now brother, brother, that which I have given you, that which you now hold"— we can imagine the men sitting together, their hands moving over the heavens depicted on their laps. Scout pointed to Polaris, perhaps feeling the bitter contrast of the star's constant station with the transience of the people who watched it from below. "The old people who are now dead use [*sic*] to call it 'the star that does not move.'" I thumb to Chamberlain's published table that lists some Skidi names and their reported meanings: *u:pirit karariwari*, "star that does not move"; *hó-pi-rit ka-wa-rí-wa-ri*, "star it does not move." "So brother," Scout said to Murie, ". . . that's what it is, the bundle that is now here. . . . Now brother you see the heavens."

The bag full of stars, the bundle of memories. I like to look at the photograph here on my desk and then go out to look at the night sky, trying to see the patterns there. One writer says the chart is three hundred years old, but I can't find substantiation for his claim. The stories themselves, however, are clearly very old, seeming to reach to some time in the Pawnees' past before they settled in Nebraska and Kansas. Chamberlain says that some of the stars clearly defined in Scout's stories cannot be seen from midcontinent latitudes; their capture into the mythic imagination of the Pawnees may date to a time when they were people of the southern plains.

Of course, I keep looking. By now both the Swimming Ducks and the Deer have caught me, and I'm looking back at them the way last week I watched a group of does move through the oak woods along Kings Creek. The animals lifted their heads to let me feel the bright intensity of their sentience as they watched me, until they decided to move away, beyond my sight. I keep thinking about them both, the living mammals, breath steaming a little in the chill, and the imagined and mnemonic creatures of the sky. Did anyone date the leather? (Evidently not.) And where is the artifact now?

The curatorial staff at the Field Museum tells me the chart is still in the collection but has been "rebundled" with the other materials, and "by agreement it is not to be opened." Not on display; not available for scholarly study. So now the images are isolated, bound firmly out of sight. When did the rebundling take place? I ask. Are there any written accounts where I could read about it? No, the curator replies, there isn't any record. It was twelve or thirteen years ago — he isn't sure of the exact date, and he doesn't seem to want to say much to me. "This is the result of a complex dialogue that I'm afraid I cannot share," he concludes.

WE KNOW THE STARS are not truly timeless, but their temporal scale dwarfs our own existence. The Swimming Ducks we see take centuries to reach us: one star lies seven hundred light-years away; the other is five hundred nineteen. But to us they appear as a matched pair, traveling the skies together through the months ungripped by winter's ice. Another pair of stars Chamberlain identifies with Pawnee names are also travelers of a sort: Hikusu', Breath; Hutu:ru´, Wind. Are these words onomatopes? Each opens with a whoosh; each whistles exhalation past the husks of its consonants. In Pawnee cosmology they're both associated with the direction of the setting sun. Along with thunder and rain, wind does come to the prairies most often from the west, from somewhere beyond the distant, perhaps even never-glimpsed mountains. The distance of the stars themselves, known in Western astronomy as Alpha and Beta Persei, has been calculated by the brightness and color of their light. While Wind lies ninety-three light-years away, Breath is nearly six hundred light-years distant. How far Breath travels before its kinetic presence can touch our skin!

I'm mingling traditions now, of course. I'm collecting stories that I've read or heard, laying them out to watch for unexpected patterns to emerge. The Pawnee elder, Scout, said that the Star that Does Not Move was sometimes represented by a feather-covered spear. A *hukawiskiria*, he called it, which translates literally as a "living-covered lance." The wooden pole "had all kinds of birds flocking around it," he explained. I like to imagine the star as the lance's tip, viewed head-on while the inspirited stars might wheel like birds around that stationary point of light. I remember in childhood trying to imagine the "poles" of the earth, seeking synthesis between the metal "axis" ("axle," I thought it must mean) around which the globe revolved and the notion of magnetic poles toward which a compass arrow would always point. I tried to combine the candy-cane poles in Santa's snowy yard from the Christmas cartoon we watched on television with the flag the European explorers sought to plant into the snowy crust—surely somewhere between the two lay the real pole, even though I'd never seen it pictured.

Look up: Polaris is the prey the Thunderbird pursues but cannot catch.

"LANGUAGE TRICKS people into believing that rises and hollows, wind and rivers, are all in some sense alive," writes cultural geographer Yi-Fu Tuan. He's interested in the ways that human beings pursue what he calls escapism: escape from nature into culture's roof and walls, escape from culture to the wilder world beyond the pale. I see his point, but I don't want to follow him into the scorched-grass circle where he's ringed himself inside his claim. "[B]ecause human beings and human speech are co-eval, there never was a time when speech did not generate this useful and reassuring illusion. Language animates," he declares; "that and human bonding are two of its most primitive and potent effects." He means here intrahuman bonding, bipedal "I" with "thou" wrapped tight in the cape of our mutual fear of death and a need for one another. Yes, I think, but . . .

It's not a trick to speak of the life-breath in the wind that drives before it rain, or fire, or animals of the hunt. It's not a trick to see Kings Creek's clean water as a kind of health or to know that light and water can engender life. Sometimes the blade-tip of attention's lance brings metaphoric plumage to enliven thought; how reductive, I think, to see this richness as outright deception. The question is how reductively one wields the figures of our language, whether we recognize the nuance of their connotations or insist that symbolism is a kind of algebraic process, rendering out each variable's singular worth. And I've always been a poor student of math.

There is a tale about a Pawnee warrior who died on the banks of the Loup River sometime in the 1830s. From the blood-and-mud of history, the story soars into the world of myth. Out hunting beaver, he was killed by a party of several Sioux, and his body was left unburied, scalped and broken. After his slaying, the Nahurac, a group of unspecified magical animals, came to restore him—all but the top of his head, which they had to replace with feathers. His allegiance, though, was still firmly rooted in the mortal world; he wished to help his tribesmen, and Ready-to-give was now his name. He could restore his mother's blind sight if she would wash grief from her eyes; with his own clairvoyance he would help the tribe defeat their enemies.

"I am in everything," he said. But he was all too human in his need for recognition. "You must never get tired of me," he told his brother. That kind of constant attention was too much to ask of even a devoted sibling. One night the brother didn't rise from his bed to keep his appointment with Ready-to-give, and that was the end of the warrior's aid.

There's some confusion as I read about the warrior. The constellation that represents him in the sky is called Pahukatawa, Knee Prints on the Stream's Bank. But the two stars said to constitute those knee prints don't lie next to one another; labeled Alpha and Beta Persei in Western star charts, they seem insignificant, mixed in with stellar scatter. Individually, they're those two star-powers the Pawnees associated with the West, Wind and Breath. I've examined planispheres and the sky itself, trying to glimpse something about those two simple star-points that could suggest the shallow mark of human weight on the grass-softened banks, but I can't. Whatever conjunction of history and metaphor is caught by those two stars, it's lost to me.

IN AUTUMN the upper Kings Creek watershed is tawny-golden, the hue of an animal warming itself in the late-season sun. "Leonine," someone once told me, and I think it's true; today my trail threads through a cougar-pelt hillside, grass rippling like muscles along the cat's resting flank. Pakstitkukek, the Mountain Lion, was a Pawnee constellation associated with Capella, known as the Yellow Star and mythically associated with the mountain lion, lightning, and the West. But this associative thinking goes only so far before I reach an impasse; the Yellow Star claimed the spring of the year, not the fall, as her season of influence.

I walk the dry creekbed, my boots clinking loose flint and limestone, rustling dry oak leaves in the autumn air. Just below a spring seeping across soft ground, a pool of clean water rests, isolate, although most of the streambed is now bone dry. I look closer: a tiny frond of watercress, its crisp stem linking each photosynthetic leaflet, a pattern almost like the shape I've memorized of Swimming Ducks and the Snake.

Several yards farther along the land rises sharply, a high cutbank dangling grass roots like vines from the dried canopy above. Here I'm scuffing my way through loose flint pebbles and sandy loam until I round a curve and am drawn up short by my own caught breath. A buck carcass, its head wrenched sideways so the antlers are positioned like an open trap poised to slam shut. Drawing closer, I can see there's still a reddish trace of flesh along the ribcage; like anklet socks, hair still clings to the bones above each hoof. Was the animal shot some distance away, then traveled as far as he could through the whispering tallgrass until he tumbled down that cliff of dirt? Was he sick, stumbling along in a search for water?

The air's so cold that there's hardly any smell. I crouch down for a moment for a better look but hesitate to touch the antlers, though I count each tine: a ten-point buck. How slender the leg bones seem in comparison to their length. How narrow the hips, how large the gristmill of the jaws. Have I ever seen a living ten-point buck along this watershed? I don't think so. Now his bones will lie here for months before spring rains will finally lift them in the creek's high current and carry them away.

Rahurahki, the Deer stars. I know I will look in the night sky to see whether the cluster of stars I've known as Orion recalls the shape of those antlers, suspended low above the horizontal streambed and the fallen buck.

It's a scramble up the lower bank, back into the wind. *Sorghastrum nutans*, Indian grass, still lifts its light-catching spikelets like feather fronds, bright in the southern-sky sun. When I turn so the tallgrass is backlit, each seed incandesces, impossibly fragile and bright. Throughout the growing season stalks lifts bits of silica skyward, then spill their mineral cargo back to earth when at last the plant degrades to soil. And what is silica? Minute ejecta from exploded stars, exhaled to the universe to fetch up in sand grains in a prairie streambed or in dead grass left standing after summer's passed.

Hikusu', Breath; Hutu:ru´, Wind. Stand very still, I tell myself. Listen. Listen.

From *Horizon's Lens: My Time on the Turning World* by Elizabeth Dodd (Lincoln: University of Nebraska Press, 2012).

Permissions

Steven I. Apfelbaum, "Getting to Know Your Neighbors," copyright © 2008, from *Nature's Second Chance: Restoring the Ecology of Stone Prairie Farm*, Boston: Beacon Press, 2009. Reprinted by permission of the author and Beacon Press, Boston.

Black Dog, "Children of the Sun and Moon" (title not in original), from *Twenty-Seventh Annual Report of the Bureau of American Ethnology to the Secretary of the Smithsonian Institution, Washington, D.C., 1905–1906*, Washington, D.C.: U.S. Government Printing Office, 1911. Public domain.

George Catlin, "Prairie Burning," from *Letters and Notes on the Manners, Customs, and Condition of the North American Indians: Written During Eight Years Travel Amongst the Wildest Tribes of Indians in North America, in 1832, 33, 34, 35, 36, 37, 38, 39*, volume 2, New York: Wiley and Putnam, 1842. Public domain.

Cindy Crosby, "Pulling Weeds: Community," from *By Willoway Brook: Exploring the Landscape of Prayer*, Brewster, Mass.: Paraclete Press, 2003. Reprinted by permission of the author.

Elizabeth B. Custer, "A Blizzard," from *Boots and Saddles*, New York: Harper and Brothers, 1885. Public domain.

Thomas K. Dean, "Finding Devonian," from *Living with Topsoil: Tending Spirits, Cherishing Land*, ed. Steve Semken, North Liberty, Iowa: Ice Cube Press, 2004. Reprinted by permission of the publisher.

Charles Dickens, "A Jaunt to the Looking-Glass Prairie and Back," from *American Notes for General Circulation*, London: Chapman and Hall, 1842. Public domain.

Elizabeth Dodd, "Constellation," from *Horizon's Lens: My Time on the Turning World*, Lincoln: University of Nebraska Press, 2012. Copyright ©2012 by the Board of Regents of the University of Nebraska. Reprinted by permission of the University of Nebraska Press.

Louise Erdrich, the essay "Big Grass," copyright © 1994 by Louise Erdrich, originally appeared in *Heart of the Land: Essays on Last Great Places*, ed. Joseph Barbato and Lisa

Weinerman, New York: Vintage, 1994. Used by permission of The Wylie Agency LLC.

Eliza Woodson Farnham, "Spring around the Prairie Lodge," from *Life in Prairie Land*, New York: Harper, 1846. Public domain.

Lance M. Foster, "*Tanji na Che*: Recovering the Landscape of the Ioway," from *Recovering the Prairie*, ed. Robert F. Sayre, Madison: University of Wisconsin Press, 1999. Copyright 1999 by the Board of Regents of the University of Wisconsin System. Reprinted by permission of the University of Wisconsin Press.

Margaret Fuller, "Chicago," from *Summer on the Lakes in 1843*, Boston: C. C. Little and J. Brown, New York: C. S. Francis, 1844. Public domain.

Hamlin Garland, "The Homestead on the Knoll," from *A Son of the Middle Border*, 1917. Public domain.

Don Gayton, "Tallgrass Dream," from *Landscapes of the Interior: Re-Explorations of Nature and the Human Spirit*, Gabriola Island, B.C.: New Society, 1996. Reprinted by permission of the publisher.

Melvin R. Gilmore, "The Plant Tribes," from *Prairie Smoke*, New York: Columbia University Press, 1929. Public domain.

Paul Gruchow, "What the Prairie Teaches Us," from *Grass Roots: The Universe of Home*, Minneapolis: Milkweed Editions, 1995. Reprinted by permission of the publisher.

William J. Haddock, "The Passing of the Prairies," from *A Reminiscence: The Prairies of Iowa and Other Notes*, Iowa City: Printed for private circulation, 1901. Public domain.

Drake Hokanson, "Habits of the Grass," from *Reflecting a Prairie Town: A Year in Peterson*, Iowa City: University of Iowa Press, 1994. Reprinted by permission of the publisher.

Bill Holm, "Horizontal Grandeur," from *Prairie Days*, San Francisco: Saybrook, 1987. Originally published in *The Music of Failure*, Plains Press, 1985, now published by the University of Minnesota Press, 2010, copyright 1985 by Bill Holm. Reprinted by permission of the author and the publisher.

Washington Irving, "Life on the Prairies," from *A Tour on the Prairies*, Paris: Baudry's European Library, 1835. Public domain.

Edwin James, "Prairie Mirage," from *Account of an Expedition from Pittsburgh to the Rocky Mountains, Performed in the Years 1819, 1820. By Order of the Hon. J. C. Calhoun*,

Secretary of War, under the Command of Maj. S. H. Long, of the U.S. Top. Engineers, London: Longman, Hurst, Rees, Orme, and Brown, 1823. Public domain.

Josephine W. Johnson, "July," from *The Inland Island*, New York: Simon and Schuster, 1971. Reprinted by permission of the Literary Estate of Josephine W. Johnson.

Teresa Jordan, "Playing God on the Lawns of the Lord," from *Heart of the Land: Essays on Last Great Places*, ed. Joseph Barbato and Lisa Weinerman, New York: Vintage, 1994. Reprinted by permission of the author.

Lisa Knopp, "Far Brought," from *The Nature of Home: A Lexicon and Essays*, Lincoln: University of Nebraska Press, 2002. Copyright 2002 by the Board of Regents of the University of Nebraska. Reprinted by permission of the University of Nebraska Press.

Francis La Flesche, "A Runaway," from *The Middle Five: Indian Boys at School*, Boston: Small, Maynard, 1901. Public domain.

William Least Heat-Moon, "Under Old Nell's Skirt," from *PrairyErth: A Deep Map*, copyright © 1991 by William Least Heat-Moon, Boston: Houghton Mifflin, 1991. Reprinted by permission of Houghton Mifflin Harcourt Publishing Company. All rights reserved.

Aldo Leopold, "July: Prairie Birthday," from *A Sand County Almanac*, New York: Oxford University Press, 1949. Reprinted by permission of the publisher.

Meridel Le Sueur, "Drought," from *North Star Country*, 1945. Reprinted by the University of Nebraska Press, 1984. Copyright 1984 by the Board of Regents of the University of Nebraska. Reprinted by permission of the University of Nebraska Press.

John Madson, "The Running Country," from *Out Home*, New York: Winchester Press, 1979. Reprinted by the University of Iowa Press, 2008. Reprinted by permission of the Estate of John Madson.

Richard Manning, "Enclosure," from *Grasslands: The History, Biology, Politics, and Promise of the American Prairie*, New York: Penguin, 1995. Reprinted by permission of the publisher and the author.

John Joseph Mathews, "Planting Moon," from *Talking to the Moon: Wildlife Adventures on the Plains and Prairies of Osage Country*, Chicago: University of Chicago Press, 1945. Reprinted by the University of Oklahoma Press, 1981. Reprinted by permission of the publisher.

John Muir, "A New World," from *The Story of My Boyhood and Youth*, Boston: Houghton Mifflin, 1913. Public domain.

Francis Parkman, "Jumping Off," from *The Oregon Trail: Sketches of Prairie and Rocky-Mountain Life*, 1849. Public domain.

John T. Price, "Man Killed by Pheasant," from *Man Killed by Pheasant and Other Kinships*, Cambridge, Mass.: Da Capo Press, 2008. Reprinted by the University of Iowa Press, 2012. Reprinted by permission of Da Capo Press.

William A. Quayle, "The Prairie," from *The Prairie and the Sea*, Cincinnati: Jennings and Graham, 1905. Public domain.

Mary Swander, "The Grasshopper Year," from *Out of This World: A Journey of Healing*, New York: Viking, 1995. Reprinted by the University of Iowa Press, 2008. Used by permission of University of Iowa Press.

Mark Twain, "My Uncle's Farm," from *The Autobiography of Mark Twain*, volume 1: *The Complete and Authoritative Edition, 1870–1909*, ed. Harriet Elinor Smith, Berkeley: University of California Press, 2010. Reprinted by permission of the publisher.

Winifred M. Van Etten, "Three Worlds," from *Growing Up In Iowa: Reminiscences of 14 Iowa Authors*, ed. Clarence A. Andrews, Ames: Iowa State University Press, 1978.

John C. Van Tramp, "Phil and the Lost Boy," from *Prairie and Rocky Mountain Adventures, Or, Life in the West*, Columbus, Ohio: Segner and Condit, 1870. Public domain.

Benjamin Vogt, "*Platanthera praeclara*," published by permission of the author.

Walt Whitman, "The Prairies, and an Undelivered Speech," from *Specimen Days*, Philadelphia: Rees Welsh, 1882. Public domain.

Zitkala-Ša, "The Great Spirit," from *American Indian Stories*, Washington, D.C.: Hayworth, 1921. Reprinted 1985 by University of Nebraska Press. Public domain.

Author and Title Index